E. coli

METHODS IN MOLECULAR MEDICINE™

John M. Walker, SERIES EDITOR

80. **Bone Research Protocols,** edited by *Stuart H. Ralston and Miep H. Helfrich,* 2003

79. **Drugs of Abuse:** *Neurological Reviews and Protocols,* edited by *John Q. Wang,* 2003

78. **Wound Healing:** *Methods and Protocols,* edited by *Luisa A. DiPietro and Aime L. Burns,* 2003

77. **Psychiatric Genetics:** *Methods and Reviews,* edited by *Marion Leboyer and Frank Bellivier,* 2003

76. **Viral Vectors for Gene Therapy:** *Methods and Protocols,* edited by *Curtis A. Machida,* 2003

75. **Lung Cancer:** *Volume 2, Diagnostic and Therapeutic Methods and Reviews,* edited by *Barbara Driscoll,* 2003

74. **Lung Cancer:** *Volume 1, Molecular Pathology Methods and Reviews,* edited by *Barbara Driscoll,* 2003

73. ***E. coli:*** *Shiga Toxin Methods and Protocols,* edited by *Dana Philpott and Frank Ebel,* 2003

72. **Malaria Methods and Protocols,** edited by *Denise L. Doolan,* 2002

71. ***Hemophilus influenzae* Protocols,** edited by *Mark A. Herbert, E. Richard Moxon, and Derek Hood,* 2002

70. **Cystic Fibrosis Methods and Protocols,** edited by *William R. Skach,* 2002

69. **Gene Therapy Protocols, 2nd ed.,** edited by *Jeffrey R. Morgan,* 2002

68. **Molecular Analysis of Cancer,** edited by *Jacqueline Boultwood and Carrie Fidler,* 2002

67. **Meningococcal Disease:** *Methods and Protocols,* edited by *Andrew J. Pollard and Martin C. J. Maiden,* 2001

66. **Meningococcal Vaccines:** *Methods and Protocols,* edited by *Andrew J. Pollard and Martin C. J. Maiden,* 2001

65. **Nonviral Vectors for Gene Therapy:** *Methods and Protocols,* edited by *Mark A. Findeis,* 2001

64. **Dendritic Cell Protocols,** edited by *Stephen P. Robinson and Andrew J. Stagg,* 2001

63. **Hematopoietic Stem Cell Protocols,** edited by *Christopher A. Klug and Craig T. Jordan,* 2002

62. **Parkinson's Disease:** *Methods and Protocols,* edited by *M. Maral Mouradian,* 2001

61. **Melanoma Techniques and Protocols:** *Molecular Diagnosis, Treatment, and Monitoring,* edited by *Brian J. Nickoloff,* 2001

60. **Interleukin Protocols,** edited by *Luke A. J. O'Neill and Andrew Bowie,* 2001

59. **Molecular Pathology of the Prions,** edited by *Harry F. Baker,* 2001

58. **Metastasis Research Protocols:** *Volume 2, Cell Behavior In Vitro and In Vivo,* edited by *Susan A. Brooks and Udo Schumacher,* 2001

57. **Metastasis Research Protocols:** *Volume 1, Analysis of Cells and Tissues,* edited by *Susan A. Brooks and Udo Schumacher,* 2001

56. **Human Airway Inflammation:** *Sampling Techniques and Analytical Protocols,* edited by *Duncan F. Rogers and Louise E. Donnelly,* 2001

55. **Hematologic Malignancies:** *Methods and Protocols,* edited by *Guy B. Faguet,* 2001

54. ***Mycobacterium tuberculosis* Protocols,** edited by *Tanya Parish and Neil G. Stoker,* 2001

53. **Renal Cancer:** *Methods and Protocols,* edited by *Jack H. Mydlo,* 2001

52. **Atherosclerosis:** *Experimental Methods and Protocols,* edited by *Angela F. Drew,* 2001

51. **Angiotensin Protocols,** edited by *Donna H. Wang,* 2001

50. **Colorectal Cancer:** *Methods and Protocols,* edited by *Steven M. Powell,* 2001

49. **Molecular Pathology Protocols,** edited by *Anthony A. Killeen,* 2001

48. **Antibiotic Resistance Methods and Protocols,** edited by *Stephen H. Gillespie,* 2001

47. **Vision Research Protocols,** edited by *P. Elizabeth Rakoczy,* 2001

46. **Angiogenesis Protocols,** edited by *J. Clifford Murray,* 2001

E. coli

Shiga Toxin Methods and Protocols

Edited by

Dana Philpott

Groupe d'Immunité Innée et Signalisation, Institut Pasteur, Paris, France

and

Frank Ebel

Max-von-Pettenkofer-Institut, Bakteriologie, Munich, Germany

Humana Press  Totowa, New Jersey

© 2003 Humana Press Inc.
999 Riverview Drive, Suite 208
Totowa, New Jersey 07512

www.humanapress.com

This publication is printed on acid-free paper. ∞
ANSI Z39.48-1984 (American Standards Institute) Permanence of Paper for Printed Library Materials.

Cover Illustration: Fig. 5A,B from Chapter 12, "Microscopic Methods to Study STEC: Analysis of the Attaching and Effacing Process," by S. Knutton.

Production Editor: Jessica Jannicelli.
Cover design by Patricia F. Cleary.

For additional copies, pricing for bulk purchases, and/or information about other Humana titles, contact Humana at the above address or at any of the following numbers: Tel.: 973-256-1699; Fax: 973-256-8341; E-mail: humana@humanapr.com; or visit our Website: www.humanapress.com

Printed in the United States of America. 10 9 8 7 6 5 4 3 2 1

Library of Congress Cataloging in Publication Data

E. coli : Shiga toxin methods and protocols / edited by Dana Philpott and Frank Ebel.
     p. ; cm. -- (Methods in molecular medicine ; 73)
   Includes bibliographical references and index.
   ISBN 0-89603-939-0 (alk. paper)
    1. Escherichia coli infections--Laboratory manuals. 2. Verocytotoxins--Laboratory manuals. I. Philpott, Dana. II. Ebel, Frank. III. Series.
    [DNLM: 1. Escherichia coli Infections--pathology. 2. Shiga Toxin--analysis. WC290 E11 2003]
   QR201.E82.E14 2003
   616.9'2--dc21

                                                                    2002068800

Preface

The study of the pathogenesis of Shiga toxin-producing *Escherichia coli* (STEC) infections encompasses many different disciplines, including clinical microbiology, diagnostics, animal ecology, and food safety, as well as the cellular microbiology of both bacterial pathogenesis and the mechanisms of toxin action. *E. coli: Shiga Toxin Methods and Protocols* aims to bring together a number of experts from each of these varied fields in order to outline some of the basic protocols for the diagnosis and study of STEC pathogenesis. We hope that our book will prove a valuable resource for the clinical microbiologist as well as the cellular microbiologist.

For the clinical microbiologist, our aim is to detail a number of current protocols for the detection of STEC in patient samples, each of which have their own advantages. Chapter 1 provides an introduction into the medical significance of STEC infections. Chapters 2–7 follow with protocols for the diagnosis and detection of STEC bacteria in patient and animal samples.

For the cellular microbiologist, we have brought together a number of experts from basic microbiologists to cell biologists to provide different protocols useful in studying the varied aspects of STEC pathogenesis. Chapters 8–13 concentrate on the cellular microbiology of STEC infections, describing protocols to study host–pathogen interactions as well as studies on the hemolysin of STEC. In Chapters 14–22, various protocols are described for studying the details of Shiga toxin (Stx) biology, from the purification of the toxin to studies of the effects of Stx on various host cell functions. Finally Chapters 23–25 provide detailed protocols for the study of STEC-mediated disease in various animal models.

The format of the chapters will be familiar to those who have used other volumes in the Methods in Molecular Medicine series. The Notes section at the end of each chapter pays particular attention to detailing the potential problems that may be encountered, as well as providing alternate methods for the protocols described.

Finally, we hope *E. coli: Shiga Toxin Methods and Protocols* will benefit those interested in both the clinical and pathological aspects of STEC infections, as well as provide a number of valuable protocols for those researchers studying host–pathogen interactions. We would like to thank the contributing authors as well as John Walker and the staff at Humana Press for their assistance in putting this volume together.

Dana Philpott
Frank Ebel

Contents

Preface v

Contributors ix

1 The Medical Significance of Shiga Toxin-Producing *Escherichia coli* Infections: *An Overview*
   Mohamed A. Karmali *1*

2 Methods for Detection of STEC in Humans: *An Overview*
   James C. Paton and Adrienne W. Paton *9*

3 Serological Methods for the Detection of STEC Infections
   Martin Bitzan and Helge Karch *27*

4 Detection and Characterization of STEC in Stool Samples Using PCR
   Adrienne W. Paton and James C. Paton *45*

5 Molecular Typing Methods for STEC
   Haruo Watanabe, Jun Terajima, Hidemasa Izumiya, and Sunao Iyoda *55*

6 STEC in the Food Chain: *Methods for Detection of STEC in Food Samples*
   Michael Bülte *67*

7 STEC as a Veterinary Problem: *Diagnostics and Prophylaxis in Animals*
   Lothar H. Wieler and Rolf Bauerfeind *75*

8 Cellular Microbiology of STEC Infections: *An Overview*
   Frank Ebel and Dana Philpott *91*

9 Analysis of Pathogenicity Islands of STEC
   Tobias A. Oelschlaeger, Ulrich Dobrindt, Britta Janke, Barbara Middendorf, Helge Karch, and Jörg Hacker *99*

10 Generation of Isogenic Deletion Mutants of STEC
   Soudabeh Djafari, Nadja D. Hauf, and Judith F. Tyczka *113*

11 Generation of Monoclonal Antibodies Against Secreted Proteins of STEC
   Kirsten Niebuhr and Frank Ebel *125*

12 Microscopic Methods to Study STEC: *Analysis of the Attaching
    and Effacing Process*
    Stuart Knutton *137*

13 Detection and Characterization of EHEC-Hemolysin
    Herbert Schmidt and Roland Benz *151*

14 Shiga Toxin Receptor Glycolipid Binding: *Pathology and Utility*
    Clifford A. Lingwood *165*

15 Methods for the Purification of Shiga Toxin 1
    **Anita Nutikka, Beth Binnington-Boyd,
        and Clifford A. Lingwood**.. *187*

16 Methods for the Identification of Host Receptors for Shiga Toxin
    **Anita Nutikka, Beth Binnington-Boyd,
        and Clifford A. Lingwood**.. *197*

17 Shiga Toxin B-Subunit as a Tool to Study Retrograde Transport
    Frédéric Mallard and Ludger Johannes......................... *209*

18 Measuring pH Within the Golgi Complex and Endoplasmic
    Reticulum Using Shiga Toxin
    Jae H. Kim.. *221*

19 Detection of Shiga Toxin-Mediated Programmed Cell Death
    and Delineation of Death-Signaling Pathways
    Nicola L. Jones *229*

20 Interaction of Shiga Toxin with Endothelial Cells
    Martin Bitzan and D. Maroeska W. M. te Loo *243*

21 Shiga Toxin Interactions with the Intestinal Epithelium
    **Cheleste M. Thorpe, Bryan P. Hurley,
        and David W. K. Acheson** *263*

22 Protocols to Study Effects of Shiga Toxin on Mononuclear
    Leukocytes
    Christian Menge *275*

23 Animal Models for STEC-Mediated Disease
    Angela R. Melton-Celsa and Alison D. O'Brien............. *291*

24 Gnotobiotic Piglets as an Animal Model for Oral Infection with O157
    and Non-O157 Serotypes of STEC
    **Florian Gunzer, Isabel Hennig-Pauka, Karl-Heinz Waldmann,
        and Michael Mengel** *307*

25 Bovine *Escherichia coli* O157:H7 Infection Model
    Evelyn A. Dean-Nystrom *329*

Index *339*

Contributors

DAVID W. K. ACHESON • *Department of Epidemiology and Preventive Medicine, University of Maryland, Baltimore, MD*

ROLF BAUERFEIND • *Institut für Hygiene und Infektionskrankheiten der Tiere, Justus-Liebig-Universität Giessen, Giessen, Germany*

ROLAND BENZ • *Lehrstuhl für Biotechnologie, Theodor-Boveri-Institut (Biozentrum) der Universität Würzburg, Würzburg, Germany*

BETH BINNINGTON-BOYD • *Departments of Laboratory Medicine and Pathobiology and Biochemistry, University of Toronto and The Research Institute, Hospital for Sick Children, Toronto, Canada*

MARTIN BITZAN • *Department of Pediatrics, Wake Forest University School of Medicine, Winston-Salem, NC*

MICHAEL BÜLTE • *Institut für Tierärztliche Nahrungsmittelkunde, Justus-Liebig-Universität Giessen, Giessen, Germany*

EVELYN A. DEAN-NYSTROM • *Pre-Harvest Food Safety and Enteric Disease Research Unit, National Animal Disease Center, Agriculture Research Service, US Department of Agriculture, Ames, IA*

SOUDABEH DJAFARI • *Institut für Medizinische Mikrobiologie, Justus-Liebig-Universität Giessen, Giessen, Germany*

ULRICH DOBRINDT • *Institut für Molekulare Infektionsbiologie, Universität Würzburg, Würzburg, Germany*

FRANK EBEL • *Bakteriologie, Max-von-Pettenkofer-Institut, Munich, Germany*

FLORIAN GUNZER • *Institut für Medizinische Mikrobiologie und Krankenhaus-hygiene, Medizinische Hochschule Hannover, Hannover, Germany*

JÖRG HACKER • *Institut für Molekulare Infektionsbiologie, Universität Würzburg, Würzburg, Germany*

NADJA D. HAUF • *Institut für Medizinische Mikrobiologie, Justus-Liebig-Universität Giessen, Giessen, Germany*

ISABEL HENNIG-PAUKA • *Klinik für kleine Klauentiere und forensische Medizin und Ambulatorische Klinik, Tierärztliche Hochschule Hannover, Hannover, Germany*

BRYAN P. HURLEY • *Department of Immunology, Tufts University School of Medicine, Boston, MA*

SUNAO IYODA • *Department of Bacteriology, National Institute of Infectious Diseases, Tokyo, Japan*

HIDEMASA IZUMIYA • *Department of Bacteriology, National Institute of Infectious Diseases, Tokyo, Japan*

BRITTA JANKE • *Institut für Molekulare Infektionsbiologie, Universität Würzburg, Würzburg, Germany*

LUDGER JOHANNES • *Laboratoire "Trafic et Signalisation–Toxines, Lipides et Vectorisation", Institut Curie, Unité Mixte de Recherche, CNRS, Paris, France*

NICOLA L. JONES • *Departments of Pediatrics and Physiology, Research Institute, The Hospital for Sick Children, University of Toronto, Toronto, Canada*

HELGE KARCH • *Institut für Hygiene, Münster, Germany*

MOHAMED A. KARMALI • *Laboratory for Foodborne Zoonoses, Health Canada, Guelph; and the Department of Pathology and Molecular Medicine, McMaster University, Hamilton, Ontario, Canada*

JAE H. KIM • *Department of Pediatrics, University of Toronto and The Research Institute, The Hospital for Sick Children, Toronto, Canada*

STUART KNUTTON • *Institute for Child Health, University of Birmingham, Birmingham, UK*

CLIFFORD A. LINGWOOD • *Departments of Laboratory Medicine and Pathobiology and Biochemistry, University of Toronto and The Research Institute, Hospital for Sick Children, Toronto, Canada*

FRÉDÉRIC MALLARD • *Laboratoire "Trafic et Signalisation–Toxines, Lipides et Vectorisation", Institut Curie, Unité Mixte de Recherche, CNRS, Paris, France*

ANGELA R. MELTON-CELSA • *Department of Microbiology and Immunology, Uniformed Services University of the Health Sciences, Bethesda, MD*

CHRISTIAN MENGE • *Institut für Hygiene und Infektionskrankheiten der Tiere, Justus-Liebig-Universität Giessen, Giessen, Germany*

MICHAEL MENGEL • *Institut für Pathologie, Medizinische Hochschule Hannover, Hannover, Germany*

BARBARA MIDDENDORF • *Institut für Molekulare Infektionsbiologie, Universität Würzburg, Würzburg, Germany*

KIRSTEN NIEBUHR • *Unité de Pathogenie Moléculaire Microbienne, Institut Pasteur, Paris, France*

ANITA NUTIKKA • *Departments of Laboratory Medicine and Pathobiology and Biochemistry, University of Toronto and The Research Institute, Hospital for Sick Children, Toronto, Canada*

ALISON D. O'BRIEN • *Department of Microbiology and Immunology, Uniformed Services University of the Health Sciences, Bethesda, MD*

TOBIAS A. OELSCHLAEGER • *Institut für Molekulare Infektionsbiologie, Universität Würzburg, Würzburg, Germany*

ADRIENNE W. PATON • *Department of Molecular Biosciences, University of Adelaide, Adelaide, Australia*

JAMES C. PATON • *Department of Molecular Biosciences, University of Adelaide, Adelaide, Australia*

DANA PHILPOTT • *Groupe d'Immunité Innée et Signalisation, Institut Pasteur, Paris, France*

HERBERT SCHMIDT • *Institut für Medizinische Mikrobiologie und Hygiene, Medizinische Fakultät Carl Gustav Carus, Technische Universität Dresden, Dresden, Germany*

D. MAROESKA W. M. TE LOO • *Department of Pediatrics, University Medical Centre St. Radboud, Nijmegen, The Netherlands*

JUN TERAJIMA • *Department of Bacteriology, National Institute of Infectious Diseases, Tokyo, Japan*

CHELESTE M. THORPE • *Division of Geographic Medicine and Infectious Disease, New England Medical Center, Boston, MA*

JUDITH F. TYCZKA • *Institut für Medizinische Mikrobiologie, Justus-Liebig-Universität Giessen, Giessen, Germany*

KARL-HEINZ WALDMANN • *Klinik für kleine Klauentiere und forensische Medizin und Ambulatorische Klinik, Tierärztliche Hochschule Hannover, Hannover, Germany*

HARUO WATANABE • *Department of Bacteriology, National Institute of Infectious Diseases, Tokyo, Japan*

LOTHAR H. WIELER • *Institut für Mikrobiologie und Tierseuchen, Freie Universität Berlin, Berlin, Germany*

1

The Medical Significance of Shiga Toxin-Producing *Escherichia coli* Infections

An Overview

Mohamed A. Karmali

1. Introduction

Shiga toxin (Stx)-producing *Escherichia coli* (STEC), also referred to as Verocytotoxin-producing *E. coli* (VTEC) *(1)*, are causes of a major, potentially fatal, zoonotic food-borne illness whose clinical spectrum includes nonspecific diarrhea, hemorrhagic colitis, and the hemolytic uremic syndrome (HUS) *(2–6)*. The occurrence of massive outbreaks of STEC infection, especially resulting from the most common serotype, O157:H7, and the risk of developing HUS, the leading cause of acute renal failure in children, make STEC infection a public health problem of serious concern *(2,5,7)*. Up to 40% of the patients with HUS develop long-term renal dysfunction and about 3–5% of patients die during the acute phase of the disease *(8–11)*. There is no specific treatment for HUS, and vaccines to prevent the disease are not yet available. The purpose of this overview is to highlight the public health impact, epidemiology, and clinicopathological features of STEC infection.

2. Public Health Impact and Epidemiology of STEC Infection

Shiga toxin-producing *E. coli* infection is usually acquired by the ingestion of contaminated food or water or by person-to-person transmission *(2,5,7)*. The natural reservoir of STEC is the intestinal tracts of domestic animals, particularly cattle and other ruminants. Sources for human infection include foods of animal origin such as meats (especially ground beef), and unpasteurized

From: *Methods in Molecular Medicine, vol. 73: E. coli: Shiga Toxin Methods and Protocols*
Edited by: D. Philpott and F. Ebel © Humana Press Inc., Totowa, NJ

milk, and other vehicles that have probably been cross-contaminated with STEC, such as fresh-pressed apple cider, yogurt, and vegetables such as lettuce, radish sprouts, alfalfa sprouts, and tomatoes *(2,5,7)*. Person-to-person transmission, facilitated by a low infectious dose, is common. Waterborne transmission and acquisition of infection in the rural setting and via contact with infected animals are becoming increasingly recognized. STEC infection occurs, typically, during the summer and fall and affects mostly young children, although the elderly also have an increased risk of infection *(2,5,7)*.

Although over 200 different OH serotypes of STEC have been associated with human illness *(5)*, the vast majority of reported outbreaks and sporadic cases in humans have been associated with serotype O157:H7 *(2,5,7)*. Other STEC serotypes that have been associated with outbreaks include O26:H11, O103:H2, O104:H21, O111:H⁻, and O145:H⁻. Outbreaks with cases of HUS have occurred almost exclusively with serotypes that exhibit the characteristic attaching and effacing (A/E) cytopathology, which is encoded for by the LEE (locus of enterocyte effacement) pathogenicity island *(2,5,7)*. However, sporadic cases of HUS have been associated with over 100 different LEE-positive and LEE-negative STEC serotypes *(5)*. In Latin America, non-O157 serotypes appear to be more commonly associated with human disease than serotype O157:H7 *(12)*.

Outbreaks of STEC infection, with some including hundreds of cases *(13–15)*, have been documented in at least 14 countries on 6 continents in a variety of settings, including households, day-care centers, schools, restaurants, nursing homes, social functions, prisons, and an isolated Arctic community *(2,16)*.

HUS, the most serious complication of STEC infection, has been reported to occur with a frequency of about 8% in several outbreaks of STEC O157:H7 infection *(2,16)*, although in one outbreak among elderly nursing home residents, it was as high as 22% *(17)*.

The frequency of sporadic HUS in North America is about 2–3 cases per 100,000 children under 5 yr of age *(2,16)*, in contrast to a roughly 10-fold higher incidence in this age group in Argentina *(12)*. In South Africa *(18)*, and in the United States *(19)*, HUS appears to be more common in white than in black children. In England, it is more common in rural than in urban areas *(10)*, and in Argentina, the syndrome occurs more commonly in upper-income than in lower-income groups *(20,21)*. The reasons for these differences between population groups are not known.

3. Clinicopathological Features and Pathophysiology of STEC Infection

After an incubation period of typically, 3–5 d, the characteristic features of STEC O157:H7 infection include a short period of abdominal cramps and nonbloody diarrhea, which may be followed, in many cases by hemorrhagic colitis, a condition distinct from inflammatory colitis that is characterized by the presence of frank hemorrhage in the stools. Fever and vomiting are not prominent features *(2,5,7)*. HUS, defined by the triad of features (acute renal failure, thrombocytopenia, and microangiopathic hemolytic anemia), develops in about one-tenth to one-quarter of the cases *(2,5,7)*. HUS may also be a complication of STEC-associated urinary tract infection *(22)*. The severity of HUS varies from an incomplete and/or a mild clinical picture to severe and fulminating disease with multiple organ involvement, including the bowel, heart, lungs, pancreas, and the central nervous system *(23)*.

The infectious dose of *E. coli* O157:H7 is very low (estimated to be less than 100 to a few hundred organisms). The organism is thought to colonize the large bowel with the characteristic A/E cytopathology mediated by components encoded by the LEE *(5)*. Pathological changes in the colon include hemorrhage and edema in the lamina propria, and colonic biopsy specimens may exhibit focal necrosis and leukocyte infiltration *(5,7)*. The pathogenesis of non-bloody diarrhea has yet to be fully elucidated.

Shiga toxin-producing *E. coli* elaborate at least four potent bacteriophage-mediated cytotoxins: Stx1 (VT 1), Stx2 (VT 2), Stx2c (VT2c), and Stx2d, which may be present alone or in combination. Stx1 is virtually identical to Shiga toxin of *Shiyolla dysenteriae*, but it is serologically distinct from the Stx2 group *(7,24)*. Among the most potent biological substances known, Stxs are toxic to cells at picomolar concentrations *(24)*.

The toxins share a polypeptide subunit structure consisting of an enzymatically active A subunit (approx 32 kDa) that is linked to a pentamer of B-subunits (approx 7.5 kDa) *(24)*. After binding to the glycolipid receptor, globotriaosylceramide (Gb3) *(25)*, on the eukaryotic cell, the toxins are internalized by receptor-mediated endocytosis and target the endoplasmic reticulum via the golgi by a process termed "retrograde transport" *(24,26)*.

The A-subunit, after it is proteolytically nicked to an enzymatically active A_1 fragment, cleaves the N-glycosidic bond at position A-4324 *(27)* of the 28S rRNA of the 60S ribosomal subunit. This blocks EF 1-dependent aminoacyl tRNA binding, resulting in the inhibition of protein synthesis *(24)*.

The development of HUS is thought to be related to the translocation of Stx into the bloodstream, although the precise mechanism for this is not known *(7)*. Histologically, HUS is characterized by widespread thrombotic microangiopathy in the renal glomeruli, gastrointestinal tract, and, other organs such as the brain, pancreas, and the lungs *(7,28,29,30)*. A characteristic swelling of glomerular capillary endothelial cells accompanied by widening of the subendothelial space is seen at the ultrastructural level, suggesting that endothelial cell damage is central to the pathogenesis of HUS *(31)*. This damage is probably mediated directly by Stx after binding to a specific receptor, globotriaosylceramide (Gb3) *(32)*, on the surface of the endothelial cell *(33)*. The toxin is internalized by a receptor-mediated endocytic process and is thought to cause cell damage by interaction with subcellular components, which result in the inhibition of protein synthesis *(24)*. Apoptosis may be another mechanism by which endothelial cells are damaged *(34)*. Although the endothelial cell appears to be the main target for Stx action, there is evidence that the toxins may also mediate biological effects by interacting with other cell types such as renal tubular cells, mesangial cells, and monocytes *(35–37)*. The blood levels of proinflammatory cytokines, especially tumor necrosis factor-α (TNF-α) and interleukin-1β (IL-1β), are elevated in HUS *(35–37)*. These cytokines have been shown, in vitro, to potentiate the action of Stx on endothelial cells by inducing expression of the receptor Gb3 *(35–37)*.

Although the injurious action of Stxs on endothelial cells appears to be crucial to the development of HUS, the precise cellular events that result in the associated pathophysiological changes, including thrombotic microangiopathy, hemolytic anemia, and thrombocytopenia, remain to be elucidated. The contributions of various host (age, immunity, receptor type and distribution, inflammatory response, and genetic factors) and parasite determinants (infectious dose, toxin types, and accessory virulence factors) to disease susceptibility and severity remain to be fully understood *(2,5,7)*. Sequencing of the genome of *E. coli* O157:H7 strain EDL 933 (in the laboratory of F. Blattner) and of its 92-kb plasmid (pO157) *(38,39)*, is expected to provide new insights into the pathogenesis of hemorrhagic colitis and the hemolytic uremic syndrome.

References

1. Konowalchuk, J., Speirs, J. I., and Stavric, S. (1977) Vero response to a cytotoxin of Escherichia coli. *Infect. Immunol.* **18,** 775–779.
2. Griffin, P. M. (1995) Escherichia coli O157:H7 and other enterohemorrhagic Escherichia coli, in *Infections of the Gastrointestinal Tract* (Blaser, M. J., Smith, P. D., Ravdin, J. I., Greenberg, H. B., and Guerrant, R. L., eds.), Raven, New York, pp. 739–761.
3. Karmali, M. A., Steele, B. T., Petric, M., and Lim, C. (1983) Sporadic cases of hemolytic uremic syndrome associated with fecal cytotoxin and cytotoxin-producing *Escherichia coli. Lancet* **1(8325),** 619–620.
4. Karmali, M. A., Petric, M., Lim, C., Fleming, P. C., Arbus, G. S., and Lior, H. (1985) The association between hemolytic uremic syndrome and infection by Verotoxin-producing *Escherichia coli. J. Infect. Dis.* **151,** 775–782.
5. Nataro, J. P. and Kaper, J. B. (1998) Diarrheagenic *Escherichia coli. Clin. Microbiol. Rev.* **11,** 142–201.
6. Riley, L. W., Remis, R. S., Helgerson, S. D., McGee, H. B., Wells, J. G., Davis, B. R., et al. (1983) Hemorrhagic colitis associated with a rare *Escherichia coli* serotype. *N. Engl. J. Med.* **308,** 681–685.

7. Karmali, M.A. (1989) Infection by Verocytotoxin-producing *Escherichia coli*. *Clin. Microbiol. Rev.* **2,** 15–38.

8. Fitzpatrick, M. M., Shah, V., Trompeter, R. S., Dillon, M. J., and Barratt, T. M. (1991) Long term renal outcome of childhood haemolytic uraemic syndrome. *Br. Med. J.* **303,** 489–492.

9. Loirat, C., Sonsino, E., Moreno, A. V., Pillion, G., Mercier, J. C., Beaufils, F., et al. (1984) Hemolytic uremic syndrome: an analysis of the natural history and prognostic features. *Acta Pediatr. Scand.* **73,** 505–514.

10. Trompeter, R.S., Schwartz, R., Chantler, C., Dillon, M.J., Haycock, G.B., Kay, R., and Barratt, T.M. (1983) Haemolytic uraemic syndrome: an analysis of prognostic features. *Arch. Dis. Child.* **58,** 101–105.

11. Siegler, R. L., Milligan, M. K., Burningham, T. H., Christofferson, R. D., Chang, S.-Y., and Jorde, L. B. (1991) Long-term outcome and prognostic indicators in the hemolytic uremic syndrome. *J. Pediatr.* **118,** 195–200.

12. Lopez, E. L., Contrini, M. M., and Rosa, M. F. D. (1998) Epidemiology of Shiga toxin-producing Escherichia coli in South America, in Escherichia coli *O157:H7 and Other Shiga Toxin-Producing* E. coli *Strains* (Kaper, J.B. and O'Brien, A.D., eds.), ASM, Washington, DC, pp. 30–37.

13. Ahmed, S. and Donaghy, M. (1998) An outbreak of Escherichia coli O157:H7 in central Scotland, in Escherichia coli *O157:H7 and Other Shiga Toxin-Producing* E. coli *Strains* (Kaper, J. B. and O'Brien, A. D., eds.), ASM, Washington, DC, pp. 59–65.

14. Centers for Disease Control and Prevention (1993) Update: multistate outbreak of *Escherichia coli* O157:H7 infections from Hamburgers—Western United States, 1992–1993. *JAMA* **269,** 2194–2196.

15. Michino, H., Araki, K., Minami, S., Nakayama, T., Ejima, Y., Hiroe, K., et al. (1998) Recent outbreaks of infections caused by *Escherichia coli* O157:H7 in Japan, in Escherichia coli *O157:H7 and Other Shiga Toxin-Producing* E. coli *Strains* (Kaper, J. B. and O'Brien, A. D., eds.), ASM, Washington, DC, pp. 73–81.

16. Griffin, P. W. and Tauxe, R. V. (1991) The epidemiology of infections caused by *Escherichia coli* O157:H7, other enterohemorrhagic *E. coli*, and the associated hemolytic uremic syndrome. *Epidemiol. Rev.* **13,** 60–98.

17. Carter, A. O., Borczyk, A. A., Carlson, J. A. K., Harvey, B., Hockin, J. C., Karmali, M. A., et al. (1987) A severe outbreak of *Escherichia coli* O157:H7-associated hemorrhagic colitis in a nursing home. *N. Engl. J. Med.* **317,** 1496–1500.

18. Kibel, M. A. and Barnard, P. J. (1968) The hemolytic uremic syndrome: a survey in Southern Africa. *S. Afr. Med. J.* **42,** 692–698.

19. Jernigan, S. M. and Waldo, F. B. (1994) Racial incidence of hemolytic uremic syndrome. *Pediatr. Nephrol.* **8,** 545–547.

20. Gianantonio, C., Vitacco, M., Mendilaharzu, F., Rutty, A., and Mendilaharzu, J. (1964) The hemolytic-uremic syndrome. *J. Pediatr.* **64,** 478–491.

21. Gianantonio, C., Vitacco, M., Mendilaharzu, F., Gallo, G. E., and Sojo, E. T. (1973) The hemolytic uremic syndrome. *Nephron* **11,** 174–192.

22. Tarr, P. I., Fouser, L. S., Stapleton, A. E., Wilson, R. A., Kim, H. H., Vary, J. C., et al. (1996) Hemolytic-uremic syndrome in a six-year old girl after a urinary tract

infection with Shiga-toxin-producing *Escherichia coli* O103:H2. *N. Engl. J. Med.* **335,** 635–660.

23. McLaine, P. N., Orrbine, E., and Rowe, P. C. (1992) Childhood hemolytic uremic syndrome, in *Hemolytic Uremic Syndrome and Thrombotic Thrombocytopenic Purpura* (Kaplan, B. S. and Moake, J. L., eds.), Marcel Dekker, New York, pp. 61–69.

24. O'Brien, A. D., Tesh, V. L., Donohue-Rolfe, A., Jackson, M. P., Olsnes, S., Sandvig, K., et al. (1992) Shiga toxin: biochemistry, genetics, mode of action and role in pathogenesis. *Curr. Topics Microbiol. Immunol.* **180,** 65–94.

25. Lingwood, C. A., Law, H., Richardson, S. E., Petric, M., Brunton, J. L., Grandis, S. D., et al. (1987) Glycolipid binding of natural and recombinant Escherichia coli produced Verotoxin in vitro. *J. Biol. Chem.* **262,** 8834–8839.

26. Sandvig, K., Prydz, K., Ryd, M., and Deurs, B. V. (1991) Endocytosis and intracellular transport of the glycolipid-binding ligand Shiga toxin in polarized MDCK cells. *J. Cell. Biol.* **113,** 553–562.

27. Endo, Y., Tsurugi, K., Yutsudo, T., Takeda, Y., Ogasawara, T., and Igarashi, K. (1988) Site of action of a Vero toxin (VT2) from *Escherichia coli* O157:H7 and of Shiga toxin on eukaryotic ribosomes; RNA N-glycosidase activity of the toxins. *Eur. J. Biochem.* **171,** 45–50.

28. Fong, J. S. C., de Chadarevian, J. P., and Kaplan, B. (1982) Hemolytic uremic syndrome. Curr. Concepts Manag. Pediatr. *Clin. North Am.* **29,** 835–856.

29. Richardson, S. E., Karmali, M. A., Becker, L. E., and Smith, C. R. (1988) The histopathology of the hemolytic uremic syndrome associated with Verocytotoxin-producing Escherichia coli infections. *Hum. Pathol.* **19,** 1102–1108.

30. Upadhyaya, K., Barwick, K., Fishaut, M., Kashgarian, M., and Segal, N. J. (1980) The importance of nonrenal involvement in hemolytic uremic syndrome. *Pediatrics* **65,** 115–120.

31. Vitsky, B. H., Suzuki, Y., Strauss, L., and Churg, J. (1969) The hemolytic uremic syndrome: a study of renal pathologic alternations. *Am. J. Pathol.* **57,** 627–647.

32. Lingwood, C. A., Mylvaganam, M., Arab, S., Khine, A. A., Magnusson, C., Grinstein, S., et al. (1998) Shiga toxin (Verotoxin) binding to its receptor glycolipid, in Escherichia coli *O157:H7 and Other Shiga Toxin-Producing* E. coli *Strains* (Kaper, J. B. and O'Brien, A. D., eds.), ASM, Washington, DC, pp. 129–139.

33. Obrig, T. (1998) Interactions of Shiga toxins with endothelial cells, in Escherichia coli *O157:H7 and Other Shiga Toxin-Producing* E. coli *Strains* (Kaper, J. B. and O'Brien, A. D., eds.), ASM, Washington, DC, pp. 303–311.

34. Inward, C. D., Williams, J., Chant, I., Crocker, J., Milford, D. V., Rose, P. E., et al. (1995) Verocytotoxin-1 induces apoptosis in Vero cells. *J. Infect.* **30,** 213–218.

35. v.d. Kar, N. C., Kooistra, T., Vermeer, M., Lesslauer, W., Monnens, L. A. H., and v. Hinsbergh, V. W. M. (1995) Tumor necrosis a induces endothelial galactosyl transferase activity and verocytotoxin receptors. Role of specific tumor necrosis factor receptors and protein kinase. C. *Blood* **85,** 734–743.

36. v.d. Kar, N. C., Sauerwein, R. W., Demacker, P. N., Grau, G. E., v. Hinsbergh, V. W., and Monnens, L. A. (1995) Plasma cytokine levels in hemolytic uremic syndrome. *Nephron* **71,** 309–313.

37. Monnens, L., Savage, C. O., and Taylor, C. M. (1998) Pathophysiology of hemolytic-uremic syndrome, in Escherichia coli *O157:H7 and Other Shiga Toxin-Producing* E. coli *Strains* (Kaper, J. B. and O'Brien, A. D., eds.), ASM, Washington, DC, pp. 287–292.

38. Burland, V., Shao, Y., Perna, N. T., Plunkett, G., Sofia, H. J., and Blattner, F. R. (1998) The complete DNA sequence and analysis of the large virulence plasmid of *Escherichia coli* O157:H7. *Nucleic Acids Res.* **26,** 4196–4204.

39. Karch, H., Schmidt, H., and Brunder, W. (1998) Plasmid-encoded determinants of *Escherichia coli* O157:H7, in Escherichia coli *O157:H7 and Other Shiga Toxin-Producing* E. coli *Strains* (Kaper, J. B. and O'Brien, A. D., eds.), ASM, Washington, DC, pp. 183–194.

2

Methods for Detection of STEC in Humans

An Overview

James C. Paton and Adrienne W. Paton

1. Introduction

Timely and accurate diagnosis of Shiga toxigenic *Escherichia coli* (STEC) disease in humans is extremely important from both a public health and a clinical management perspective. In the outbreak setting, rapid diagnosis of cases and immediate notification of health authorities is essential for effective epidemiological intervention. Early diagnosis also creates a window of opportunity for therapeutic intervention. Agents capable of adsorbing and neutralizing free Shiga toxin (Stx) in the gut lumen have been described *(1,2)*, and these are likely to be most effective when adminstered early in the course of disease, before serious systemic sequelae develop. Also, the clinical presentation of STEC disease can sometimes be confused with other bowel conditions; thus, early definitive diagnosis may prevent unnecessary invasive and expensive surgical and investigative procedures or administation of antibiotic therapy, which may be contraindicated *(3)*. However, detection of STEC is fraught with difficulty, particularly for strains belonging to serogroups other than O157. In the early stages of infection, there may be very high numbers of STEC in feces (the STEC may constitute >90% of aerobic flora), but as disease progresses, the numbers may drop dramatically. In cases of hemolytic uraemic syndrome (HUS), the typical clinical signs may become apparent as much as 2 wk after the onset of gastrointestinal symptoms, by which time the numbers of the causative STEC may be very low indeed. Also, in some cases, diarrhea is no longer present and only a rectal swab may be available at the time of admission to the

From: *Methods in Molecular Medicine, vol. 73: E. coli: Shiga Toxin Methods and Protocols*
Edited by: D. Philpott and F. Ebel © Humana Press Inc., Totowa, NJ

hospital, limiting the amount of specimen available for analysis. For these reasons, STEC detection methods need to be very sensitive and require minimal specimen volumes.

Shiga toxigenic *E. coli* diagnostic methods are based on the detection of the presence of either Stx or *stx* genes in fecal extracts or fecal cultures, and/or isolation of the STEC (or other Stx-producing organism) itself (reviewed in **refs. 4–7**). These procedures differ in complexity, speed, sensitivity, specificity, and cost, and so diagnostic strategies need to be tailored to the clinical circumstances and the resources available.

2. Detection of Stx

2.1. Tissue Culture Cytotoxicity Assays

Cytotoxicity for Vero (African green monkey kidney) cells remains the "gold standard" for the demonstration of the presence of Stx-related toxins in a fecal sample. Vero cells have a high concentration of Gb_3 receptors in their plasma membranes as well as Gb_4 (the preferred receptor for Stx2e) and thus are highly sensitive to all known Stx variants. In a typical assay, Vero monolayers (usually in 96-well trays) are treated with filter-sterilized fecal extracts or fecal culture filtrates and examined for cytopathic effect after 48 to 72 h incubation. Historically, this assay has played an important role in establishing a diagnosis of STEC infection, particularly where subsequent isolation of the causative organism has proven to be difficult (*4*). The sensitivity is influenced by the abundance of STEC in the fecal sample, as well as the total amount and potency of the Stx produced by the organism itself, and the degree to which the particular Stx is released from the bacterial cells. Karmali et al. (*8*) found that treating mixed fecal cultures with polymyxin B to release cell-associated Stx improved the sensitivity of the Vero cell assay, such that it could reliably detect STEC when present at a frequency of 1 CFU (Colony-forming unit) per 100. Clearly, some STEC produce very high levels of toxin and these can be detected at even lower frequencies; however, the converse also applies.

Although detection of Stx by tissue culture cytotoxicity is a valuable diagnostic method, it is labor intensive, time-consuming and cumbersome. Not all microbiological diagnostic laboratories are appropriately set up for tissue culture work, with Vero cell monolayers available on demand. Moreover, speed of diagnosis is important and the results of cytotoxicity assays are generally not available for 48–72 h. Also, the presence of cytoxicity in a crude filtrate could be the result of the effects of other bacterial products or toxins; thus, positive samples should always be confirmed (and typed) by testing for neutralization of cytotoxicity by specific (preferably monoclonal) antibodies to Stx1 or Stx2.

2.2. ELISA Assays for the Direct Detection of Stx

A number of enzyme-linked immunosorbent assays (ELISA) have been developed for direct detection of Stx1 and Stx2 in fecal cultures and extracts. Like Vero cytotoxicity, these have a potentially important role in diagnosis, because they are capable of detecting the presence of STEC (or other Stx-producing species) regardless of serogroup. However, ELISA assays are more rapid, permitting a result within 1 d. Most of the published ELISA methods involve a sandwich technique using immobilized monoclonal or affinity-purified polyclonal antibodies to the toxins as catching ligands. Purified Stx receptor (Gb$_3$) or hydatid cyst fluid (containing P$_1$ glycoprotein, which also binds Stx) have also been used to coat the solid phase. After incubation with cultures (or direct fecal extracts), bound toxin is detected using a second Stx-specific antibody followed by an appropriate anti-immunoglobulin–enzyme conjugate. Some assays employ a Stx detection antibody directly conjugated to the enzyme or a biotinylated detection antibody that is used with a streptavidin–enzyme conjugate (*5*).

Importantly, Stx ELISA assays are now commercially available in kit form (e.g., Premier EHEC from Meridian Diagnostics; LMD from LMD Laboratories, Carlsbad, CA). Reported specificities for both the in-house and commercial ELISA assays, determined by testing reference isolates and by comparing ELISA results for fecal extracts with culture and Vero cytotoxicity, have generally been very high. The sensitivity of the various ELISA assays is affected by a number of variables, including the avidity of the antibodies employed as well as the type and amount of Stx produced by a given strain. Early in-house ELISAs were generally less sensitive than the Vero cytotoxicity assay and sensitivity was inadequate to reliably detect low levels of Stx found in direct fecal extracts. However, the amount of free Stx present in primary fecal cultures is generally higher, particularly when broths are supplemented with polymyxin B and/or mitomycin C to enhance the production and release of Stx. Under such circumstances, ELISAs have been reported to be capable of detecting the presence of STEC comprising as little as 0.1% of total flora (*9,10*). Moreover, in two large studies, the Premier EHEC ELISA has been shown to be at least as sensitive as Vero cytotoxicity for detection of STEC in fecal culture extracts (*11,12*). Such assays will be of considerable utility for routine clinical laboratories without access to more specialized diagnostic procedures, particularly for detection of non-O157 STEC.

2.3. Reverse Passive Latex Agglutination

A reverse passive latex agglutination (RPLA) assay for detection of Stx production is also commercially available in kit form (VTEC-RPLA from Oxoid,

Unipath Limited, Basingstoke, UK; Verotox-F from Denka Seiken, Tokyo, Japan). This test involves incubation of serially diluted polymyxin B extracts of putative STEC cultures, or culture filtrates, with Stx1- and Stx2-specific antibody-coated latex particles and examining agglutination after 24 h. Beutin et al. *(13)* detected toxin production (of the appropriate type) in strains containing stx_1, stx_2, and stx_2c, but it did not detect toxin produced by the strains carrying stx_2e. However, a number of Stx2 and Stx2c producers gave positive reactions only when undiluted extracts were tested, which suggested that sensitivity might be inadequate for testing primary fecal culture extracts. More promising results have since been reported by Karmali et al. *(14)*, who demonstrated 100% sensitivity and specificity with respect to the Vero cytotoxicity assay when testing culture filtrates of reference STEC isolates, as did the previous study. However, analysis of dilutions of purified toxins demonstated that the end-point sensitivity of Verotox-F was comparable to Vero cytotoxicity. Although data on the performance of these assays using mixed fecal culture extracts are not yet available, it appears that they are simple, rapid, and accurate and may enable widespread screening for STEC by clinical laboratories.

3. Detection of *stx* Genes
3.1. Hybridization with DNA and Oligonucleotide Probes

Access to cloned stx_1 and stx_2 genes and their respective nucleotide sequences enabled the development of DNA and oligonucleotide probes for the detection of STEC (reviewed in **ref. 5**). The introduction of non-radioactive labels such as digoxigenin (DIG) and biotin has overcome many of the disadvantages associated with ^{32}P- or ^{35}S- labeled probes, which were used in earlier studies. Typically, these probes have been used for testing large numbers of fecal *E. coli* isolates, or for the direct screening of colonies on primary isolation plates for the presence of *stx* genes by colony hybridization *(15)*. These procedures are both sensitive and specific, and when stringent washing conditions are used, strains carrying stx_1, stx_2, or both can be differentiated. Although hybridization with DNA or oligonucleotide probes is not a particularly sensitive means for screening broth cultures or fecal extracts for the presence of STEC, it is a powerful tool for distinguishing colonies containing *stx* genes from commensal organisms, as discussed later.

3.2. Polymerase Chain Reaction

Access to sequence data for the various *stx* genes has permitted design of a variety of oligonucleotide primer sets for amplification of *stx* genes using polymerase chain reaction (PCR) (reviewed in **ref. 5**). Crude lysates or DNA extracts from single colonies, mixed broth cultures, colony sweeps, or even direct extracts of feces or foods can be used as templates for PCR. *Stx*-specific PCR products are usually detected by ethidium bromide staining after separation of the reaction mix by agarose gel electrophoresis. Some of the *stx* PCR assays described to date combine different primer pairs for *stx$_1$* and *stx$_2$*, and, in some cases, *stx$_2$* variants, in the same reaction. These direct the amplification of fragments which differ in size for each gene type *(16–19)*. Other *stx* PCR assays use single pairs of primers based on consensus sequences, which are capable of amplifying all *stx* genes, with subsequent identification of gene type requiring a second round of PCR, or hybridization with labeled oligonucleotides directed against type-specific sequences within the amplified fragment *(20,21)*. Apart from the added sensitivity, secondary hybridization steps act as independent confirmation of identity of the amplified product. Restriction analysis of amplified portions of *stx$_2$* genes has also been used to discriminate between *stx$_2$* and *stx$_2$* variants *(22–24)*.

Polymerase chain reaction technology is ideally suited to the detection of *stx* genes in microbiologically complex samples such as feces and foodstuffs, and it is potentially extremely sensitive. However, such samples may contain inhibitors of *Taq* polymerase, and sensitivity is often suboptimal when direct extracts are used as template. For both feces and food samples, the sensitivity of PCR assays is vastly increased if template DNA is extracted from broth cultures *(18,21)*. Broth enrichment (which can involve as little as 4 h incubation) serves two purposes. Inhibitors in the sample are diluted and bacterial growth increases the number of copies of the target sequence. Optimization of sensitivity is of paramount importance, because the numbers of STEC in the feces of patients with serious Stx-related diseases or in suspected contaminated foodstuffs may be very low indeed. Another consideration that may impact upon performance of some *stx*-specific PCR assays is the DNA sequence polymorphisms that are known to exist. This is particularly so for *stx$_2$*-related genes, for which significant variation has been reported (reviewed in **ref.** *5*). Sequence divergence between the primer and its target (particularly at the 3' end of the primer) will significantly reduce the efficiency of annealing with potentially dramatic effects on sensitivity of the PCR reaction. When selecting or designing primers, care must be taken to avoid regions where sequence heterogeneity has already been reported. PCR assays that use a single primer pair to amplify both *stx$_1$* and *stx$_2$* may be less susceptible to this potential complication. Target sequences that are conserved between otherwise widely divergent genes are likely to encode structurally important domains; thus, random mutations will be strongly selected against.

Speed of diagnosis of STEC infection is also an important consideration in the clinical setting. The precise time required for a PCR assay varies with the amplification protocol itself (number of cycles and incubation times at each temperature), the method used for DNA extraction, and the procedure for detection of the PCR products. The minimum time required for direct PCR analysis of an unenriched fecal sample analyzed by agarose gel electrophoresis could be as little as 4 h. Inclusion of a broth enrichment step and use of a more sophisticated DNA purification procedure would increase this time to 12–24 h, whereas hybridization of PCR products with *stx* probes could add a further day. The cumulative increase in sensitivity resulting from each additional step needs to be balanced against the increase in time, and this equation will vary in accordance with the particular clinical or epidemiological context.

It has often been argued that PCR is a technique that should be confined to reference laboratories, because it is labor intensive and requires highly skilled staff. However, an increasing number of clinical laboratories are now routinely using PCR for a range of applications. Unlike the Stx-specific antibodies and other specialized reagents needed for ELISA assays, custom-made oligonucleotide primers are inexpensive and universally available and have a very long shelf life. Modern versatile PCR thermal cyclers are no more expensive than ELISA plate readers and can handle assays in the 96-well format for laboratories that have a high specimen throughput. Moreover, a variety of alternatives to agarose gel electrophoresis have been developed for high-volume, sensitive, semiautomatable detection of PCR products (e.g., the TaqMan and AmpliSensor fluorogenic PCR assay systems) *(25,26)*.

3.3. PCR for Detection of Other STEC Markers

Polymerase chain reaction has also been used for the detection of genes encoding accessory virulence factors of STEC, such as *eae*, a component of the locus of enterocyte effacement (LEE) pathogenicity island, which encodes the capacity to form attaching/effacing lesions on enterocytes, and EHEC-*hlyA*, which encodes an enterohemolysin and is located on a large (approx 60 MDa) plasmid present in may STEC strains *(27,28)*. This information may be of clinical significance, because there appears to be a link between the presence of these genes and the capacity of an STEC isolate to cause serious human disease *(29,30)*. PCR assays exploiting sequence variation in the 3' portion of the *eae* gene have been used as a basis for distinguishing O157 STEC strains from certain other common serogroups *(27,31)*. However, availability of sequence data for the genetic loci (*rfb* regions) encoding O-antigen biosynthesis in *E. coli* serogroups such as O157, O111, and O113 *(32,33)* have enabled development of more reliable serogroup-specific PCR assays. Two other genetic markers associated with O157:H7 STEC strains have also been used as the basis of PCR assays. These are the *fliCh7* gene, which encodes the H7 antigen *(34)*, and a single base mutation in the *uidA* gene (detected by mismatch amplification mutation assay), which is responsible for the β-glucoronidase-negative phenotype of O157:H7 strains *(35)*.

Polymerase chain reaction primers specific for the various STEC markers referred to here are typically deployed as components of multiplex PCR assays, which also detect *stx* genes, enabling simultaneous detection and partial genetic characterization of STEC in a sample. However, the increased complexity of these assays renders them less suitable for routine, high-volume screening of fecal samples or foods. In our laboratory, we have adopted a two-tiered approach in which fecal culture extracts are initially screened using a *stx* PCR assay yielding a single PCR product for all *stx* types *(21)*. Any positive samples are then subjected to multiplex PCR analysis using two primer sets. The first utilizes four primer pairs and detects the presence of stx_1, stx_2 (including variants of stx_2), *eae*, and EHEC-*hlyA* *(32)*. The second assay uses three primer pairs directed against serogroup-specific sequences in the *rfb* regions of *E. coli* O157, O111, and O113 *(33)*. These two multiplex assays provide independent confirmation of the initial *stx* screening assay, as well as information on the serogroup and virulence traits of the STEC strain or strains present in a sample. Details of these assays are provided in a later chapter in this volume.

4. Isolation of STEC

Although a substantial amount of information on the causative STEC can be obtained by molecular analysis of mixed cultures, isolation of the STEC itself must be considered as the definitive diagnostic procedure. Apart from confirming the molecular data, isolation permits additional characterization of the STEC by a variety of methods, including O:H serotyping, phage typing, restriction fragment length polymorphism (RFLP), pulsed-field gel electrophoresis (PFGE), amplification-based DNA typing, and so forth. Although such characterization may have limited clinical application, it is of great importance from an epidemiological point of view, particularly in the outbreak setting, and methods for this are presented in a later chapter in this volume.

4.1. Culture for O157 STEC

Culture on sorbitol–MacConkey agar (SMAC) has been the most commonly used method for isolation of O157 STEC. This is because unlike the majority of fecal *E. coli* strains, most O157:H7 and O157:H⁻ STEC, which are the most common causes of human STEC disease in many parts of the world, are unable to ferment sorbitol *(36)*. SMAC plates are inoculated with the fecal specimen and examined after 18–24 h incubation for the presence of colorless, sorbitol-negative colonies. Individual colonies can then be tested by slide or tube agglutination with (commercially available) O157- and H7-specific antisera or latex reagents. It should, of course, be remembered that not all O157 *E. coli* produce Stx, thus toxigenicity needs to be confirmed by tissue culture, ELISA, or RPLA, as discussed earlier.

The sensitivity of SMAC is limited by the capacity to recognize nonfermenting colonies against the background of other organisms on the plate, and this is difficult when the O157 strain comprises less than 1% of the flora. Isolation rates can be improved by incorporation of cefixime to inhibit *Proteus* sp. and rhamnose, which is fermented by most sorbitol-negative non-O157 *E. coli* (O157 strains generally do not ferment rhamnose) *(37)*, or cefixime and potassium tellurite (CT-SMAC) *(38)*. Although screening fecal cultures on SMAC and its variants is inexpensive and involves minimal labor and equipment, it will principally detect STEC belonging to serogroup O157. Serious STEC disease has been associated with many other serogroups, and although some of these can also be sorbitol-negative, the majority are sorbitol-positive *(4)*. Furthermore, Stx2-producing, sorbitol-positive *E. coli* O157:H⁻ isolates have been associated with cases of HUS in Germany and the Czeck Republic *(39,40)*. These strains were also very sensitive to tellurite, which mitigates against the use of CT-SMAC for isolation of STEC in these regions.

E. coli O157:H7 can also be distinguished from other *E. coli* strains by failure to produce β-D-glucuronidase *(41)*, an enzyme that can be readily detected fluorigenically using the substrate 4-methylumbelliferyl-β-D-glucuronide or colorimetrically on plates supplemented with 5-bromo-6-chloro-3-indolyl-β-D-glucuronide *(42)*. Again, this criterion is not useful for the detection of non-O157 STEC or the sorbitol-positive O157 STEC isolates referred to earlier, as these are generally glucuronidase-positive.

Various specialized commercial agar media for isolation of O157 STEC are now available. Rainbow Agar O157 (Biolog Inc., Hayward, CA), for example, contains selective agents for *E. coli* and chromogenic substrates for β-D-glucuronidase and β-galactosidase. Glucuronidase-negative, galactosidase-positive O157 strains appear as black or gray colonies on this medium, whereas commensal *E. coli* strains are pink. It has also been claimed that some non-O157 STEC strains overproduce β-galactosidase relative to β-D-glucuronidase on this medium, giving the colonies a distinctive intermediate color. To date, analyses of the efficacy of this medium for detection of either O157 or non-O157 STEC in fecal samples are limited, but at least one study has shown that Rainbow Agar O157 is clearly superior to SMAC *(43)*. CHROMagar O157 (Becton Dickinson Microbiology Systems) also distinguishes O157 on the basis color; O157 colonies are mauve, and other bacteria are either blue or colorless. For both of these media, the manufacturers suggest incorporation of additional selective agents (novobiocin and tellurite, respectively) to improve isolation rates. Again, it should be emphasized that isolation of a putative O157 strain from either of these chromogenic selective media is not a definitive diagnosis in itself, and as for SMAC, isolates must be tested to confirm Stx production.

4.1.1. Direct Detection of O157 Antigen in Fecal Samples

Direct immunofluorescent staining of fecal specimens using polyclonal anti-O157-FITC is a potential alternative to SMAC for detection of *E. coli* O157 involving only about a 2-h turn-around time *(44)*. Commercial ELISAs for rapid (less than 1 h) detection of the presence of O157 antigen in fecal specimens are also available (LMD from LMD Laboratories, Carlsbad, CA; Premier *E. coli* O157 from Meridian Diagnostics, Inc., Cincinnati, OH). Both the immunofluorescence and ELISA tests have similar or superior sensitivity to SMAC *(12,44,45)*, and, importantly, are capable of detecting sorbitol-fermenting O157 STEC, should they be present. A number of other O157 immunoassay detection kits are commercially available (e.g., Ampcor *E. coli* O157:H7 [Ampcor]; Tecra *E. coli* O157 [Tecra]; EHEC-TEK [Organon Teknika]), but data on their utility for detection of *E. coli* O157 in human fecal cultures or extracts are not available. Again, all of these assays require confirmation either by culture or by demonstration of Stx in the sample.

4.2. Culturing for Non-O157 STEC

The high dependence of most clinical laboratories on SMAC culture for screening fecal samples from patients with suspected STEC infection has undoubtedly led to an over-estimation of the relative importance of O157 STEC as a cause of human disease. However, it has been known for many years that *E. coli* strains belonging to a large range of serotypes as well as certain strains of other bacterial species are capable of producing Stx and causing serious disease in humans *(4)*. Regrettably, there is no definitive biochemical characteristic that distinguishes STEC belonging to serogroups other than O157 from commensal fecal *E. coli* strains, a fact that significantly complicates isolation of such organisms. However, nearly all O157 STEC, as well as a significant proportion of non-O157 STEC strains, produce the plasmid-encoded enterohemolysin EHEC-Hly. Such strains are not hemolytic on standard blood agar, but produce small, turbid hemolytic zones on washed sheep erythrocyte agar (supplemented with Ca^{2+}) after 18–24 h incubation at 37°C. Production of EHEC-Hly is highly indicative that a given isolate is an STEC, but the predictive value of a negative result is low *(30,46)*. As a consequence, hemolytic phenotype on washed sheep erythrocyte agar is a useful means of identifying colonies for further analysis, but nonhemolytic colonies should also be tested.

The only comprehensive means of isolating STEC or other Stx-producing organisms involves direct analysis of colonies on nonselective agar plates using either *stx*-specific nucleic acid probes or antibodies to Stx, and a variety of protocols for this purpose have been described (reviewed in **ref. 5**). This is a labor-intensive process and can only be justified for specimens that have tested positive in screens for Stx (by cytotoxicity or ELISA) or for *stx* (by PCR). Colonies from agar plates can be directly blotted onto a suitable membrane (e.g., nitrocellulose or polyvinylidene difluoride [PVDF] for immunoblots, or positively charged nylon for hybridization). A carefully aligned replicate of the filter must be made and then it can be processed and reacted with antibody or nucleic acid probe by standard procedures. Theoretically, up to several hundred discrete colonies can be tested on a single filter, although this may require dilution and replating of primary cultures. Alternatively, colonies from primary isolation plates can be picked off and inoculated into 96-well microtiter trays containing broth. This is a time-consuming step (15–20 min per tray), but the 96-well format enables the subsequent use of semiautomated machinery to make replicate copies of trays and, after incubation, to transfer aliquots onto appropriate filters; the trays are also convenient for preservation of the isolates at –80°C. Comparisons of the sensitivity and specificity of immunoblotting and DNA probing for the detection of STEC colonies indicate that the latter is probably a more reliable method. Immunoblot techniques have the further disadvantage of having to grow colonies on special media in order to optimize production and/or release of Stx *(47)*.

4.3. Immunomagnetic Separation for Isolation of STEC

Immunomagnetic separation (IMS) is a potentially powerful enrichment technique for the isolation of STEC from low-abundance specimens. The procedure involves coating magnetic beads with anti-LPS (lipopolysaccharide) and mixing these with broth cultures or suspensions of feces or suspect food homogenates. Beads and bound bacteria are then trapped in a magnetic field, the unbound suspension is decanted, and the beads are washed. After additional binding/washing cycles, the beads are plated and resultant colonies tested for reactivity with the appropriate O-antiserum and more importantly for Stx production. The principal drawback of IMS is, of course, its serogroup specificity, and, at present, only O157-specific magnetic beads are available commercially (Dynabeads anti-*E. coli* O157 from Dynal, Oslo; Captivate O157 from Lab M). Notwithstanding this, it is an extremely valuable enrichment technique in circumstances where deliberate and exclusive targeting of this serogroup is justifiable (e.g., analysis of samples epidemiologically linked to proven cases of O157 STEC disease). Several studies have shown that IMS enrichment using the commercial O157-specific beads prior to plating on CT-SMAC significantly increases the isolation rate of *E. coli* O157 from fecal samples *(48,49)*. Also, during the investigation of an outbreak of HUS caused by an O111:H– STEC strain, enrichment using an in-house O111-specific IMS reagent enabled isolation of O111 STEC from a suspected food source after direct plating and colony hybridization had yielded negative results *(50)*.

5. Serological Diagnosis of STEC Infection

Diagnosis of STEC-related disease can be particularly problematic when patients present late in the course of disease, because the numbers of STEC in feces may be extremely low and hence undetectable even by PCR analysis of enrichment broths. However, STEC infection often elicits humoral antibody responses to a range of bacterial products, which may permit elucidation of the etiology of infection by serological means, as discussed in a subsequent chapter in this volume.

Several previous studies have examined immune responses of patients with STEC disease to Stx1, Stx2, and LPS (reviewed in **ref.** *5*) and, more recently, to products of the LEE locus such as intimin, Tir, EspA and EspB *(51)*. Theoretically, Stx should be the preferred target because all STEC, by definition, produce Stx1 and/or an Stx2-related toxin. However, previous studies have shown that only a minority of patients with proven STEC disease mount detectable serum antibody responses to the respective toxin type, as judged by either ELISA, cytotoxicity neutralization, or Western blotting *(52–55)*. Moreover, an appreciable proportion of healthy individuals may have detectable serum antibodies to Stx1, particularly in rural populations *(54)*. This would complicate interpretation of results obtained using single serum specimens unless geographically- and age-matched baseline data for the healthy population were available. Ideally, acute and convalescent sera should be tested for rising or falling antibody titres.

More encouraging results have been obtained by testing for antibodies to LPS, although this diagnostic approach suffers from the disadvantage of being able to target only specified serogroups. Not surprisingly, the majority of these studies have focused on serodiagnosis of O157 STEC infections. A high proportion of patients infected with this STEC serogroup have elevated acute-phase serum antibody levels to O157 LPS, as measured by ELISA or passive hemagglutination assay, and the background seropositivity rate in healthy controls is generally low *(52,56–60)*. In several of these studies, anti-LPS titers fell rapidly during the immediate post-acute phase, and so elevated titers in a single specimen may, indeed, be a reliable indicator of current or very recent infection. Clinical laboratory testing, at least for O157 antibodies, is also facilitated by the availability of a commercial latex agglutination test kit, which has been shown to be both sensitive and specific *(61)*. Although data on serological responses to infections caused by other STEC-associated serogroups are more limited, such analyses have been shown to be helpful in determing the etiology in a number of sporadic cases of HUS *(62,63)* and in the investigation of at least three outbreaks of HUS caused by non-O157 STEC strains *(50,64,65)*.

Diagnosis of STEC infection on the basis of serological responses to LEE-encoded proteins has also been advocated. This has the advantage of targeting a wider range of STEC types, although not all strains associated with serious human disease are LEE-positive. Antibody responses to intimin (the *eae* gene product) were more frequent among HUS patients than responses to other LEE proteins, but the frequency of intimin seroconversion was lower than for O157 LPS *(51)*. It should also be remembered that other enterobacterial pathogens, including enteropathogenic *E. coli*, are LEE-positive and so would be expected to elicit anti-intimin responses in humans. Problems of interpretation may also arise with anti-LPS responses, as even for O157, the association between Stx-production and serogroup is not absolute, and for all serogroups, highly purified LPS antigens are required to minimize crossreactions. Thus, caution should be exercised when interpreting serological data, particularly in the absence of coroborating evidence (e.g., Stx production or the presence of *stx* genes in fecal cultures).

6. Strategies for STEC Detection

Selection of the most appropriate methodology for detection of STEC will involve striking a balance among speed, specificity, sensitivity, and cost of the alternatives. Ideally, clinical microbiology laboratories should screen all fecal samples from patients with acute diarrhea (not just those that are bloody) for the presence of STEC, using methods which are not serogroup restricted. PCR analysis of primary fecal cultures is probably the most sensitive and specific means of screening for the presence of STEC. However, for those laboratories that lack PCR capability, direct screening of fecal cultures for the presence of Stx using one of the commercially available ELISA (or possibly RPLA) kits is recommended. Verocytotoxicity, although slower, is a highly satisfactory alternative. Methods targeted specifically at O157 STEC (e.g., CT-SMAC culture, O157 antigen detection, etc.) are suboptimal stand-alone primary screens, but if comprehensive screening is not possible, it is better to use these methods than not to screen at all. It would be prudent, however, for such laboratories to refer negative specimens from cases of severe bloody diarrhea or suspected HUS to a reference laboratory.

All samples and cultures that test positive after screening should be sent to a reference laboratory for confirmation and attempted isolation of STEC if adequate resources are not available on site. Given the widespread instability of *stx* genes during subculture *(66)*, it is important that initial samples and primary cultures are referred in addition to putative STEC isolates. It is at the isolation stage where the specialized plate media referred to earlier may save time by directing attention to suspect colonies, particularly where they are in low abundance. However, if using such media rather than nonselective plates, it is essential to test a range of colony types and not just those with the STEC-associated phenotype. Given the sensitivity of PCR screens, a proportion of genuine STEC-positive specimens may not yield an isolate even after heroic attempts. It may still be possible to obtain meaningful additional information about the causative organism in such circumstances. PCR analysis will indicate toxin type and whether virulence-related genes, or genes associated with important serogroups are also present in the sample. However, the interpretation of this information is complicated by the possibility that the composite genotypic profile may represent the sum of genotypes of more than one STEC organism. At least in cases of HUS, information on the likely infecting serogroup can also be obtained by serological tests for anti-LPS.

References

1. Armstrong, G. D., Rowe, P. C., Goodyer, P., Orrbine, E., Klassen, T. P., Wells, G., et al. (1995) A phase I study of chemically synthesized verotoxin (Shiga-like toxin) Pk-trisaccharide receptors attached to chromosorb for preventing hemolytic-uremic syndrome. *J. Infect. Dis.* **171,** 1042–1045.
2. Paton, A. W., Morona, R., and Paton, J. C. (2000) A new biological agent for treatment of Shiga toxigenic *Escherichia coli* infections and dysentery in humans. *Nat. Med.* **6,** 265–270.
3. Tarr, P. I. (1995) *Escherichia coli* O157:H7: clinical, diagnostic, and epidemiological aspects of human infection. *Clin. Infect. Dis.* **20,** 1–8.
4. Karmali, M. A. (1989) Infection by verocytotoxin-producing *Escherichia coli.* *Clin. Microbiol. Rev.* **2,** 15–38.
5. Paton, J. C. and Paton, A. W. (1998) Pathogenesis and diagnosis of Shiga toxin-producing *Escherichia coli* infections. *Clin. Microbiol. Rev.* **11,** 450–479.
6. Strockbine, N. A., Wells, J. G., Bopp, C. A., and Barrett, T.J. (1998) Overview of detection and subtyping methods, in Escherichia coli *O157:H7 and Other Shiga Toxin-Producing* E. coli *Strains* (Kaper, J. B. and O'Brien, A. D. eds.), American Society for Microbiology, Washington, DC, pp. 331–356.
7. Karch, H., Bielaszewska, M., Bitzan, M., and Schmidt, H. (1999) Epidemiology and diagnosis of Shiga toxin-producing *Escherichia coli* infections. *Diagn. Microbiol. Infect. Dis.* **34,** 229–243.
8. Karmali, M. A., Petric, M., Lim, C., Cheung, R., and Arbus, G. S. (1985) Sensitive method for detecting low numbers of verotoxin-producing *Escherichia coli* in mixed cultures by use of colony sweeps and polymyxin extraction of verotoxin. *J. Clin. Microbiol.* **22,** 614–619.
9. Ashkenazi, S. and Cleary, T. G. (1990) A method for detecting Shiga toxin and Shiga-like toxin-I in pure and mixed culture. *J. Med. Microbiol.* **32,** 255–261.
10. Law, D., Ganguli, L. A., Donohue-Rolfe, A., and Acheson, D. W. (1992) Detection by ELISA of low numbers of Shiga-like toxin-producing *Escherichia coli* in mixed cultures after growth in the presence of mitomycin C. *J. Med. Microbiol.* **36,** 198–202.
11. Kehl, K. S., Havens, P., Behnke, C. E., and Acheson, D. W. K. (1997) Evaluation of the Premier EHEC assay for detection of Shiga toxin-producing *Escherichia coli. J. Clin. Microbiol.* **35,** 2051–2054.

12. Mackenzie, A. M. R., Lebel, P., Orrbine, E., Rowe, P. C., Hyde, L., Chan, F., et al., and the Synsorb PK Study Investigators (1998) Sensitivities and specificities of Premier *E. coli* O157 and Premier EHEC enzyme immunoassays for diagnosis of infection with verotoxin (Shiga-like toxin)-producing *Escherichia coli. J. Clin. Microbiol.* **36,** 1608–1611.

13. Beutin, L., Zimmermann, S., and Gleier, K. (1996) Rapid detection and isolation of Shiga-like toxin (Verocytotoxin)-producing *Escherichia coli* by direct testing of individual enterohemolytic colonies from washed sheep blood agar plates in the VTEC-RPLA assay. *J. Clin. Microbiol.* **34,** 2812–2814.

14. Karmali, M. A., Petric, M., and Bielaszewska, M. (1999) Evaluation of a microplate latex agglutination method (Verotox-F assay) for detecting and characterizing verotoxins (Shiga toxins) in *Escherichia coli. J. Clin. Microbiol.* **37,** 396–399.

15. Thomas, A., Smith, H. R., Willshaw, G. A., and Rowe, B. (1991) Non-radioactively labelled polynucleotide and oligonucleotide DNA probes, for selectively detecting *Escherichia coli* strains producing Vero cytotoxins VT1, VT2 and VT2 variant. *Mol. Cell. Probes* **5,** 129–135.

16. Begum, D., Strockbine, N. A., Sowers, E. G., and Jackson, M. P. (1993) Evaluation of a technique for identification of Shiga-like toxin-producing *Escherichia coli* by using polymerase chain reaction and digoxigenin-labeled probes. *J. Clin. Microbiol.* **31,** 3153–3156.

17. Brian, M. J., Frosolono, M., Murray, B. E., Miranda, A., Lopez, E. L., Gomez, H. F., and Cleary, T. G. (1992) Polymerase chain reaction for diagnosis of enterohemorrhagic *Escherichia coli* infection and hemolytic-uremic syndrome. *J. Clin. Microbiol.* **30,** 1801–1806.

18. Gannon, V. P. J., King, R. K., Kim, J. Y., and Thomas, E. J. (1992) Rapid and sensitive method for detection of Shiga-like toxin-producing *Escherichia coli* in ground beef using the polymerase chain reaction. *Appl. Environ. Microbiol.* **58,** 3809–3815.

19. Pollard, D. R., Johnson, W. M., Lior, H., Tyler, S. D., and Rozee, K. R. (1990) Rapid and specific detection of verotoxin genes in *Escherichia coli* by the polymerase chain reaction. *J. Clin. Microbiol.* **28,** 540–545.

20. Karch, H. and Meyer, T. (1989) Single primer pair for amplifying segments of distinct Shiga-like toxin genes by polymerase chain reaction. *J. Clin. Microbiol.* **27,** 2751–2757.

21. Paton, A. W., Paton, J. C., Goldwater, P. N., and Manning, P. A. (1993) Direct detection of *Escherichia coli* Shiga-like toxin genes in primary fecal cultures by polymerase chain reaction. *J. Clin. Microbiol.* **31,** 3063–3067.

22. Lin, Z., Kurazono, H., Yamasaki, S., and Takeda, Y. (1993) Detection of various variant verotoxin genes in *Escherichia coli* by polymerase chain reaction. *Microbiol. Immunol.* **37,** 543–548.

23. Russmann, H., Schmidt, H., Heesemann, J., Caprioli, A., and Karch, H. (1994) Variants of Shiga-like toxin II constitute a major toxin component in *Escherichia coli* O157 strains from patients with haemolytic uraemic syndrome. *J. Med. Microbiol.* **40,** 338–343.

24. Tyler, S. D., Johnson, W. M., Lior, H., Wang, G., and Rozee, K. R. (1991) Identification of verotoxin type 2 variant B subunit genes in *Escherichia coli* by the polymerase chain reaction and restriction fragment length polymorphism analysis. *J. Clin. Microbiol.* **29,** 1339–1343.

25. Chen, S., Xu, R., Yee, A., Wu, K. Y., Wang, C. N., Read, S., et al. (1998) An automated fluorescent PCR method for detection of shiga toxin-producing *Escherichia coli* in foods. *Appl. Environ. Microbiol.* **64,** 4210–4216.

26. Sharma, V. K., Dean-Nystrom E. A., and Casey, T. A. (1999) Semi-automated fluorogenic PCR assays (TaqMan) for rapid detection of *Escherichia coli* O157:H7 and other Shiga toxigenic *E. coli. Mol. Cell. Probes* **13,** 291–302.

27. Gannon, V.P.J., Rashed, M., King, R.K., and Thomas, E.J.G. (1993) Detection and characterization of the eae gene of Shiga-like toxin-producing *Escherichia coli* using polymerase chain reaction. *J. Clin. Microbiol.* **31,** 1268–1274.

28. Schmidt, H., Beutin, L., and Karch, H. (1995) Molecular analysis of the plasmid-encoded hemolysin of *Escherichia coli* O157:H7 strain EDL 933. *Infect. Immun.* **63,** 1055–1061.

29. Barrett, T. J., Kaper, J. B., Jerse, A. E., and Wachsmuth, I. K. (1992) Virulence factors in Shiga-like toxin-producing *Escherichia coli* isolated from humans and cattle. *J. Infect. Dis.* **165,** 979–980.

30. Schmidt, H. and Karch, H. (1996) Enterohemolytic phenotypes and genotypes of Shiga toxin-producing *Escherichia coli* O111 strains from patients with diarrhea and hemolytic-uremic syndrome. *J. Clin. Microbiol.* **34,** 2364–2367.

31. Louie, M., De-Azavedo, J., Clarke, R., Borczyk, A., Lior, H., Richter, M., et al. (1994) Sequence heterogeneity of the eae gene and detection of verotoxin-producing Escherichia coli using serotype-specific primers. *Epidemiol. Infect.* **112,** 449–461.

32. Paton, A. W. and Paton, J. C. (1998) Detection and characterization of Shiga toxigenic Escherichia coli using multiplex PCR assays for *stx1, stx2, eaeA,* Enterohemorrhagic *E. coli hlyA, rfbO111* and *rfbO157. J. Clin. Microbiol.* **36,** 598–602.

33. Paton, A. W. and Paton, J. C. (1999) Direct detection of Shiga toxigenic *Escherichia coli* strains belonging to serogroups O111, O157, and O113 by multiplex PCR. *J. Clin. Microbiol.* **37,** 3362–3365.

34. Gannon, V. P. J., D'Souza, S., Graham, T., King, R. K., Rahn, K., and Read, S. (1997) Use of the flagellar H7 gene as a target in multiplex PCR assays and improved specificity and identification of enterohemorrhagic *Escherichia coli* strains. *J. Clin. Microbiol.* **35,** 656–662.

35. Cebula, T. A., Payne, W. L., and Feng, P. (1995) Simultaneous identification of strains of Escherichia coli serotype O157:H7 and their Shiga-like toxin type by mismatch amplification mutation assay-multiplex PCR. *J. Clin. Microbiol.* **33,** 248–250. [Published erratum appears in *J. Clin. Microbiol.* **33,** 1048.]

36. March, S. B. and Ratnam, S. (1986) Sorbitol-MacConkey medium for detection of *Escherichia coli* O157:H7 associated with hemorrhagic colitis. *J. Clin. Microbiol.* **23,** 869–872.

37. Chapman, P. A., Siddons, C. A., Zadik, P. M., and Jewes, L. (1991) An improved selective medium for the isolation of *Escherichia coli* O157. *J. Med. Microbiol.* **35**, 107–110.

38. Zadik, P. M., Chapman, P. A., and Siddons, C. A. (1993) Use of tellurite for the selection of verocytotoxigenic *Escherichia coli* O157. *J. Med. Microbiol.* **39**, 155–158.

39. Gunzer, F., Bohm, H., Russmann, H., Bitzan, M., Aleksic, S., and Karch, H. (1992) Molecular detection of sorbitol-fermenting *Escherichia coli* O157 in patients with hemolytic-uremic syndrome. *J. Clin. Microbiol.* **30**, 1807–1810.

40. Bielaszewska, M., Janda, J., Blahova, K., Karch, H., Karmali, M. A., Preston, M. A., et al. (1997) Isolation of sorbitol-fermenting (SF) verocytotoxin-producing *Escherichia coli* O157:H- in the Czech Republic. 13rd International Symposium and Workshop on Shiga Toxin (Verotoxin)-producing *Escherichia coli* Infections, Baltimore, MD.

41. Ratnam, S., March, S. B., Ahmed, R., Bezanson, G. S., and Kasatiya, S. (1988) Characterization of *Escherichia coli* serotype O157:H7. *J. Clin. Microbiol.* **26**, 2006–2012.

42. Thompson, J. S., Hodge, D. S., and Borczyk, A. A. (1990) Rapid biochemical test to identify verocytotoxin-positive strains of *Escherichia coli* serotype O157. *J. Clin. Microbiol.* **28**, 2165–2168.

43. Novicki, T.J., Daly, J.A., Mottice, S.L., and Carroll, K.C. (2000) Comparison of sorbitol MacConkey agar and a two-step method which utilizes enzyme-linked immunosorbent assay toxin testing and a chromogenic agar to detect and isolate enterohemorrhagic *Escherichia coli*. *J. Clin. Microbiol.* **38**, 547–551.

44. Park, C.H., Hixon, D.L., Morrison, W.L., and Cook, C.B. (1994) Rapid diagnosis of enterohemorrhagic *Escherichia coli* O157:H7 directly from fecal specimens using immunofluorescence stain. *Am. J. Clin. Pathol.* **101**, 91–94.

45. Park, C. H., Vandel, N. M., and Hixon, D. L. (1996) Rapid immunoassay for detection of *Escherichia coli* O157 directly from stool specimens. *J. Clin. Microbiol.* **34**, 988–990.

46. Beutin, L., Montenegro, M. A., Orskov, I., Orskov, F., Prada, J., Zimmermann, S., et al. (1989) Close association of verotoxin (Shiga-like toxin) production with enterohemolysin production in strains of *Escherichia coli*. *J. Clin. Microbiol.* **27**, 2559–2564.

47. Hull, A. E., Acheson, D. W., Echeverria, P., Donohue-Rolfe, A., and Keusch, G. T. (1993) Mitomycin immunoblot colony assay for detection of Shiga-like toxin-producing Escherichia coli in fecal samples: comparison with DNA probes. *J. Clin. Microbiol.* **31**, 1167–1172.

48. Chapman, P. A. and Siddons, C. A. (1996) A comparison of immunomagnetic separation and direct culture for the isolation of verocytotoxin-producing *Escherichia coli* O157 from cases of bloody diarrhoea, non-bloody diarrhoea and asymptomatic contacts. *J. Med. Microbiol.* **44**, 267–271.

49. Karch, H., Janetzki-Mittmann, C., Aleksic, S., and Datz, M. (1996) Isolation of enterohemorrhagic *Escherichia coli* O157 strains from patients with hemolytic–uremic syndrome by using immunomagnetic separation, DNA-based methods and direct culture. *J. Clin. Microbiol.* **34**, 516–519.

50. Paton, A. W., Ratcliff, R., Doyle, R. M., Seymour-Murray, J., Davos, D., Lanser, J. A., et al. (1996) Molecular microbiological investigation of an outbreak of hemolytic uremic syndrome caused by dry fermented sausage contaminated with Shiga-like toxin-producing *Escherichia coli*. *J. Clin. Microbiol.* **34**, 1622–1627.
51. Jenkins, C., Chart, H., Smith, H.R., Hartland, E.L., Batchelor, M., Delahay, R.M., et al. (2000) Antibody response of patients infected with verocytotoxin-producing *Escherichia coli* to protein antigens encoded on the LEE locus. *J. Med. Microbiol.* **49**, 97–101.
52. Barrett, T. J., Green, J. H., Griffin, P. M., Pavia, A. T., Ostroff, S. M., and Wachsmuth, I. K. (1991) Enzyme-linked immunosorbent assays for detecting antibodies to Shiga-like toxin I, Shiga-like toxin II, and *Escherichia coli* O157:H7 lipopolysaccharide in human serum. *Curr. Microbiol.* **23**, 189–195.
53. Chart, H., Law, D., Rowe, B., and Acheson, D. W. (1993) Patients with haemolytic uraemic syndrome caused by *Escherichia coli* O157: absence of antibodies to Vero cytotoxin 1 (VT1) or VT2. *J. Clin. Pathol.* **46**, 1053,1054.
54. Karmali, M. A., Petric, M., Winkler, M., Bielaszewska, M., Brunton, J., van-de-Kar, N., et al. (1994) Enzyme-linked immunosorbent assay for detection of immunoglobulin G antibodies to *Escherichia coli* Vero cytotoxin 1. *J. Clin. Microbiol.* **32**, 1457–1463.
55. Reymond, D., Karmali, M. A., Clarke, I., Winkler, M., and Petric, M. (1997) Comparison of the western blot assay with the neutralizing-antibody and enzyme-linked immunosorbent assays for measuring antibody to verocytotoxin 1. *J. Clin. Microbiol.* **35**, 609–613.
56. Bitzan, M. and Karch, H. (1992) Indirect hemagglutination assay for diagnosis of *Escherichia coli* O157 infection in patients with hemolytic-uremic syndrome. *J. Clin. Microbiol.* **30**, 1174–1178.
57. Bitzan, M., Moebius, E., Ludwig, K., Muller-Wiefel, D. E., Heesemann, J., and Karch, H. (1991) High incidence of serum antibodies to *Escherichia coli* O157 lipopolysaccharide in children with hemolytic-uremic syndrome. *J. Pediatr.* **119**, 380–385.
58. Chart, H., Smith, H. R., Scotland, S. M., Rowe, B., Milford, D. V., and Taylor, C. M. (1991) Serological identification of *Escherichia coli* infection in haemolytic uraemic syndrome. *Lancet* **337**, 138–140.
59. Greatorex, J. S. and Thorne, G.M. (1994) Humoral immune responses to Shiga-like toxins and *Escherichia coli* O157 lipopolysaccharide in hemolytic-uremic syndrome patients and healthy subjects. *J. Clin. Microbiol.* **32**, 1172–1178.
60. Yamada, S., Kai, A., and Kudoh, Y. (1994) Serodiagnosis by passive hemagglutination test and verotoxin enzyme-linked immunosorbent assay of toxin-producing Escherichia coli infections in patients with hemolytic–uremic syndrome. *J. Clin. Microbiol.* **32**, 955–959.
61. Chart, H. (1999) Evaluation of a latex agglutination kit for the detection of human antibodies to the lipopolysaccharide of *Escherichia coli* O157, following infection with verocytotoxin-producing *E. coli* O157. *Lett. Appl. Microbiol.* **29**, 434–436.
62. Bielaszewska, M., Janda, J., Blahova, K., Feber, J., Potuznik, V., and Souckova, A. (1996) Verocytotoxin-producing *Escherichia coli* in children with hemolytic uremic syndrome in the Czech Republic. *Clin. Nephrol.* **46**, 42–44.

63. Chart, H. and Rowe, B. (1990) Serological identification of infection by Vero cytotoxin producing *Escherichia coli* in patients with haemolytic uraemic syndrome. *Serodiagn. Immunother. Infect. Dis.* **4,** 413–418.

64. Caprioli, A., Luzzi, I., Rosmini, F., Resti, C., Edefonti, A., Perfumo, F., et al. (1994) Community-wide outbreak of hemolytic-uremic syndrome associated with non-O157 verocytotoxin-producing *Escherichia coli. J. Infect. Dis.* **169,** 208–211.

65. Paton, A. W., Woodrow, M. C., Doyle, R. M., Lanser, J. A. and Paton, J. C. (1999) Molecular characterization of a Shiga-toxigenic *Escherichia coli* O113:H21 strain lacking *eae* responsible for a cluster of cases of hemolytic-uremic syndrome. *J. Clin. Microbiol.* **37,** 3357–3361.

66. Karch, H., Meyer, T., Russmann, H., and Heesemann, J. (1992) Frequent loss of Shiga-like toxin genes in clinical isolates of *Escherichia coli* upon subcultivation. *Infect. Immun.* **60,** 3464–3467.

3

Serological Methods for the Detection of STEC Infections

Martin Bitzan and Helge Karch

1. Introduction

The detection of antibodies to Shiga toxin (Stx)-producing *Escherichia coli* (STEC) antigens serves varied purposes: (1) the etiologic diagnosis of acute hemolytic uremic syndrome (HUS) and hemorrhagic colitis (HC) in the clinical laboratory; (2) epidemiological investigations; (3) the study of immune responses in STEC-mediated diseases, immunization trials, and animal models. Although the isolation of STEC from the feces of a patient with HC or HUS is generally sufficient evidence for its etiological role in these diseases, it may fail because of a number of circumstances. For example, a timely stool specimen may not be available, the primary laboratory may be unaware of the clinical diagnosis or apply inadequate isolation methods, or the patient may have received suppressive antibiotics. Moreover, when patients present with HUS, usually 5–7d after the onset of diarrhea, the excretion rate of STEC organisms is already substantially diminished. Among *E. coli* isolates from patients with HUS and HC, STEC O157:H7 predominates. However, so-called non-O157:H7 STEC serotypes are emerging both as causes of outbreaks and sporadic HUS and diarrhea, especially in Europe, Australia and South America. The clinical features of non-O157 STEC infections closely resemble those of prototypic *E. coli* O157:H7 disease *(1,2)*. The microbiological diagnosis of non-O157:H7 STEC strains is complicated by the lack of easily detectable biochemical or growth characteristics and large serotype diversity. Serological techniques offer a complementary, culture-independent diagnostic approach. They are indispensable for epidemiological and immunization studies.

From: *Methods in Molecular Medicine, vol. 73: E. coli: Shiga Toxin Methods and Protocols*
Edited by: D. Philpott and F. Ebel © Humana Press Inc., Totowa, NJ

Shiga toxin-producing *E. coli* display a dazzling array of potentially immunogenic antigens and/or virulence factors, which include at least four serologically discernible Shiga toxins, lipopolysaccharide (LPS), secreted and membrane proteins. Karmali first reported the detection of antibodies to Stx in sera of patients with STEC-mediated HUS, using a Vero cell toxicity neutralization assay *(3)*. This approach is appealing considering that Shiga toxins are the primary cause of HUS and HC. Most clinical STEC isolates express Stx2, often in combination with Stx1 and/or Stx2 variants. The practical usefulness of the neutralization assay for the acute diagnosis, however, is limited: many patients exhibit only modest titer changes between acute and convalescent samples *(3–7)*. Furthermore, serum samples from virtually all patients and healthy individuals neutralize Stx2 *(8,9)*. The neutralizing principle does not reside in the immunoglobulin fraction *(8)*, but in other serum compartments *(9a)*. Its clinical significance is uncertain. The correlation between the genotype of the STEC isolate and the presence of Stx-specific antibodies is poor. For example, using an Stx type-specific enzyme-linked immunosorbent assay (ELISA), Barrett et al. found Stx1-specific but not Stx2-specific antibodies in patient's sera during an outbreak by a sole Stx2-producing *E. coli* strain *(4)*.

Western blot assays using Stx1 and Stx2 offer the advantage of higher specificity compared to toxin neutralization test and ELISA, because it allows the visualization of antigen band(s) of specific molecular size and the identification of the reactive serum component as immunoglobulin. Karmali's group reported an excellent agreement between neutralization test, ELISA (IgG) and immunoblot (IgG) using Stx1 as antigen in paired serum samples from patients with infection by Stx1 producing *E. coli* *(10)*. Interestingly, about 50% of the samples had antibodies to both the A- and B-subunits, whereas the remaining samples had detectable antibodies to either the A- or subunit. This study also demonstrated that the Stx1 IgG Western blot assay can discriminate true from spurious neutralization assay and ELISA results *(10)*. Hence, the authors considered Western blotting the "gold standard" for Stx-based serological assays *(10)*. However, in a previous study, Chart et al. failed to detect Stx1- or Stx2-specific antibodies by Western blot in sera from patients with HUS *(11)*. Although the lack of a detectable immune response to Stx1 may be explained by the predominance of Stx2 producing *E. coli* strains in the United Kingdom *(12)*, the above findings contrast with those of Ludwig et al., who detected Stx2-specific antibodies by Western blotting in the majority of children with HUS in Germany *(12a)*. Taken together, Stx-based serological tests are potentially valuable seroepidemiological tools, but unsatisfactory for the acute serological diagnosis *(2,10,13)*.

More encouraging results for the acute serological diagnosis of STEC infection were obtained with assays using LPS as antigen. Most laboratories investigating the immune response of patients with HUS to STEC O157 LPS used Western blotting, ELISA or indirect hemagglutination (IHA) (for review, *see* **refs.** *2* and *12*). All three techniques yield similar results. ELISA and Western blot allow differentiation of the immunoglobulin class of anti-LPS antibodies, whereas the IHA allows the quantitation of the immune response as antibody titer. The agglutination is largely the result of the presence of LPS-specific IgM antibodies *(14)*. The majority of children with HUS indeed mount a brisk IgM and IgA antibody response to STEC O157 LPS as determined by immunoblot and ELISA *(5,15,16)*. IgM/IHA anti-O157 LPS antibodies are detectable over a period of 2–3 mo, but occasionally considerably longer *(5,14–16)*. IgG antibodies to O157 LPS may be detected early in the course of HUS and support the diagnosis, especially when IgM class antibodies are only marginally elevated. More often their rise is delayed or not observed at all, even in follow-up samples *(7,16,17)*. The usefulness of O157 LPS-specific IgG antibodies for seroepidemiological purposes *(13)* has still to be established. IgA anti-O157 LPS antibodies decline rapidly after the manifestation of the HUS, but they are occasionally found in controls without documented or suspected STEC infection *(5,17)*. *E. coli* O157-induced LPS-specific antibodies are also found in patients with uncomplicated diarrhea or HC, but less consistently than in patients with HUS *(4,16,17a)*.

Few studies have explored the immune response to STEC LPS other than O157 *(14,17–20)*. It emerges from these reports that patients with HUS associated with non-O157:H7 STEC strains mount an *E. coli* serogroup-specific immune response similar to that observed in patients with *E. coli* O157:H7 infection *(17)*. Classical enteropathogenic *E. coli* (EPEC) strains and non-O157 STEC strains share O:H serotypes, predominantly those belonging to serogroups O26, O55, O111, and O126 *(12)*. Occasionally, patients with a non-O157 STEC isolate exhibit antibodies to *E. coli* O157 LPS, in addition to the homologous LPS type *(2,17)*. Possible explanations include the occurrence of true double infections by *E. coli* O157 and non-O157 strains, the secondary transmission (in the gut) of Stx carrying bacteriophages to other susceptible *E. coli* serotypes, and crossreactivity between antigenic epitopes or spurious nonspecific stimulation *(2,17)*.

The immune response to STEC protein antigens encoded by genes clustering on the locus of enterocyte attachment and effacement (LEE) has recently been evaluated independently by two groups *(21,22,22a)*. Purified recombinant intimin, *E. coli*-secreted protein (Esp) A filament structural protein, translocated EspB, and intercellular and extracellular domains of the translocated intimin receptor (Tir) were used as antigens in Western blotting and/or ELISA.

Although the majority of patients with documented *E. coli* O157 infection demonstrated reactivity with the extracellular domains of the examined proteins, their diagnostic usefulness has yet to be established. Similarly, Schmidt et al. demonstrated that most patients with *E. coli* O157-induced HUS have antibodies to the plasmid-encoded EHEC hemolysin by Western blotting *(23)*. Technical details relating to antigen preparation, reagents, and assay technology are beyond the scope of this chapter; the interested reader is referred to the original publications *(21–23)*.

At present, LPS-based techniques, primarily those detecting IgM class antibodies, appear to be of greatest diagnostic values *(12)*. In the clinical diagnostic setting, it is recommended that sera from patients with HUS or HC be first tested for IgM or IgM+IgG antibodies to the *E. coli* O157 O-antigen and subsequently, if negative, for antibodies to candidate non-O157 serogroups, such as O26, O55, O91, O103, O111, O113; O128, and O145 *(2,12)*. STEC outbreaks by rare STEC serotypes offer the opportunity to extract LPS from the outbreak strain for specific antibody screening.

This chapter focuses on the methodology of assays to detect antibodies towards LPS and Stx.

2. Materials

1. General equipment: minigel (slab gel) electrophoresis chamber and transfer (electro blot) apparatus; 96-well microtiter (ELISA) plate reader; X-ray film developer; autorad cassettes.
2. Tissue culture facility with class II biological safety cabinet, inverted microscope.
3. Disposable plastics:

    Flat-bottom or C-shaped 96-well microtiter plates (e.g., Immunoplates; MaxiSorp, NUNC A/S, Roskilde, Denmark/Nalge Nunc International) for LPS ELISA; round-bottom microtiter plates (e.g., Nunc), for the indirect hemagglutination assay; incubation trays for immunoblot strips (e.g., disposable eight-well Mini-incubation trays from Bio-Rad).

4. Tissue culture flasks; 96-well, flat-bottom microtiter plates (e.g. Corning; Nunc).
5. Dialysis tubing, exclusion size approx 10 kDa (Visking[R], MWCO 12-14,000; Spectra/Por[R], MWCO 6-8000; from Serva Electrophoresis GmbH, Heidelberg, Germany, and Spectrum Laboratories, Rancho Dominguez, CA). Boil tubing in 1 m*M* sodium-EDTA, pH 7.0, for 15 min before use.
6. Membranes for immunoblotting: nitrocellulose; polyvinylidene difluoride (PVDF; e.g., Bio-Rad Laboratories).
7. Tryptic soy (TSB) or Luria Bertani (LB) broth (Trypton/Yeast extract) for bacterial cultures.
8. STEC strains (primarily *E. coli* O157) for lipopolysaccharide (LPS) extraction.
9. LPS from non-O157 STEC O-groups. For this purpose, LPS from enteropathogenic (EPEC) strains can be purchased (e.g., from *E. coli* O26:B6, O55:B5,

O111:B4, and O128: B12 [Sigma Chemicals, St. Louis, MO]).

10. O-Group antigen-specific *E. coli* test sera for *E. coli* O157 and enteropathogenic (non-O157) *E. coli* strains, such as O26:K60 (B6), O55:K59 (B5), O111:K58 (B4), O113:K75 (B19), O119:K69 (B14), O126:K71 (B16), O128:K67 (B12) (Dade-Behring, Liederbach, Germany [not available in the United States]; Difco Laboratories [via Lee Labs, Grayson, GA, a subsidiary of Becton Dickinson]; SA Scientific, Inc., San Antonio, TX).

11. Sheep red blood cells (SRBC).

12. Purified (or crude) Stx1 and Stx2 (for a detailed protocol, compare Chapter 15).

13. Stx1- and Stx2-specific monoclonal (purified, hybridoma) and polyclonal antibodies (e.g., ATCC, or Toxin Technology [Sarasota, FL],); defined human (patient) immune sera.

14. Tissue culture cell lines (American Type Culture Collection): Vero (ATCC CCL-81), ACHN cell (CRL-1611), or HeLa S3 cells (ATCC CCL-2); HeLa cells are less susceptible to the cytotoxic effect of Stx2c.

15. Tissue culture media (suitable for all three cell lines): minimal essential medium (MEM) with Earle's salts, supplemented with 2 m*M* glutamine and 5% (Vero and HeLa cells) or 10% (ACHN cells) fetal bovine serum (FBS). For specific media recommendations, see ATCC information.

16. Crystal violet (0.13%) for tissue culture staining (mix 26 mL of 0.5% crystal violet stock solution with 5 mL absolute ethanol, 5.4 mL of 37% formaldehyde, and phosphate-buffered saline (PBS) to a total volume of 100 mL).

17. General buffers and solutions (final concentrations):
    a. Phosphate-buffered saline (PBS), pH 7.5: 10 m*M* Na_2HPO_4, 1.5 m*M* KH_2PO_4, 137 m*M* NaCl.
    b. Tris-buffered saline (TBS), pH 7.4–7.6: 10 or 50 m*M* Tris-HCl, 137 m*M* NaCl.

18. Sodium dodecyl sulfate–polyacrylamide gel electrophoresis (SDS-PAGE) buffers:
    a. Laemmli sample buffer: 62.5 m*M* Tris-HCl, pH 6.8, 2% (w/v) SDS, 10% glycerol (v/v), 0.001% (w/v) bromophenol blue, 5% (v/v) 2-mercaptoethanol *(24)*.
    b. Electrode (running) buffer *(24)*: 25 m*M* Tris-HCl, 192 m*M* glycine, 0.1% SDS.
    c. Transfer (blotting) buffer *(25)*: 192 m*M* glycine, 25 m*M* Tris, 20% methanol (v/v).

19. Stains for polyacrylamide gels: silver stain reagents (e.g., Bio-Rad Silver stain [Bio-Rad, Hercules, CA]; Coomassie Brilliant Blue R-250 or GelCode Blue Stain Reagent [Pierce, Rockford, IL]; Serva Blue G [Serva Electrophoresis; Boehringer Ingelheim]).

20. Tricine–SDS-polyacrylamide electrophoresis buffers *(26)*:
    a. Sample buffer (1X): 50 m*M* Tris-HCl (pH 6.8), 4% SDS, 12% (v/v) glycerol, 2% β-mercaptoethanol, Serva Blue G, tip of small spatula (Serva Electrophoresis).
    b. Anode (running) buffer: 0.2 *M* Tris-HCl, pH 8.9.
    c. Cathode (running) buffer: 0.1 *M* Tris-HCl, 0.1 *M* Tricine, 0.1% SDS (pH ~ 8.25).

21. Conjugate antibodies: peroxidase or alkaline phosphatase (AP)-conjugated anti-human (IgM, IgG), anti-sheep (IgG), or anti-rabbit (IgG) antibody.

22. Enhanced, luminol-based chemiluminescence assay system (for immunoblot; e.g. ECL from Amersham Pharmacia Biotech).
23. Substrate buffer for AP conjugate antibody (immunoblotting):
    a. AP substrate buffer: diethanolamine, 96 mL/L dH_2O, pH 9.8 (with HCl), 200 mg/L $MgCl_2 \cdot X\ 6H_2O$. Dilute 1 : 5 (v/v) with 0.85% NaCl ("Ready-to-use" AP solution).
    b. NBT: 0.1 g 4-nitrotetrazolium chloride blue hydrate in 100 mL distilled water (dH_2O); store at 4°C in the dark.
    c. Indolyl phosphate: 0.05 g 5-bromo-4-chloro-3-indolyl phosphate (BCIP) *p*-toluidine salt in 10 mL *N*,*N*-dimethylformamide (DMF; e.g., Sigma); store 1-mL aliquots at –20°C. BCIP and NBT are also available as simple-to-use tablets (e.g., Sigma).
24. Substrates for conjugated enzymes (ELISA): *o*-phenylenediamine di-hydrochloride (OPD) or *p*-nitrophenylphosphate, disodium (pNPP) (also available as substrate/buffer tablets [Sigma *Fast* OPD; Sigma *Fast* pNPP; Sigma]).
25. Other reagents and chemicals: acetic acid, acrylamide, ammonium persulfate (APS), bovine serum albumin (BSA), fetal bovine serum, formaldehyde (37%), methylenebisacrylamide, phenol, proteinase K, sodium dodecyl sulfate, standard protein molecular-weight marker (prestained), *N*,*N*,*N'*,*N'*-tetramethylenediamine (TEMED), Tween-20.

3. Methods

3.1. Lipopolysaccharide Extraction

1. Grow *E. coli* test strain in 300 mL tryptic soy or LB broth (may upscale to 1.5 L) overnight at 37°C in shaking incubator (200–300 rpm). To start culture, inoculate prewarmed medium with 2 mL of 6 h preculture in same broth (*see* **Note 1**).
2. Collect bacteria by centrifugation at 6000*g* for 20 min.
3. Lyse bacterial pellet with 7 mL of warm (68°C) double distilled water (ddH_2O), transfer lysate into 25- or 50-mL glass centrifuge tube, and add the same volume of 68°C phenol (90% phenol in ddH_2O). Mix vigorously for 1 min (*see* **Note 2**).
4. Incubate for 20 min at 68°C with repeated agitation.
5. Chill on ice for 5 min. When phases are separated, spin at 6000*g* for 15 min.
6. Collect watery (upper) phase, which contains the LPS (5–7 mL), and transfer into conditioned dialysis tubing of approx 10 kDa exclusion size. Secure ends well.
7. Dialyze extract against at least 500 mL of distilled H_2O at 4°C for 48 h. Replace the water every 12 h.
8. Collect LPS extract and digest with proteinase K (1 mg per 5–7 mL) for 5 h at 42°C.
9. Repeat phenol extraction and dialysis as above.
10. Lyophilize LPS extract and store in sealed glass tube in desiccator at 4°C (*see* **Note 3**).
11. Check quality and purity of LPS preparation by SDS-PAGE with subsequent silver and protein staining.

3.2. LPS Fractionation by SDS-PAGE and Characterization of Purified LPS

1. Fractionate LPS by SDS-PAGE. Separating gel: 11% acrylamide in 0.037 *M* Tris-HCl, pH 8.8; stacking gel: 6% acrylamide in 0.125 *M* Tris-HCl, pH 6.8 (*see* **Note 4**).

2. Dilute LPS stock solution in sample buffer to a final concentration of 1 mg LPS/mL. Boil 3 min and load 6- to 10-μg per lane, along with a protein molecular-weight marker (*see* **Note 5**).
3. Separate LPS in electrode (running) buffer for approx 60 min at 30–100 mA (minigel).
4. Stain polyacrylamide gel with periodic acid–silver (*see* **Note 6**). The Bio-Rad silver staining involves gel fixation in 40% methanol/10% acetic acid (v/v), oxidization, washes in deionized dH_2O, addition of silver nitrate-containing reagent, and development with carbonate and paraformaledehyde-containing solution. The gel can be kept overnight in the first fixative. Stop staining when polysaccharide ladders appear as yellow to brown bands (*see* **Fig. 1** and **Notes 7–9**).
5. Assess for proteins by staining gel with Coomassie Brilliant Blue R-250 or comparable protein stains or staining kits. Coomassie blue-stained gels can be destained in 40% methanol/10% acetic acid prior to silver staining.
6. The purified LPS is characterized serologically by Western blotting using LPS (O-antigen)-specific antibodies (*see* **Subheading 3.3.5.**).

3.3. Immunoblot for the Detection of Serum Antibodies to STEC LPS

1. Fractionate LPS by SDS-PAGE as described in **Subheading 3.2.**. Part of the gel can be stained to ensure good separation and quality of the LPS. For the preparation of blot strips, pour stacking gel without comb and load 100–400 μg of LPS per mini slab gel.
2. Electroblot LPS onto nitrocellulose membrane in blotting buffer, using a wet-transfer system. Detection of protein molecular-size markers on the blot does not reliably indicate the completeness of LPS transfer. If in doubt, silver stain the gel for remaining LPS after transfer.
3. Block membrane in 10 mM PBS, pH 7.5, containing 0.01–0.05% Tween-20 (PBS-T) and 5% BSA at 37°C for 2 h or at 4°C overnight in covered container (on rocking platform) or tube (on rotator). Decant blocking buffer.
4. For later use, store blots in sealed plastic bags at 4°C or –20°C. Membrane may be cut in 2.5- to 4-mm-wide blot strips.
5. Dilute human (patient) serum (1 : 100) or test (immune) serum in PBS-T 0.05% with 1% BSA, add to blot, and incubate for 2 h at room temperature or overnight at 4°C. Optimize dilutions of immune sera according to titer, usually between 1 : 100 and 1:1000. Include human or rabbit immune serum as positive control in each assay (*see* **Note 10**)
6. Rinse and wash blot three times with 10 mM PBS-T 0.05%.
7. Add conjugate (secondary) antibody at optimized dilution in PSB-T for 1–2 h at room temperature (start with 1:1000 for AP conjugate and indolyl/NBT substrate; start with 1:20,000 for peroxidase conjugate and enhanced chemiluminescence). Examples for the detection of antibodies to *E. coli* O157 and O26 LPS are shown in **Figs. 2** and **3**.

3.4. Lipopolysaccharide ELISA

1. Coat microtiter plates with 0.5–1 μg LPS/per well in 10 mM PBS (pH 7.4). Dry overnight at room temperature. It is convenient to use 50 μL LPS at a concentra-

kDa

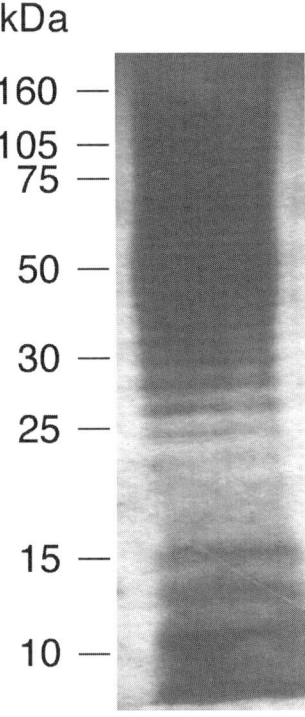

E.coli 0111

Fig. 1. *E. coli* O111:B4 lipopolysaccharide (Sigma), 6 µg/lane, fractionated by denaturing SDS-PAGE and silver stained, with protein molecular-weight markers of the indicated sizes.

    tion of 10 or 20 µg/mL PBS. Wash plates five times with 10 m*M* PBS containing 1% Tween. LPS-Coated microtiter plates can be stored at room temperature for several months.

2. Add 0.1 mL of patient serum per well, diluted 1:100 to 1:500 in 10 m*M* PBS-T 0.1% containing 2% FBS (v/v). Use triplicate wells. Incubate at 37°C overnight. Rinse five times with PBS-T 0.05%. Add 100 µL of secondary (conjugate) antibody diluted in PBS-T 0.1% to wells. Optimize dilutions (1:1000 to 1:10,000). Incubate for 2 h at room temperature

3. Rinse five times with PBS-T 0.05%; remove all fluid. Add 200 µL of enzyme substrate solution (e.g., OPD or pNPP) (*see* **Note 11**).

4. Incubate in the dark at room temperature. Both substrate reactions result in yellow color. Read after 30 min at 450 nm (OPD) or 402 nm (pNPP). For delayed reading, the reaction can be stopped by adding 50 µL of 3 *M* H$_2$SO$_4$, 3 *M* HCl (OPD), or 3 *M* NaOH (pNPP). Read at 492 nm (OPD) or 405 nm (pNPP) using a standard ELISA plate reader.

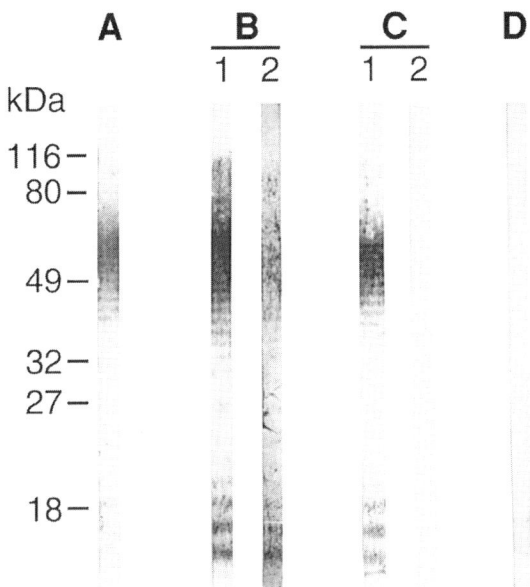

Fig. 2. Detection of antibodies to *E. coli* O157 LPS by immunoblotting. Nitrocellulose blot strips were developed with rabbit immune serum or serum samples from children with enteropathic HUS and alkaline phosphatase-conjugated goat anti-rabbit IgG or rabbit anti-human IgM and with 5-bromo-4-chloro-3-indolyl phosphate (BCIP)/ nitro blue tetrazolium (NBT) as substrate. Sample **A**: rabbit immune serum to *E. coli* O157; Samples **B** and **C**: serum samples from children with HUS due to *E. coli* O157; Sample **D**: normal control serum; 1, acute-phase serum; 2, convalescent phase serum (collected 1 and 3 mo, respectively, after the onset of diarrhea). Samples B1, B2 and C1 are positive for IgM antibodies to *E. coli* O157 LPS and recognize high- and low-molecular-weight bands; samples D and C2 are negative.

5. Include the following controls: blank (substrate and stop solution only), omission of the primary antibody, positive and negative serum samples at same dilutions as test samples (to ensure assay consistency). Breakpoints (cutoffs) are defined as mean plus three standard deviations and determined for each assay using control sera from age-matched healthy individuals. Actual readings can be normalized for the positive control and reported as ELISA (OD) units or as multiples of standard deviations above the mean (SD units) (*17*).

3.5. Indirect Hemagglutination Assay

1. Alkali treatment of LPS: suspend 5 mg of LPS in 2 mL of dH_2O. Add 0.4 mL of 1 *M* NaOH. Incubate mixture at 56°C for 60 min. Neutralize with 1 *M* acetic acid (final pH 7.0) and add dH_2O to a total volume of 5 mL. Use for IHA or lyophilize for serum absorption.

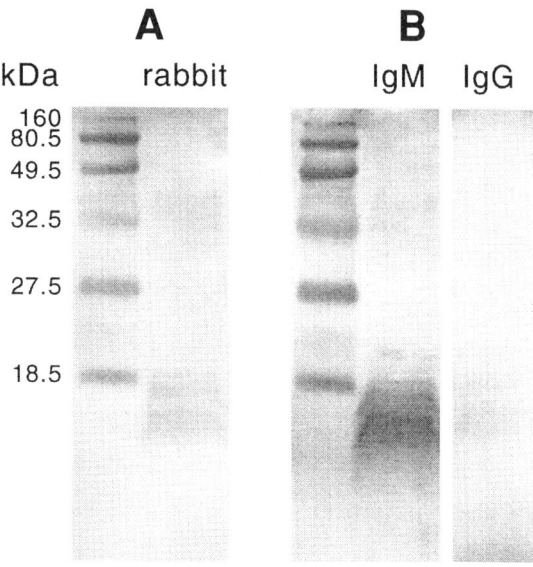

E.coli 026 LPS

Fig. 3. Detection of antibodies to *E. coli* O26 LPS, developed with rabbit immune serum (**A**) or serum from a patient with HUS secondary to STEC O26 infection (**B**). The blots were developed with anti-rabbit IgG, anti-human IgM or IgG alkaline phosphatase conjugates as indicated.

2. Wash SRBC in 0.85% saline until supernatant is clear. Resuspend pellet to 0.5% (v/v) in saline.

3. Add 1.2 mL of alkali-treated LPS dropwise to 100 mL of 0.5% SRBC. Incubate mixture at 37°C for 30 min with gentle agitation.

4. Wash sensitized SRBC three times in saline. Resuspend pellet to 0.6% (v/v) in saline. Control SRBCs are alkali-treated as above, except that the LPS is omitted.

5. Heat-inactivate patient serum (56°C, 30 min) and dilute fourfold, starting at 1:2, 1:8, 1:32, and so on to 1:32,768. Transfer 30 µL of each serum dilution per well of a round-bottom microtiter plate or dilute directly in the microtiter plate.

6. Include known positive and negative control samples in each assay.

7. Add 30 µL of LPS-sensitized or control SRBC, yielding final serum dilutions of 1:4 to 1:65,536. Use 30 µL saline and 30 µL LPS-sensitized SRBC to control for spontaneous agglutination.

8. Incubate plates at room temperature for 3 h and read results. The highest dilution giving a clear agglutination pattern is considered the end point (IHA titer) (*see* **Note 14**).

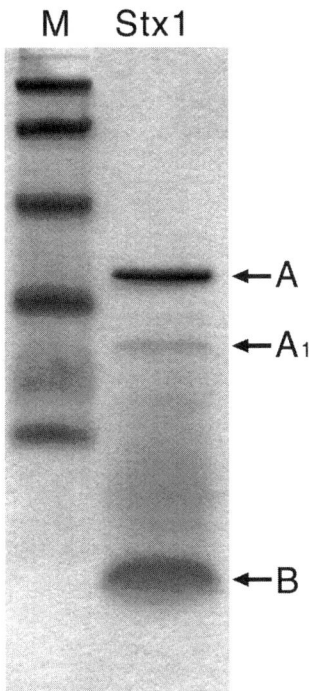

Fig. 4. Purified Shiga toxin 1 (10 µg), fractionated by tricine SDS–polyacrylamide gel electrophoresis (10% separating gel) and stained with Serva Blue G. The major bands correspond to the A- and B-subunits, and the faint band represents the nicked A1-subunit. Molecular-size markers: phosphorylase B (111 kDa), bovine serum albumin (73 kDa), ovalbumin (47.5 kDa), carbonic anhydrase (33.9 kDa), soybean trypsin inhibitor (28.8 kDa), and lysozyme (20.5 kDa) (*see* **Note 21**).

3.6. Stx Neutralization Assay

1. Determine the 50% cytotoxic dose ($CD_{50\%}$) of Stx stock solution (*see* **Notes 1 and 15**). Grow test cells (Vero, ACHN, or HeLa cells) in 96-well microtiter plate for 24–48 h so that they become just confluent (*see* **Note 16**). Use of 5×10^4 cells/well is a good starting point; use exactly 100 µL cell suspension per well. Add logarithmic (10-fold) dilutions of toxin in 100 µL tissue culture medium and incubate at 37°C in 5% CO_2 for 24–48 h (HeLa cells) or 72 h (Vero, ACHN cells) (*see* **Note 17**). The $CD_{50\%}$ is the reciprocal of the highest dilution, killing 50% of the cells. Quantitate the cytotoxic effect by crystal violet staining (*see* below and **Note 18**).

2. For the cytotoxicity neutralization assay, grow cell monolayer as above.
3. Dilute test samples (rabbit immune serum, patient serum, immunoglobulin preparations etc.) geometrically in complete tissue culture medium, beginning with 1:5. It is convenient to dilute samples in the microtiter plate in the same order as for the assay. Final volume is 50 µL/well. Prepare all dilutions in duplicate or triplicate.
4. Dilute Stx in complete tissue culture medium to a concentration of 40–80 $CD_{50\%}$/mL (2–4 $CD_{50\%}$/50 µL).
5. Mix constant amounts of Stx (2–4 $CD_{50\%}$ in 50 µL) with equal volumes of sample dilutions (test) or tissue culture medium alone (toxin control) in wells of a microtiter plate. Final volume is 100 µL, yielding a starting dilution of 1:10.
6. Prepare identical sample dilutions and mix with vehicle (tissue culture medium) without Stx to control for nonspecific stimulatory, inhibitory or toxic effects on cell monolayer. Incubate mixtures at 37°C for 1 h in a 5% CO_2 incubator.
7. Transfer mixtures to the tissue culture monolayer, conveniently with a multi-channel pipettor. Observe for 24–48 (HeLa) or 72 h (Vero, ACHN) using an inverted microscope. Record any damage to monolayers.
8. Rinse wells with PBS, aspirate all liquid. Fix cells with 70 µL of 2% formalin per well for 1 min. Remove formalin (dump into bleach with vigorous shaking). Add 70 µL of crystal violet stain for at least 20 min. Rinse microtiter plates generously with tap water until no more dye is flowing off and air-dry (*see* **Note 19**).
9. Elute bound stain from cells with 50% (v/v) ethanol in water (200 µL). Tap plate or use orbital shaker.
10. Read absorbance at 490 or 550 nm in microplate reader. The optical density directly correlates with the number of attached cells. If OD exceeds the linear range, because of high cell density, transfer 50-µL aliquots from each well to a new 96-well plate and add 150 µL PBS or stain with less concentrated crystal violet.
11. Determine protective (toxin-neutralizing) effect of each sample dilution as a fraction of the corresponding (toxin-free) control dilution, expressed as the ratio of their OD values after correction for nonspecific sample effects: (Sample – Stx)/ (Control – Stx). The highest dilution protecting 50% or the cells is taken as the end point (50% neutralization titer). The OD ratios can be plotted against the sample dilutions and the titer determined graphically or calculated using least-square regression analysis (8).

3.7. Stx Immunoblot Assay

1. Fractionate Stx by denaturing discontinuous tricine–SDS-PAGE (*see* **Note 20**):
   Separating gel (final concentrations): 10% acrylamide in 1 *M* Tris-HCl, pH 8.45 (adjust pH with HCl), and 13.3% glycerol (separating gel stock (w/v): acrylamide 46.5%, bisacrylamide 3%). Let completely polymerize.
   Stacking gel: 4% acrylamide in 1 *M* Tris-HCl, pH 8.45, without glycerol (stacking gel stock (w/v): acrylamide 48%, bisacrylamide 1.5%).
2. Mix Stx sample (v/v) with tricine sample buffer (*see* **Subheading 2.**). Boil for 5 min, and load gel.

3. Fractionate samples at 50 mA until Serva Blue reaches the bottom of the separating gel.
4. Fix gel with 10% (v/v) acetic acid in 50% ethanol for 30 min. Stain with Serva Blue G in same fixative for 60 min. Destain in 10% acetic acid in water until background is clear (*see* **Fig. 4** and **Note 21**).
5. Immunoblotting: Electroblot proteins from the unstained gel onto PVDF or nitrocellulose membrane in transfer (blotting) buffer, using a current of 0.15–0.25 A for 1 h or 30–40 mA overnight in the coldroom (wet transfer; pre-cool the buffer, insert an ice pack).
6. Cut membrane in 2.5-mm-wide strips. Stain one strip with Coomassie blue to ascertain the presence and location of the bands representing the A/A1- and B-subunit.
7. All subsequent steps are performed at room temperature and with 50 mM TBS, pH 7.4, or TBS–Tween (TBS-T).
8. Place strips in separate trays. Block membranes with TBS-T 0.1% containing 5% BSA for 2 h. Wash three times with TBS-T 0.05%.
9. Incubate strips with (human) test sera diluted 1:100 in TBS-T 0.05% containing 1% BSA. Wash three times with TBS-T 0.05%.
10. Incubate strips with peroxidase- (or AP)-conjugated goat anti-human IgG at the optimal dilution (start with 1:10,000 for ECL).
11. Place blot on cellophane wrap and add sufficient luminol-based substrate and enhancer solutions to cover blots according to the manufacturer's instructions. Remove all extra fluid, envelope blots in cellophane wrap, place in cassette, and expose to X-ray film. Start with 10–30 s, repeat with longer or shorter exposure time as necessary.
12. For AP-conjugate secondary antibody develop with BCIP/NBT as described in **Subheading 3.3.**).

4. Notes

1. Work with STEC isolates and Stx requires safety precautions for handling, storage, and removal according to local and national regulations.
2. The LPS extraction is based on the "hot phenol" method of Westphal and Jann *(27)*. Use only glass or phenol-resistant polypropylene tubes.
3. Lyophilized LPS is extremely light, fluffy, and sticky and difficult to handle. Use face mask.
4. Sodium dodecyl sulfate-PAGE of LPS is based on **refs.** *28* and *29*. Unpolymerized acrylamide is neurotoxic; use face mask and gloves.
5. Alternatively, mix 2X Laemmli sample buffer (v/v) with diluted LPS (in dH$_2$O).
6. Silver staining of LPS gels was originally described in **ref.** *28* and subsequently in **ref.** *29* (for LPS from *E. coli*). Simplified methods and test kits, based on the method of **ref.** *30*, are available (e.g., Bio-Rad Silver Stain). Silver stain may cause cancer by inhalation.
7. The stained gel can be stored in a zip-locked plastic bag with a few drops of water or dried on filter paper. If the gel darkens during drying, change the stop solution several times to remove all developer or dry gel between two pieces of cellophane. For additional details and troubleshooting, *see* the manufacturer's instructions.

8. The silver stain-reactive component of the LPS is the polysaccharide portion *(28)*. The many orderly spaced bands represent LPS molecules with varying numbers of repeating units in their O side chains (*see* **Fig. 1**).

9. The electrophoretic mobility of LPS in the SDS–polyacrylamide gel is determined by its lipid content and the number of repeating units *(28,31)*. On 11% polyacrylamide gels, larger-size LPS bands migrate between 30 and 70 kDa and smaller bands between 10 and 15 kDa relative to protein molecular weight markers (*see* **Figs. 1** and **2**).

10. O-Group (LPS)-specific antisera are marketed for slide or tube agglutination of *E. coli* isolates. Titers given by the manufacturer refer to standard tube agglutination and not to immunoblot or ELISA. Note that most antisera are raised in rabbits, but some are raised in other hosts (e.g., sheep).

11. The preparation of substrate and buffer solutions for AP and peroxidase conjugates for immunoblotting and ELISA can be simplified by various commercial ready-to-use reagents (e.g., from Sigma). OPD and pNPP are carcinogens. Avoid direct contact.

12. To remedy high-background signals, further dilute (titrate) the secondary antibody, change the blocking reagent(s) (low-fat dried milk, casein, or BSA) or the percentage of Tween (0.01–0.1%), or add 5% nonimmune serum to the blocking buffer (1% to the primary antibody) of the host species used to raise the secondary antibody (e.g., goat for goat anti-human IgG conjugate).

13. If space allows, perform the last step in the darkroom area close to the developer.

14. Rarely, crossreactions between *E. coli* O157, *Salmonella* and *Brucella* spp. and *Yersinia enterocolitica* O:9 may cause false-positive results in LPS-based assays. Absorption of the serum with homologous and heterologous LPS or with whole-cell bacterial suspensions helps clarify the specificity of the reaction *(14)*.

15. The Stx neutralization assay can be performed with purified or crude Stx. Divide toxin preparations in small aliquots and store at –80°C. Preferably, purified Stx is adjusted to a concentration of 1 mg/mL. Avoid repeated thawing and freezing cycles. Crude toxin is especially sensitive to degradation and inactivation; discard aliquot after each use. For consistent results, avoid varying cell density, toxin dose, or incubation times. Ensure that all reagents and plastic ware are of tissue-culture-grade quality and kept sterile.

16 All human material including cell lines (tissue cultures) and serum may contain viral agents and has to be handled with care and discarded safely (bleach; autoclaving).

17. Cytotoxicity test ($CD_{50\%}$) and toxin neutralization assay can be further simplified by adding freshly trypsinized cells directly to the wells containing preincubated mixtures of the test sample and/or Stx.

18. Alternative methods for the quantitation of Stx-induced cytotoxicity (or residual cell viability) include the 3-(4,5-dimethylthiazol-2-yl)-2,5-diphenyltetrazolium bromide (MTT) assay, neutral red incorporation, and lactate dehydrogenase release *(32)*.

19. Use of a suitable container and running tap water have proved effective in rinsing the crystal violet-stained tissue culture plates. Use gloves and protect your clothes!

20. For Stx immunoblot assays, purified Stx can also be separated by conventional SDS-PAGE with 15% separating gel and 9% stacking gel *(10,33)*. However, the small B-subunit monomers forms a considerably shaper band on the tricine gel.

21. The migration rates of the molecular-weight markers from various vendors were found to differ relative to the Stx A and B subunits using tricine–SDS-PAGE. The cause of the observed discrepancies is not known. The same Stx preparation migrates with an apparent molecular weight of approx 32 (A-subunit) and 10 kDa (B-subunit) by conventional SDS-PAGE through a 15% gel.

Acknowledgments

The authors thank Olga Böhler and Dr. Kerstin Ludwig and Dr. Martina Bielaszewska for sharing protocols and technical advice.

References

1. Karmali, M. A. (1989) Infection by verocytotoxin-producing *Escherichia coli. Clin. Microbiol. Rev.* **2**, 15–38.
2. Paton, J. C. and Paton, A. W. (1998) Pathogenesis and diagnosis of Shiga toxin-producing *Escherichia coli* infections. *Clin. Microbiol. Rev.* **11**, 450–479.
3. Karmali, M. A., Petric, M., Lim, C., Fleming, P. C., Arbus, G. S., and Lior, H. (1985) The association between idiopathic hemolytic uremic syndrome and infection by verotoxin-producing *Escherichia coli. J. Infect. Dis.* **151**, 775–782.
4. Barrett, T., Green, J., Griffin, P., Pavia, A., Ostroff, S., and Wachsmuth, I. (1991) Enzyme-linked immunosorbent assays for detecting antibodies to Shiga-like toxin I, Shiga-like toxin II, and *Escherichia coli* O157:H7 lipopolysaccharide in human serum. *Curr. Microbiol.* **23**, 189–195.
5. Bitzan, M., Ludwig, K., Klemt, M., Konig, H., Buren, J., and Muller-Wiefel, D. E. (1993) The role of *Escherichia coli* O 157 infections in the classical (enteropathic) haemolytic uraemic syndrome: results of a Central European, multicentre study. *Epidemiol. Infect.* **110**, 183–196.
6. Karmali, M. A., Petric, M., Winkler, M., Bielaszewska, M., Brunton, J., van de Kar, N., et al. (1994) Enzyme-linked immunosorbent assay for detection of immunoglobulin G antibodies to *Escherichia coli* Vero cytotoxin 1. *J. Clin. Microbiol.* **32**, 1457–1463.
7. Greatorex, J. S. and Thorne, G. M. (1994) Humoral immune responses to Shiga-like toxins and *Escherichia coli* O157 lipopolysaccharide in hemolytic–uremic syndrome patients and healthy subjects. *J. Clin. Microbiol.* **32**, 1172–1178.
8. Bitzan, M., Klemt, M., Steffens, R., and Muller-Wiefel, D. E. (1993) Differences in verotoxin neutralizing activity of therapeutic immunoglobulins and sera from healthy controls. *Infection* **21**, 140–145.
9. Caprioli, A., Luzzi, I., Seganti, L., Marchetti, M., Karmali, M. A., Clarke, I., et al. (1994) Frequency and nature of verocytotoxin 2 (VT2) neutralizing activity (NA) in human and animal sera, in *Recent advances in verocytotoxin-producing Escherichia coli Infections* (Karmali, M. A. and Goglio, A., eds.) Elsevier, Amsterdam, pp. 353–356.
9a. Kimura, T., Tani, S., Matsumoto, Y. Y., and Takeda, T. (2001) Serum amyloid P component is the Shiga toxin 2-neutralizing factor in human blood. *J. Biol. Chem.* **276**, 41,576–41,579.

10. Reymond, D., Karmali, M. A., Clarke, I., Winkler, M., and Petric, M. (1997) Comparison of the western blot assay with the neutralizing-antibody and enzyme-linked immunosorbent assays for measuring antibody to verocytotoxin 1. *J. Clin. Microbiol.* **35,** 609–613.

11. Chart, H., Law, D., Rowe, B., and Acheson, D. W. (1993) Patients with haemolytic uraemic syndrome caused by *Escherichia coli* O157: absence of antibodies to Vero cytotoxin 1 (VT1) or VT2. *J. Clin. Pathol.* **46,** 1053–1054.

12a. Ludwig, K., Karmali, M. A., Sarkim, V., Bobrowski, C., Petric, M., Karch, H., et al. (2001) Antibody response to Shiga toxins Stx2 and Stx1 in children with enteropathic hemolytic-uremic syndrome. *J. Clin. Microbiol.* **39,** 2272–2279.

12. Karch, H., Bielaszewska, M., Bitzan, M., and Schmidt, H. (1999) Epidemiology and diagnosis of Shiga toxin-producing *Escherichia coli* infections. *Diagn. Microbiol. Infect. Dis.* **34,** 229–243.

13. Reymond, D., Johnson, R. P., Karmali, M. A., Petric, M., Winkler, M., Johnson, S., et al. (1996) Neutralizing antibodies to *Escherichia coli* Vero cytotoxin 1 and antibodies to O157 lipopolysaccharide in healthy farm family members and urban residents. *J. Clin. Microbiol.* **34,** 2053–2057.

14. Bitzan, M. and Karch, H. (1992) Indirect hemagglutination assay for diagnosis of *Escherichia coli* O157 infection in patients with hemolytic-uremic syndrome. *J. Clin. Microbiol.* **30,** 1174–1178.

15. Chart, H., Smith, H. R., Scotland, S. M., Rowe, B., Milford, D. V., and Taylor, C. M. (1991) Serological identification of *Escherichia coli* O157:H7 infection in haemolytic uraemic syndrome. *Lancet* **337,** 138–140.

16. Ludwig, K., Ruder, H., Bitzan, M., Zimmermann, S., and Karch, H. (1997) Outbreak of *Escherichia coli* O157:H7 infection in a large family. *Eur. J. Clin. Microbiol. Infect. Dis.* **16,** 238–241.

17. Ludwig, K., Bitzan, M., Zimmermann, S., Kloth, M., Ruder, H., and Muller-Wiefel, D. E. (1996) Immune response to non-O157 Vero toxin-producing *Escherichia coli* in patients with hemolytic uremic syndrome. *J. Infect. Dis.* **174,** 1028–1039.

17a. Ludwig, K., Sarkim, V., Bitzan, M., Karmali, M. A., Bobrowski, C., Ruder, H., et al. (2002) Shiga toxin-producing *Escherichia coli* infection and antibodies against Stx2 and Stx1 in household contacts of children with enteropathic hemolytic-uremic syndrome. *J. Clin. Microbiol.* **40,** 1773–1782.

18. Caprioli, A., Luzzi, I., Rosmini, F., Resti, C., Edefonti, A., Perfumo, F., et al. (1994) Community-wide outbreak of hemolytic⁄uremic syndrome associated with non-O157 verocytotoxin-producing *Escherichia coli*. *J. Infect. Dis.* **169,** 208–211.

19. Morooka, T., Matano, H., Umeda, A., Oda, T., Amako, K., and Karmali, M. A. (1995) Indirect hemagglutination assay for antibodies to *Escherichia coli* lipopolysaccharides O157, O111 and O26 in patients with hemolytic uremic syndrome. *Acta Paediatr. Jpn.* **37,** 469–473.

20. Luzzi, I., Tozzi, A. E., Rizzoni, G., Niccolini, A., Benedetti, I., Minelli, F., et al. (1995) Detection of serum antibodies to the lipopolysaccharide of *Escherichia coli* O103 in patients with hemolytic-uremic syndrome. *J. Infect. Dis.* **171,** 514–515.

21. Jenkins, C., Chart, H., Smith, H. R., Hartland, E. L., Batchelor, M., Delahay, R. M., et al. (2000) Antibody response of patients infected with verocytotoxin-producing *Escherichia coli* to protein antigens encoded on the LEE locus. *J. Med. Microbiol.* **49,** 97–101.

22. Li, Y., Frey, E., Mackenzie, A. M., and Finlay, B. B. (2000) Human response to *Escherichia coli* O157:H7 infection: antibodies to secreted virulence factors. *Infect. Immun.* **68,** 5090–5095.

22a. Karpman, D., Bekassy, Z. D., Sjogren, A. C., Dubois, M. S., Karmali, M. A., Mascarenhas, M., et al. (2002) Antibodies to intimin and *Escherichia coli* secreted proteins A and B in patients with enterohemorrhagic *Escherichia coli* infections. *Pediatr. Nephrol.* **17,** 201–211.

23. Schmidt, H., Beutin, L., and Karch, H. (1995) Molecular analysis of the plasmid-encoded hemolysin of *Escherichia coli* O157:H7 strain EDL 933. *Infect. Immun.* **63,** 1055–1061.

24. Laemmli, U. K. (1970) Cleavage of structural proteins during the assembly of the head of bacteriophage T4. *Nature* **227,** 680–685.

25. Towbin, H., Staehelin, T., and Gordon, J. (1979) Electrophoretic transfer of proteins from polyacrylamide gels to nitrocellulose sheets: procedure and some applications. *Proc. Natl. Acad. Sci. USA* **76,** 4350–4354.

26. Schagger, H. and von Jagow, G. (1987) Tricine–sodium dodecyl sulfate–polyacrylamide gel electrophoresis for the separation of proteins in the range from 1 to 100 kDa. *Anal. Biochem.* **166,** 368–379.

27. Westphal, O. and Jann, K. (1965) Bacterial lipopolysaccharides: extraction with phenol–water and further applications of the procedure. *Methods Carbohydr. Chem.* **5,** 83–91.

28. Tsai, C. M. and Frasch, C. E. (1982) A sensitive silver stain for detecting lipopolysaccharides in polyacrylamide gels. *Anal. Biochem.* **119,** 115–119.

29. Karch, H., Leying, H., and Opferkuch, W. (1984) Analysis of electrophoretically heterogeneous lipopolysaccharides of Escherichia coli by immunoblotting. *FEMS Microbiol. Lett.* **22,** 193–196.

30. Merril, C. R., Goldman, D., Sedman, S. A., and Ebert, M. H. (1981) Ultrasensitive stain for proteins in polyacrylamide gels shows regional variation in cerebrospinal fluid proteins. *Science* **211,** 1437–1438.

31. Jann, B., Reske, K., and Jann, K. (1975) Heterogeneity of lipopolysaccharides. Analysis of polysaccharide chain lengths by sodium dodecylsulfate–polyacrylamide gel electrophoresis. *Eur J. Biochem.* **60,** 239–246.

32. Williams, J. M., Boyd, B., Nutikka, A., Lingwood, C. A., Barnett Foster, D. E., Milford, D. V., et al. (1999) A comparison of the effects of verocytotoxin-1 on primary human renal cell cultures. *Toxicol. Lett.* **105,** 47–57.

33. Bielaszewska, M., Clarke, I., Karmali, M. A., and Petric, M. (1997) Localization of intravenously administered verocytotoxins (Shiga-like toxins) 1 and 2 in rabbits immunized with homologous and heterologous toxoids and toxin subunits. *Infect. Immun.* **65,** 2509–2516.

4

Detection and Characterization of STEC in Stool Samples Using PCR

Adrienne W. Paton and James C. Paton

1. Introduction

Production of a member of the Shiga toxin (Stx) family is a *sine qua non* of virulence for Shiga toxigenic *Escherichia coli* (STEC), and therefore polymerase chain reaction (PCR) amplification of *stx* genes is, unquestionably, a definitive diagnostic procedure. Moreover, PCR is extremely sensitive, which is an important feature, because as STEC disease progresses from the initial diarrheal phase to the more serious complications, the numbers of STEC in the feces often diminish markedly *(1,2)*. Rapid PCR assays also permit the timely diagnosis of index cases, which is important for the recognition and subsequent management of outbreaks. Although direct extracts of feces can be used as a template for PCR, sensitivity has often been suboptimal because of the presence of inhibitors of *Taq* polymerase. For this reason, it is strongly recommended that fecal samples be first cultured (even for as little as 4 h) in a suitable enrichment broth. This has the advantage of diluting any inhibitors present and increasing the number of target STEC organisms. Crude DNA extracts from such cultures can then be subjected to subsequent PCR analysis.

Sequence data for *stx₁* and *stx₂* genes have been published for more than a decade *(3,4)* and since then, a number of variant members of the *stx* family have been described (reviewed in **ref. 2**). Numerous PCR primer sets have been described that are capable of detecting either or both of the two major toxin gene types *(2)*. In addition, several putative STEC accessory virulence genes, such as *eae* and EHEC-*hlyA* have been described *(5,6)*. These serve as markers for the locus of enterocyte effacement (LEE) pathogenicity island and

From: *Methods in Molecular Medicine, vol. 73: E. coli: Shiga Toxin Methods and Protocols*
Edited by: D. Philpott and F. Ebel © Humana Press Inc., Totowa, NJ

the large STEC virulence plasmid, respectively. Furthermore, sequence data from the O-antigen-encoding *rfb* regions of important STEC serogroups (e.g., O157, O111, etc.) have been reported *(7,8)*. This has enabled the development of multiplex PCR assays capable of providing a degree of genetic characterization of STEC strains present in a sample. In our laboratory we have adopted a two-tiered approach to PCR analysis of fecal samples from patients with suspected STEC infection. Fecal culture extracts are initially screened for the presence of *stx* genes using a pair of redundant oligonucleotide primers capable of directing the amplification of a 212- to 215-base pair (bp) product from either stx_1 or stx_2 (including all known stx_2 variants associated with human disease) *(9)*. PCR products are detected by agarose gel electrophoresis and ethidium bromide staining. The fact that the stx_1 and stx_2 products are essentially the same size can be an advantage, because if both genes are present, a more intense PCR signal will be obtained, thereby maximizing sensitivity of the assay. Any extracts yielding a positive result are subjected to a second round of analysis using two multiplex PCR assays. The first utilizes four primer pairs specific for stx_1, stx_2 (including variants of stx_2), *eae*, and EHEC-*hlyA*, generating amplification products of 180, 255, 384, and 534 bp, respectively (**Fig. 1A**). The second uses three primer pairs specific for portions of the *rfb* regions of *E. coli* serogroups O157, O111, and O113 generating PCR products of 259, 406, and 593 bp, respectively (**Fig. 1B**) *(10,11)*. The initial *stx* screening assay establishes the diagnosis quickly and with maximal sensitivity. The subsequent multiplex assays provide independent confirmation of this, because the stx_1- and stx_2-specific primer pairs direct amplification of distinct regions of their respective target genes compared to the redundant screening primers. Information concerning toxin gene type and presence of serogroup-specific or accessory virulence genes is also clinically useful. Capacity of a given STEC strain to produce severe human disease appears to be enhanced by the production of Stx2, the presence of LEE, and carriage of the large virulence plasmid *(2,12–15)*. Also, LEE-positive STEC belonging to serogroups O157 and O111 are known to be of high human virulence and have caused numerous outbreaks of STEC disease *(2)*. Thus, a patient presenting with diarrhea caused by an organism carrying such markers should be carefully monitored for signs of complications. However, LEE is not essential for pathogenesis, and a significant minority of sporadic cases and at least one outbreak of severe disease have been caused by LEE-negative STEC strains *(2,16)*. One of the most common LEE-negative STEC serogroups associated with human disease is O113 (particularly serotype O113:H21) *(2,16,17)* and so this serogroup has also been targeted. A fur-

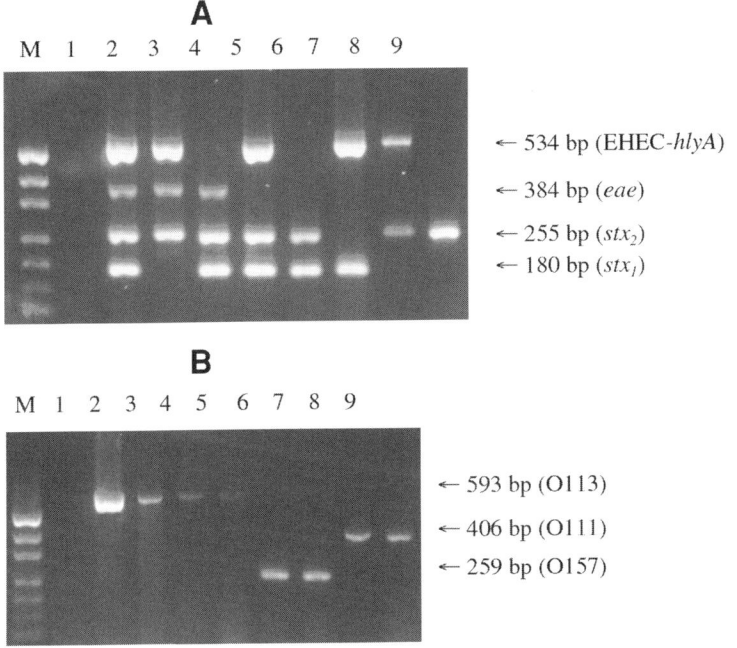

Fig. 1. Multiplex PCR analysis. (**A**) Characterization of reference STEC strains by multiplex PCR Assay 1. Lanes: M, DNA size markers (pUC19 DNA digested with *Hpa*II, fragment sizes visible are 501/489, 404, 331, 242, 190, 147, and 111 bp); 1, negative control; 2, O157:H⁻ strain 96/2998 (*stx₁⁺*, *stx₂⁺*, *eae⁺*, EHEC-*hlyA⁺*); 3, O157 strain 94-8628 (*stx₂⁺*, *eae⁺*, EHEC-*hlyA⁺*); 4, O157 strain 96/0052 (*stx₁⁺*, *stx₂⁺*, *eae⁺*); 5, O48:H21 strain 94CR (*stx₁⁺*, *stx₂⁺*, EHEC-*hlyA⁺*); 6, O128 strain 95AS1 (*stx₁⁺*, *stx₂⁺*); 7, O91 strain 95HE4 (*stx₁⁺*, EHEC-*hlyA⁺*); 8, O113 strain MW10 (*stx₂⁺*, EHEC-*hlyA⁺*); 9, OX3:H21 strain 031 (*stx₂⁺*). The expected mobilities for the various specific PCR products are also indicated. (**B**) Crude DNA extracts of *stx*-positive primary fecal cultures analyzed using multiplex PCR Assay 2. Lanes: M, DNA size markers (pUC19 DNA digested with *Hpa*II, fragment sizes visible are 501/489, 404, 331, 242, 190, 147, and 111 bp); 1, negative control; 2–4, extracts from three patients with culture-proven O113:H21 STEC infection; 5, extract from a *stx₂*-positive but culture-negative HUS patient with serological evidence of O113 infection; 6 and 7, extracts from patients with culture-proven O157:H⁻ STEC infection; 8 and 9, extracts from patients with culture-proven O111:H⁻ STEC infection. Reproduced from **refs. *10*** and ***11*** with permission from the *Journal of Clinical Microbiology.*

ther argument for the deployment of multiplex PCR analysis is that the detection of similar multiplex PCR profiles in two or more patients may be an early indicator of epidemiological linkage.

2. Materials

2.1. Specimen Preparation

1. Micropipets and sterile tips (preferably plugged).
2. Reference STEC strains.
3. Luria–Bertani (LB) broth: 1% Bacto-tryptone (Difco), 0.5% Bacto yeast extract (Difco), 1% NaCl, adjusted to pH 7.5, dispensed in 5-mL aliquots in 10-mL tubes and autoclaved at 121°C for 10 min.
4. 1.5-mL Screw-capped microcentrifuge tubes.
5. TS buffer: 25% sucrose, 50 mM Tris-HCl, pH 8.0 (sterilized by autoclaving at 121°C for 10 min and stored at 4°C).
6. Lysis solution: 10 mg/mL lysozyme (Roche Molecular Biochemicals) in 0.25 M EDTA (store in aliquots at –15°C and do not freeze–thaw more than three times).
7. 20 mg/mL Proteinase K (Roche Molecular Biochemicals) (store in aliquots at –15°C).
8. 95% Ethanol (prechilled at –15°C).
9. TE buffer: 10 mM Tris-HCl, 1 mM EDTA, pH 8.0 (sterilized by autoclaving).
10. Heating block.
11. Microcentrifuge.

2.2. PCR

1. 0.5-mL Thin-walled PCR tubes.
2. 10X PCR buffer: 100 mM Tris-HCl (pH 8.3), 500 mM KCl, 20 mM MgCl$_2$, 1% gelatin, 1% Tween-20, 1% Nonidet P-40 (*see* **Note 1**).
3. dNTP Mix: 1.25 mM dATP, 1.25 mM dCTP, 1.25 mM dGTP, 1.25 mM dTTP. This is prepared by diluting molecular biology grade 100 mM stock solutions (Roche Molecular Biochemicals) in sterile water. Store in aliquots at –15°C and avoid freeze-thawing more than five times.
4. *Taq* polymerase, 5 U/µL (Roche Molecular Biochemicals) (stored at –15°C).
5. Oligonucleotide primers, 1 mg/mL dissolved in TE and stored at 4°C; *see* **Table 1** for sequences.
6. Ultrapure sterile water.
7. Thermal cycler.

2.3. PCR Product Detection

1. Agarose: molecular biology grade.
2. TBE buffer: 90 mM Tris-base, 90 mM boric acid, 2 mM EDTA, supplemented with 0.5 µg/mL ethidium bromide (N.B. EtBr is carcinogenic, avoid skin contact).
3. Loading buffer: TE supplemented with 0.05% bromophenol blue and 50% glycerol.
4. DNA size markers; for example, pUC19 DNA restricted with *Hpa*II (fragment sizes 501, 489, 404, 331, 242, 190, 147, 111, 110, 67, 34, and 26 bp) (store at 4°C).
5. Horizontal electrophoresis apparatus and power supply.
6. Ultraviolet (UV) transilluminator.
7. Polaroid MP-4 camera and type 667 film, or other gel documentation system.

Table 1
PCR Primers

Primer	Sequence (5'–3')	Specificity	Amplicon size (bp)
stx Screen			
stxUF	ATACAGAG[GA]G[GA]ATTTCGT[a]	nt 586–797/800 of A-subunit coding region of stx_1 and stx_2	212/215
stxUR	TGATGATG[AG]CAATTCAGTAT		
Multiplex assay 1 (stx_1, stx_2, *eae*, and EHEC-*hlyA*)			
stx1F	ATAAATCGCCATTCGTTGACTAC	nt 454–633 of A-subunit coding region of stx_1	180
stx1R	AGAACGCCCACTGAGATCATC		
stx2F	GGCACTGTCTGAAACTGCTCC	nt 603–857 of A-subunit coding region of stx_2 (including stx_2 variants)	255
stx2R	TCGCCAGTTATCTGACATTCTG		
eaeF	GACCCGGCACAAGCATAAGC	nt 27–410 of *eae* (conserved between EPEC and STEC)	384
eaeR	CCACCTGCAGCAACAAGAGG		
hlyAF	GCATCATCAAGCGTACGTTCC	nt 70–603 of EHEC-*hlyA*	534
hlyAR	AATGAGCCAAGCTGGTTAAGCT		
Multiplex assay 2 (*E. coli* serogroups O157, O111, and O113)			
O157F	CGGACATCCATGTGATATGG	nt 393–651 of $rfbE_{O157:H7}$	259
O157R	TTGCCTATGTACAGCTAATCC		
O111F	TAGAGAAATTATCAAGTTAGTTCC	nt 24–429 of orf 3.4 of *E. coli* O111 *rfb* region	406
O111R	ATAGTTATGAACATCTTGTTTAGC		
O113F	AGCGTTTCTGACATATGGAGTG	O113 *wzy* gene (nt 3690–4282 of O113 *rfb* region)	593
O113R	GTGTTAGTATCAAAAGAGGCTCC		

[a]Bracketed nucleotides denote positions at which two alternative nucleotides were incorporated to accommodate known sequence variations between *stx* types.

3. Methods

3.1. Specimen Preparation

1. Inoculate 5 mL LB broth with feces (do not use more than approximately one rice grain of specimen per tube, otherwise inhibition of subsequent PCR reactions may occur) or reference STEC strain and incubate at 37°C for 4–16 h. An uninoculated broth should be included in each run as a negative control.
2. Transfer 1.0 mL of each culture to 1.5 mL tubes and microfuge for 1 min at 15,000g (*see* **Note 2**). Tip off the supernatant and then carefully remove all of the remaining fluid with a micropipet.
3. Resuspend each pellet in 95 µL of TS buffer, add 5 µL of lysis solution, and incubate at 37°C for 20 min.
4. Add 3 µL of proteinase K and incubate at 65°C for 60 min and then at 95°C for 20 min (to inactivate the proteinase K).
5. Add 500 µL of 95% ice-cold ethanol and precipitate the DNA at –70°C for 30 min.
6. Microfuge samples at 15,000g for 15 min, remove the supernatant fractions, and dry the pellets either under vacuum or by heating at 65°C for 10 min.
7. Resuspend the dried pellets in 300 µL of TE (*see* **Note 3**).

The above method is somewhat cumbersome and labor intensive, but inclusion of the ethanol precipitation step is effective at removing inhibitors, and the DNA extract is stable for many months (even >1 yr) at 4°C. However, for laboratories with high throughput, we have provided an alternative rapid method (*see* **Note 4**) employing Chelex 100 ion-exchange resin (Bio-Rad Laboratories) to remove inhibitory substances from proteinase K digests. Sensitivity is comparable to the above method when fresh extracts are tested, but we have found that the Chelex extracts are less stable, and false-negative results are occasionally obtained when extracts are stored for more than a few days at 4°C before analysis.

3.2. Screening PCR for Presence of stx Genes

1. For each sample to be tested combine 5 µL 10X PCR buffer, 8 µL dNTP mix, 0.25 µL each of primers stxUF and stxUR (*see* **Table 1**), 1 U *Taq* polymerase, and sterile water to a total volume of 47 µL. It is best to prepare this as a separate master reagent mix sufficient for the total number of assays required (allowing one additional assay volume) and then dispense 47-µL aliquots into 0.5-mL PCR reaction tubes.
2. Add 3 µL of DNA extract to the reaction tube, flick mix, and pulse microfuge (or shake) to ensure that the reaction mix is at the bottom of the tube (*see* **Notes 5** and **6**). Each run should include analysis of DNA extracts from a reference STEC strain(s) (positive control), as well as the uninoculated broth extract (negative control) and a PCR reaction mix with 3 µL TE instead of a DNA extract (reagent blank) (*see* **Note 7**).
3. Place tubes in a thermal cycler and subject to 35 PCR cycles, each consisting of 1-min denaturation at 95°C, 1-min annealing at 50°C, and 1-min elongation at 72°C (*see* **Note 8**).

4. Prepare a 2% agarose gel in TBE/EtBr and when set place in a horizontal electro-phoresis apparatus with the buffer covering the gel.
5. Combine 20 µL of each PCR reaction with 5 µL loading buffer and load into the wells of the agarose gel. Also, load DNA size markers in at least one track.
6. Electrophorese at 5–8 V/cm until tracker dye approaches the end of the gel.
7. Visualize the DNA bands in the EtBr-stained gel on a UV transilluminator and photograph with a Polaroid camera or another gel documentation system.
8. A given sample is deemed positive for *stx* if a PCR product of the appropriate size (212–215 bp) is visible (*see* **Note 9**). The assay is invalid if such a product appears in either of the negative control or reagent blank tracks.

3.3. Multiplex PCR Analysis of stx-Positive Samples

A positive result in the *stx* screening assay indicates unequivocally that the patient is infected with an Stx-producing organism, thereby establishing a diagnosis. The following multiplex PCR assays can be used to independently confirm the positive screening result and to provide additional information on the genotype of the infecting strain (*see* **Note 10**).

1. Prepare and dispense PCR master reagent mixes as in **Subheading 3.2.**, but replace the screening primers with stx1F, stx1R, stx2F, stx2R, eaeF, eaeR, hlyAF, and hlyAR for Assay 1 or with O157F, O157R, O111F, O111R, O113F, and O113R for Assay 2 (0.25 µL of each primer per assay) (*see* **Table 1**).
2. Add 3 µL of DNA extract, mix, and place in the thermal cycler as in **Subheading 3.2.** (*see* **Note 11**).
3. Subject samples to 35 PCR cycles, each consisting of 1-min denaturation at 95°C, 2-min annealing at 65°C for the first 10 cycles, decrementing to 60°C by cycle 15, and 1.5-min elongation at 72°C, incrementing to 2.5 min from cycles 25 to 35 (*see* **Note 12**).
4. Analyze PCR products by agarose gel electrophoresis as in **Subheading 3.2.**
5. For each sample, score as positive or negative for each gene tested in accordance with the size of the PCR products (expected sizes are listed in **Table 1**) and by comparison with the mobility of PCR products obtained for the reference STEC samples (*see* **Note 13**).

3.4. Procedures for Minimizing PCR Contaminations

A frequent criticism of diagnostic PCR assays is the possibility of false posi-tives caused by contamination of PCR reactions. In the overwhelming majority of cases, this contamination emanates from previous PCR reactions. Contami-nation of a batch of reagents is easily detected by the presence of a PCR prod-uct in the negative control or the reagent blank included in each PCR run, but spot contamination of individual tubes is harder to identify. Nevertheless, care-ful adherence to appropriate work practices and physical separation of the areas where preamplification and postamplification steps are performed can minimize these problems. Ideally, specimens and reagents should be prepared and reactions set up in a dedicated room, located as far away as possible from the place where the thermal cycler is housed and the PCR products are detected. This "PCR clean" room should be fitted with an UV light to facilitate decontamination, and work surfaces and equipment should be regularly cleaned with hypochlorite or dilute HCl. Staff should wash their hands and remove potentially contaminated gowns or lab coats worn elsewhere in the laboratory before entering, and then put on clean gowns and gloves once inside. Items of equipment and consumables brought into the clean room should be decontaminated beforehand (by exposure to UV or swabbing with hypochlorite) wherever practicable. Equipment items such as micropipets and microfuges should be dedicated to the facility. Use of plugged micropipet tips is also recommended to prevent aerosol carryover, although these are more expensive than conventional tips.

4. Notes

1. Gelatin is included in the PCR buffer to minimize degradation of the *Taq* polymerase by residual proteinase K activity; the detergents also appear to improve the performance of the *Taq* polymerase, particularly with crude DNA preps.
2. Retain the remainder of the culture so that attempts can be made to isolate STEC from any positive samples and/or for other analyses.
3. A gentle flick-mix is all that is required. This will resuspend the DNA, but other precipitated cell debris may remain pelleted.
4. Pellet the broth culture and remove the supernatant as in **step 3.1.2.**, and then resuspend in 75 μL of sterile water, 3 μL of proteinase K (20 mg/mL), and 25 μL of Chelex 100 (prepared as a 20% [w/v] suspension in sterile water [pH should be ≥ 9.5]). Vortex the tubes and then incubate at 65°C for 60 min and then at 95°C for 20 min. Vortex again and store at 4°C. Microfuge samples for 3 min to pellet the Chelex resin immediately before withdrawing 3 μL of the supernatant for PCR template.
5. Do not set up reactions on ice, as this may promote spurious annealing.
6. If the thermal cycler used does not have a heated lid facility, overlay each reaction with 50 μL mineral oil to prevent evaporation.
7. In the event of a contamination, this will enable one to determine whether the DNA extraction solutions or the PCR reagents are the source.
8. Optimal amplification parameters may vary from machine to machine and so it may be necessary to adjust the annealing temperature to maintain specificity.
9. Bands of the incorrect size may be seen in occasional samples, presumably a consequence of mispriming. These can be eliminated by increasing the annealing temperature, although sensitivity may be somewhat lower.
10. More than one STEC type may be present in a given sample and so the multiplex PCR profiles may represent the sum of these strains.
11. Include extracts of a sufficient range of STEC reference strains to cover all genes tested.

12. The reason for decrementing annealing temperatures from 65°C to 60°C is to enable only high-specificity annealing during the early PCR cycles (65°C is above the T_m for the primers), but then reduce annealing to below the T_m for the latter part of the assay to improve efficiency of amplification. Elongation time is incremented to maximize efficiency during the latter stages of the cycling.

13. The multiplex assays are not quite as sensitive as the *stx*-screening assay, because annealing temperatures have to be kept high in order to maintain specificity. Thus, if the initial *stx* PCR screen is only weakly positive, it may not be possible to confirm this by multiplex analysis.

References

1. Karmali, M. A. (1989) Infection by verocytotoxin-producing *Escherichia coli*. *Clin. Microbiol. Rev.* **2**, 15–38.
2. Paton, J. C. and Paton, A. W. (1998) Pathogenesis and diagnosis of Shiga toxin-producing *Escherichia coli* infections. *Clin. Microbiol. Rev.* **11**, 450–479.
3. Jackson, M. P., Newland, J. W., Holmes, R. K., and O'Brien, A. D. (1987) Nucleotide sequence analysis of the structural genes for Shiga-like toxin I encoded by bacteriophage 933J from *Escherichia coli*. *Microb. Pathogen.* **2**, 147–153.
4. Jackson, M. P., Neill, R. J., O'Brien, A. D., Holmes, R. K., and Newland, J. W. (1987) Nucleotide sequence analysis and comparison of the structural genes for Shiga-like toxin I and Shiga-like toxin II encoded by bacteriophages from *Escherichia coli*. *FEMS Microbiol. Lett.* **44**, 109–114.
5. Yu, J. and Kaper, J.B. (1992) Cloning and characterization of the eae gene of enterohaemorrhagic *Escherichia coli* O157:H7. *Mol. Microbiol.* **6**, 411–417.
6. Schmidt, H., Beutin, L., and Karch, H. (1995) Molecular analysis of the plasmid-encoded hemolysin of *Escherichia coli* O157:H7 strain EDL 933. *Infect. Immun.* **63**, 1055–1061.
7. Bilge, S. S., Vary, J. C., Dowell, S. F., and Tarr, P. I. (1996) Role of the *Escherichia coli* O157:H7 O side chain in adherence and analysis of an *rfb* locus. *Infect. Immun.* **64**, 4795–4801.
8. Bastin, D. A. and Reeves, P. R. (1995) Sequence analysis of the O antigen gene (*rfb*) cluster of *Escherichia coli* O111. *Gene* **164**, 17–23.
9. Paton, A. W., Paton, J. C., Goldwater, P. N., and Manning, P. A. (1993) Direct detection of *Escherichia coli* shiga-like toxin genes in primary fecal cultures using the polymerase chain reaction. *J. Clin. Microbiol.* **31**, 3063–3067.
10. Paton, A. W. and Paton, J. C. (1998) Detection and characterization of Shiga toxigenic *Escherichia coli* using multiplex PCR assays for *stx1*, *stx2*, *eaeA*, Enterohemorrhagic *E. coli hlyA*, *rfbO111* and *rfbO157*. *J. Clin. Microbiol.* **36**, 598–602.
11. Paton, A. W. and Paton, J. C. (1999) Direct detection of Shiga toxigenic *Escherichia coli* strains belonging to serogroups O111, O157, and O113 by multiplex PCR. *J. Clin. Microbiol.* **37**, 3362–3365.
12. Ostroff, S. M., Tarr, P. I., Neill, M. A., Lewis, J. H., Hargrett-Bean, N., and Kobayashi, J. M. (1989) Toxin genotypes and plasmid profiles as determinants of systemic sequelae in *Escherichia coli* O157:H7 infections. *J. Infect. Dis.* **160**, 994–999.

13. Kleanthous, H., Smith, H. R., Scotland, S. M., Gross, R. J., Rowe, B., Taylor, C. M., et al. (1990) Haemolytic uraemic syndromes in the British Isles, 1985–8: association with Verocytotoxin producing *Escherichia coli*. Part 2: Microbiological aspects. *Arch. Dis. Child.* **65,** 722–727.

14. Barrett, T. J., Kaper, J. B., Jerse, A. E., and Wachsmuth, I. K. (1992) Virulence factors in Shiga-like toxin-producing *Escherichia coli* isolated from humans and cattle. *J. Infect. Dis.* **165,** 979–980.

15. Schmidt, H. and Karch, H. (1996) Enterohemolytic phenotypes and genotypes of Shiga toxin-producing *Escherichia coli* O111 strains from patients with diarrhea and hemolytic-uremic syndrome. *J. Clin. Microbiol.* **34,** 2364–2367.

16. Paton, A. W., Woodrow, M. C., Doyle, R. M., Lanser, J. A., and Paton, J. C. (1999) Molecular characterization of a Shiga-toxigenic *Escherichia coli* O113:H21 strain lacking *eae* responsible for a cluster of cases of hemolytic-uremic syndrome. *J. Clin. Microbiol.* **37,** 3357–3361.

17. Karmali, M. A., Petric, M., Lim, C., Fleming, P. C., Arbus, G. S., and Lior, H. (1985) The asociation between hemolytic uremic syndrome and infection by Verotoxin-producing *Escherichia coli*. *J. Infect. Dis.* **151,** 775–782.

5

Molecular Typing Methods for STEC

Haruo Watanabe, Jun Terajima, Hidemasa Izumiya, and Sunao Iyoda

1. Introduction

In order to investigate the relatedness of Shiga toxin-producing *Escherichia coli* STEC strains isolated from outbreaks or sporadic cases, traditional methods of strain typing, such as serotyping and bacteriophage typing, have been used. However, newer molecular typing methods have been recently introduced in many laboratories, which include plasmid fingerprinting *(1)*, ribotyping *(2)*, polymerase chain reaction (PCR)-based methods such as RAPD (random-amplified polymorphic DNA)-PCR *(3)*, and AFLP (amplified fragment length polymorphism) *(4)*, and analysis of chromosomal DNA restriction patterns by pulsed-field gel electrophoresis (PFGE) *(5)*. In this chapter, we describe three types of method for investigating relatedness of STEC strains: PFGE, RAPD-PCR, and AFLP.

In ordinary agarose gel electrophoresis, DNA molecules larger than a certain size migrate at about the same rate, called the "limiting mobility," and cannot be sieved by the gel. By cyclically varying the orientation of the electric field in the gel during the run, it was shown that molecules that would otherwise run at the limiting mobility could be resolved *(6)*. There are several systems for alternating-angle electrophoresis, but the most commonly used is contour-clamped homogeneous electric fields *(7)*. Because the size of molecules to be resolved in the gel is less than 1000 kb in the case of STEC, several kinds of PFGE apparatus are commercially available. PFGE subtyping has been successfully applied to the subtyping of many pathogenic bacteria, including STEC O157:H7 *(8)*. We have investigated STEC isolates from outbreaks and sporadic cases in Japan since 1996 *(9,10)*. The method has worked effectively

From: *Methods in Molecular Medicine, vol. 73: E. coli: Shiga Toxin Methods and Protocols*
Edited by: D. Philpott and F. Ebel © Humana Press Inc., Totowa, NJ

to identify STEC isolates from multiple sporadic cases that were epidemiologically related *(11)*. A set of guidelines for interpreting DNA restriction patterns by PFGE has been recently published *(12)*.

Random-amplified polymorphic DNA–PCR, also called AP (arbitrarily primed)-PCR, is one of the molecular typing methods applicable for bacterial isolates *(13,14)*. In this method, DNA fragments with sequences sandwiched between inverted repeat sequences homologous to a primer are randomly amplified *(15)*. Patterns of bands of amplified fragments after electrophoresis vary among isolates if their sequences of genome are different from each other. The detection of such a difference allows one to distinguish between strains.

The AFLP™ (amplified fragment length polymorphism) technique is based on the selective PCR amplification of restricted DNA fragments. It has been applied to various organisms, including several bacteria *(16–18)*, and has been evaluated as one of the highly sensitive DNA subtyping methods. We summarize here the AFLP procedure using the AFLP Microbial Fingerprinting kit (PE Applied Biosystems). As the supplied selective primers complementary to the restriction site are fluorescently labeled, amplified fragments can be analyzed automatically on an automated DNA sequencer along with in-lane size standards. This enables standardization of fragment sizes and facilitates identification of polymorphic bands.

2. Materials
2.1. PFGE

1. Disposable plastics: Falcon 2054 tubes; sample plug mold (Bio-Rad); 15-mL tubes with screw cap (Sarsstedt); Eppendorf tubes (1.5 mL, 2.0 mL); inoculating loops (Greiner).
2. STEC O157:H7 strains for analysis.
3. Low-melt preparative-grade agarose (Bio-Rad); pulsed-field certified agarose (Bio-Rad).
4. Tris-EDTA (TE): 10 mM Tris-HCl, 1 mM EDTA, pH 8.0.
5. Tris-borate–EDTA buffer (TBE): 45 mM Tris-borate, 1 mM EDTA.
6. Enzyme buffer: Sure/Cut Buffer H, Roche.
7. Lysozyme solution: 1 mg/mL lysozyme in 0.5 M EDTA, pH 8.0.
8. Lysis buffer: 1 mg/mL proteinase K (Boehringer Mannheim), 1% *N*-lauroylsarcosine (Sigma) in 0.5 M EDTA, pH 8.0.
9. Restriction enzyme (*Xba*I, Roche).
10. Reagents: Pefabloc SC (Roche), ethidium bromide.
11. DNA size standard (Lambda ladder, Bio-Rad).
12. CHEF-DRIII system (Bio-Rad), including power module, electrophoresis box, variable-speed pump, 14-cm × 13-cm gel casting stand, 20-well comb and comb holder, tygon tubing, leveling bubble.

2.2. RAPD-PCR

1. Disposable plastics: 90-mm Petri dishes; inoculating loop (Greiner, Germany); 0.2-mL thin-wall plastic tube; 1.5-mL polypropylene tube; plastic pipet tips.
2. Thermal cycler PCR2400/9600/9700 (Perkin-Elmer)
3. Electrophoresis apparatus.
4. Nutrient agar (Difco).
5. Lysis mixture: 10% (w/v) dextrose, 25 mM Tris-HCl, 1 mM EDTA (pH 8.0), 5 mg/mL lysozyme. Add lysozyme prior to use.
6. 10% Sodium dodecyl sulfate (SDS).
7. Tris-saturated phenol.
8. Chloroform.
9. 3 M Sodium acetate (pH 5.3).
10. 70% and 99% Ethanol.
11. TE: 10 mM Tris-HCl, 1 mM EDTA (pH 8.0).
12. *Taq* DNA polymerase (5 U/µL).
13. 10X amplification buffer: 100 mM Tris-HCl (pH 8.3), 500 mM KCl, 15 mM MgCl$_2$, 0.01% gelatin.
14. dNTP solution containing each 2.5 mM of dATP, dTTP, dCTP, and dGTP.
15. Primer solution: 10 µM of the primer (e.g., G11; 5'-TGCCCGTCGT-3') in TE.
16. 1X TAE electrophoresis buffer: 0.04 M Tris-acetate, 1 mM EDTA.
17. Agarose.
18. Gel-loading buffer: 0.25% xylene cyanol, 0.25% bromophenol blue, 30% glycerol.
19. DNA size marker: lambda DNA(30 ng/µL) digested with *Hin*dIII.
20. Gel-staining solution: 0.1 µg/mL ethidium bromide.
21. Ultraviolet (UV) transilluminator.

2.3. AFLP

All of the materials for AFLP are essentially the same as that describing on the protocol of the AFLP Microbial Fingerprinting kit (PE Applied Biosystems). This kit consists of the following three individual modules:

1. AFLP *Eco*RI Ligation/Amplification Module (consists of *Eco*RI adaptor, *Eco*RI core sequence, and nine selective primers).
2. AFLP *Mse*I Ligation/Amplification Module (consists of *Mse*I adaptor, *Mse*I core sequence, and nine selective primers).
3. AFLP Amplification Core Mix Module (consists of preselective primers, buffer, nucleotides, and AmpliTaq DNA polymerase®).

Other materials, which are not supplied from the kit are as follows:

4. Disposable plastic tubes: sterile 0.5-mL microcentrifuge tubes for sample preparation; sterile 0.2-mL thin-walled tubes for PCR reaction.
5. Thermal cycler (GeneAmp PCR systems 9600, 2400, or DNA Thermal Cycler 480 [PE Applied Biosystems]).
6. Total DNA of STEC for analysis.
7. *Eco*RI and *Mse*I restriction endonuclease.

8. T4 DNA ligase.
9. 10X T4 DNA ligase buffer containing 10 mM ATP.
10. 0.5 M NaCl (autoclaved).
11. TE buffer (autoclaved)
12. ABI 373, 377 or 310 DNA Sequencer (PE Applied Byosystems) with necessary reagents and software for analysis (all of which are available from PE Applied Biosystems).

3. Methods

3.1. PFGE

3.1.1. Embedding of Bacteria Into Agarose–Gel Block

1. By using an inoculating loop, transfer appropriate amount of STEC O157:H7 from the plate into the Falcon 2054 tube containing 200 µL of sterile water, so that the optical density (OD) at 600 nm of the suspension will be approx 0.2 (*see* **Note 1**).
2. Add equal amount of melted 1% of low-melt agarose solution into the tube and after mixing dispense the mixture into two wells of plug mold per sample, which were prechilled on ice. Leave the plug molds on ice at least for 20 min to solidify the agarose (*see* **Note 2**).

3.1.2. Lysis of the Samples

1. Remove the tape on the bottom of the plug mold and carefully push out the hardened gel block into a 15-mL tube containing 1 mL lysozyme solution.
2. Incubate, with gentle agitation, for more than 1 h at 37°C.
3. Pour off the lysozyme solution from the tube and add 1 mL of lysis buffer.
4. Incubate, with gentle agitation, overnight at 50°C. After the incubation, samples can be stored in the same buffer at 4°C for a few years.

3.1.3. Digestion of Samples by Restriction Enzyme

1. Take a piece of gel from the lysis buffer using a spatula and cut it into half with a cover glass.
2. Transfer the cut piece to a 2.0 mL U-bottom Eppendorf tube containing 0.5 mL of 4 mM Pefabloc SC in TE (*see* **Note 3**).
3. Incubate tubes at 50°C with gentle agitation at least for half an hour.
4. After the second incubation, take as much as possible out of the buffer from the tube and add 1 mL of TE into a tube (*see* **Note 4**).
5. Incubate the tube on ice with gentle agitation for half an hour.
6. Change TE to 0.2 mL of Sure/Cut H buffer and incubate the tube on ice at least for half an hour.
7. Change the buffer to 0.1 mL of freshly prepared Sure/Cut H buffer containing 30 U of restriction enzyme *Xba*I.
8. Incubate the tube at 37°C with gentle agitation at least for 4 h.
9. After the incubation, change the buffer containing *Xba*I to 0.5X TBE and leave the tubes on ice until loading samples into wells of the gel.

3.1.4. Loading of Restriction Digest

1. Prepare 100 mL of 1% of pulsed-field certified agarose gel for a horizontal gel apparatus (14 cm × 13 cm) using 0.5X TBE buffer.
2. Dissolve the agarose completely in a microwave oven and the melted agarose should be cooled down to approximately 55°C before pouring onto the gel apparatus.
3. Let the agarose solidify.
4. Transfer a sample onto the gel and insert it into a well using a spatula or a Pasteur pipet that is made to have L-shaped edge on its end. Gently push plugs to bottom and be sure that there are no bubbles (*see* **Note 5**).
5. Insert any molecular size markers such as Lambda DNA ladder in both end wells of the gel.

3.1.5. Electrophoresis

1. Set the gel box horizontally on a bench. Pour 2 L of 0.5X TBE buffer that was made the previous day and kept at 4°C overnight. Place the gel into the gel box.
2. Adjust the circulation pump for an appropriate flow.
3. Make sure that surface of the gel is set in horizontal and covered with the buffer.
4. Set the electrophoretic apparatus with the following conditions: 6 V/cm, buffer temperature 12°C, included angle at 120°, a linearly ramped switching time from 4 s to 8 s for 9 h and then a linearly ramped switching time from 8 s to 50 s for 13 h (*see* **Note 6**).
5. After electrophoresis, take the gel out of the gel box and stain with 0.2 µg/mL of ethidium bromide at room temperature for half an hour with gentle shaking.
6. Destain with distilled water at least for half an hour and photograph the gel under illumination with UV light (*see* **Notes 7** and **8**).

3.2. RAPD-PCR

3.2.1. Preparation of total DNA

1. Streak an STEC strain onto a nutrient agar plate and incubate at 37°C overnight.
2. Suspend colonies into a microtube containing 200 µL of lysis mixture at approx OD 1 of 600 nm.
3. Incubate at 37°C for 30 min.
4. Add 10 µL of 10% SDS and mix the solution well. Heat at 65°C for 30 min.
5. Spin down briefly in a microfuge.
6. Add 100 µL of Tris-saturated phenol and vortex vigorously for 30 s (*see* **Note 9**).
7. Spin down briefly.
8. Add 100 µL of chloroform and vortex vigorously for 30 s.
9. Centrifuge at 12,000g for 10 min in a microfuge.
10. Transfer the aqueous phase (approx 200 µL) to a fresh microtube. Discard the interface and organic phase.
11. Add 20 µL of 3 *M* sodium acetate and 500 µL of ice-cold ethanol and mix the solution well (*see* **Note 10**).
12. Store the solution at −20°C for 30 min.

13. Recover the DNA by centrifugation at 12,000 rpm at 4°C for 10 min.
14. Remove the supernatant carefully.
15. Add 300 µL of ice-cold 70% ethanol and centrifuge at 12,000 rpm at 4°C for 5 min.
16. Remove the supernatant.
17. Evaporate the last traces of fluid by drying.
18. Dissolve the DNA in 300 µL of TE.

3.2.2. PCR and electrophoresis

1. In a sterile 0.2-mL tube, mix the reaction on ice as follows:

Sterile water	18 µL
10X Amplification buffer	2.5 µL
dNTP solution	2 µL
Primer solution	1 µL
Taq DNA polymerase	0.5 µL
Template DNA	1 µL

2. Carry out PCR amplification as following program:
   95°C for 3 min
   35 cycles of 93°C for 30 s
   37–55°C for 30 s
   72°C for 60 s
   72°C for 3 min (*see* **Note 11**).
3. Prepare a 1.2% agarose gel in 1X TAE buffer.
4. Load a total of 10 µL of the PCR reaction, which is mixed with 2 µL of gel-loading buffer, onto the gel, and carry out electrophoresis.
5. Stop electrophoresis when the dye of bromophenol blue reach about three-fourths of the gel.
6. Soak the gel in gel-staining solution for 30 min at room temperature.
7. Soak the gel in water for destaining.
8. Take a photograph under UV illumination.

3.3. AFLP (see Note 12)

The AFLP procedure consists of the following four major steps:

1. Digestion of total DNA with two restriction endonucleases (*Eco*RI and *Mse*I).
2. Ligation of adaptors that have complementary sequences to *Eco*RI or *Mse*I.
3. Polymerase chain reaction reaction with preselective primers having complementary sequences to adapters. Next, PCR reaction using selective primer set, which has additional nucleotides (one or two out of A, T, G, or C) at the 3'-end.
4. Analysis of PCR products by automated DNA sequencer.

We follow the protocol of the AFLP Microbial Fingerprinting kit except for preparing the total DNA of STEC. For more details of each step, *see* the manufacturer's protocol. We describe here the important considerations for analysis.

3.3.1. Selection for the Primer Set

Before starting to analyze actual samples, it is important to select the best selective primer combination among available primer sets supplied from the AFLP Microbial Fingerprinting kit. For this purpose, subtyping data obtained from PFGE- or RAP -analysis should be very helpful. Even if they are not available, epidemiological data with tester strains should be helpful also. In this case, a criterion to be taken into account is that the isolates derived from the apparently same origin should show the same AFLP pattern, wheras epidemiologically unlinked isolates should show different pattern. We choose the *Eco*RI+A/*Mse*I+C primer set as a best for STEC O157:H7 analysis *(18)*.

3.3.2. Evaluation of AFLP Results

Because there has been no gold standard for the AFLP analysis, evaluation of AFLP results should be done in comparison with that of PFGE or RAPD. In our analysis for STEC O157:H7, AFLP data have been comparable to that of PFGE in most cases *(18)*. However, interpretation of AFLP results should be done with caution when the AFLP pattern is the same between the two isolates. Results from epidemiological analysis should be considered for linkage of the isolate to the outbreak.

3.4. Concluding Remarks

We have used these three methods and phage typing for the analysis of STEC O157:H7 strains isolated in epidemiologically related outbreaks in Osaka, Wakayama, and Kyoto and unrelated outbreaks and sporadic cases of STEC infection. The results summarizing the ability of each of these methods to discriminate the different strains is shown in **Table 1**. PFGE is most powerful for the discrimination of strains isolated from epidemiologically unrelated outbreaks and sporadic cases. However, it takes one or more days to get a reliable result. AFLP seems to give a result comparable to PFGE; however, it requires a DNA sequences for the analysis. RAPD-PCR is rapid for obtaining results, however, it is less sensitive for discrimination when there are little differences among strains. Phage typing, which is a more classical method, is also a sensitive method comparable to PFGE. Each method has some advantages and disadvantages. One can choose one of these methods according to each circumstance. However, a combination of these methods may increase the ability to discriminate between strains and help to determine the relatedness of strains *(18,20)*.

4. Notes
1. The amount of DNA (e.g., amount of bacteria) is adjusted appropriately by the density of the suspension so that half of the gel block is suitable for further analy-

Table 1
PFGE, RAPD-PCR, AFLP, and Phage Typings of STEC O157:H7 Isolates

Origin, city isolated	Date isolated	Types of pattern by			
		PFGE	RAPD-PCR	AFLP	Phage
Human isolates					
Hiroshima	June 1996	Ia	I	A	21
Gifu	June 1996	Ib	I	A	40
Okayama (Oku)	May 1996	Ic	I	A	40
Aichi	June 1996	Ib	I	A	21
Osaka (Sakai)	July 1996	IIa	IIe	C	32
Wakayama (Hashimoto)	July 1996	IIa	IIe	C	32
Kyoto	July 1996	IIa	IIe	C	32
Tokyo	June 1996	IId	IIe	D	40
Iwate	Sept. 1996	IIj	IIe	E	14
Gunma	June 1996	IV	IIe	F	2
Hokkaido	Aug 1996	VI	IIe	F	8
Nonhuman isolates					
Kanagawa	June 1996	Va	IIe	G	26
Okinawa	Nov. 1996	IIIb	IIm	H	23

sis by PFGE. The use of low-melt agarose for preparing sample gel blocks makes it fragile to handle but certainly it increases the exchange rate of reagents in and out of the gel.

2. Chromosomal-grade agarose can be substituted for low-melt agarose because it is easier to handle the gel. When chromosomal-grade agarose is used, the incubation period of washing gel blocks should be extended at least to 1 h for each step.

3. Pefablock SC in TE can be prepared as 100 mM stock solution in TE and stored at −20°C. Phenylmethylsulfonyl fluoride (PMSF) can also be used at the concentration of 1 mM, but the incubation time should be extended to 1 h. PMSF is both toxic and volatile and should be handled with precaution. It is unstable in aqueous solution, so solutions should always be freshly prepared.

4. Be sure to remove all of the liquid during this and subsequent wash steps because appropriate buffer condition will be required when the plugs are digested with restriction enzymes.

5. Loading the plug slices can be tedious. Cutting a plug in an appropriate size (i.e., cutting a plug a little smaller than the width of a well) may help to put a plug into a well easily.

6. It is important that the temperature of the gel (buffer) should remain at the same temperature during the run to maintain sharpness of the bands.

7. The addition of destaining process after staining will certainly decrease the background and give better contrast in the photograph.
8. We have so far examined more than 3000 strains by PFGE, and we have found that PFGE patterns always show some difference between epidemiologically unrelated isolates. Genomic DNA of STEC O157:H7 seems to be unstable and variable. When you find the strains for which PFGE patterns are identical to epidemiologically indeterminate cases, you should consider that the isolates come from epidemiologically linked cases, namely the same outbreak. On the other hand, when you find strains with some difference in PFGE patterns among epidemiologically linked isolates, you should consider the relatedness following the criteria proposed by Tenover et al. *(12)*.
9. Use Tris-saturated phenol but not phenol : chloroform (and isoamyl alcohol) for the total DNA preparation because the yield of DNA is affected.
10. The yield of DNA is not sufficient when the threadlike precipitate cannot be seen after adding ethanol.
11. The cause of inefficient amplification is mainly the result of the following reasons: The yield of total DNA is low or the activity of polymerase is low. They can be overcome by increasing the amount of the DNA or polymerase. It will help the increase of the efficiency to keep all reagents on ice while preparing the reaction mixture of PCR.
    The short length of the primer (10–12 bases) or low annealing temperature (about 37°C) is usually used in order to produce a higher number of amplified DNA fragments. However, some weak bands appearing are not reproducible. Application of a higher annealing temperature (usually 45–50°C, sometimes 55°C) may help to amplify more specific and reproducible DNA fragments.
    Several primers have been reported to classify STEC O157 isolates *(9,19)*. Primers should be chosen dependent on cases. We usually apply the G11 primer (Operon Technologies, Inc.). The combination of a primer and annealing temperature should be optimized to obtain a good resolution among isolates.
12. Pay attention to the DNA preparation as in the case of RAPD-PCR.

References

1. Tenover, F. C. (1985) Plasmid fingerprinting. A tool for bacteria strain identification and surveillance of nosocomial and community-acquired infections. *Clin. Lab. Med.* **5,** 413–436.
2. Stull, T. L., LiPuma, J. J., and Edlind, T. D. (1988) A broad-spectrum probe for molecular epidemiology of bacteria: ribosomal RNA. J. Infect. Dis. 157, 280–286.
3. van Belkum, A. (1994) DNA fingerprinting of medically important microorganism by use of PCR. *Clin. Microbiol. Rev.* **7,** 174–184.
4. Vos, P., Hogers, R., Bleeker, M., Reijans, M., van der Lee, T., Hornes, M., et al. (1995) AFLP: a new technique for DNA fingerprinting. *Nucleic Acids Res.* **23,** 4407–4427.
5. Arbeit, R. D., Arthur, M., Dunn, R. D., Kim, C., Selander, R. K., and Goldstein, R. (1990) Resolution of recent evolutionary divergence among *Escherichia coli*

from related lineages: the application of pulsed-field gel electrophoresis to molecular epidemiology. *J. Infect. Dis.* **161,** 230–235.

6. Schwarts, D, C., Saffran, W., Welsh, J., Hass, R., Goldenberg, M., and Cantor, C. R. (1983) New techniques for purifying large DNAs and studying their properties and packaging. *Cold Spring Harbor Symp. Quant. Biol.* **47(Pt 1),** 189–195.

7. Chu, G., Vollrath, D., and Davis, R. W. (1986) Separation of large DNA molecules by contour-clamped homogeneous electric fields. *Science* **234,** 1582–1585.

8. Barrett, T. J., Lior, H., Green, J. H., Khakhria, R., Wells, J. G., Bell, B. P., et al. (1994) Laboratory investigation of a multistate food-borne outbreak of *Escherichia coli* O157:H7 by using pulsed-field gel electrophoresis and phage typing. *J. Clin. Microbiol.* **32,** 3013–3017.

9. Watanabe, H., Wada, A., Inagaki, Y., Itoh, K., and Tamura, K. (1996) Outbreaks of enterohaemorrhagic *Escherichia coli* O157:H7 infection by two different genotype strains in Japan. *Lancet* **348,** 831–832.

10. Izumiya, H., Terajima, J., Wada A., Inagaki, Y., Itoh, K., Tamura, K., et al. (1998) Molecular typing of enterohemorrhagic *Escherichia coli* O157:H7 isolates in Japan by using pulsed-field gel electrophoresis. *J. Clin. Microbiol.* **35,** 1675–1680.

11. Terajima, J., Izumiya, H., Iyoda, S., Tamura, K., and Watanabe, H. (1999) Detection of a multi-prefectural *E. coli* O157:H7 outbreak caused by contaminated Ikura-Sushi ingestion. *Jpn. J. Infect. Dis.* **52,** 52–53.

12. Tenover, F. C., Arbeit, R. D., Goering, R. V., Mickelsen, P. A., Murray, B. E., Persing, D. H., et al. (1995) Interpreting chromosomal DNA restriction patterns produced by pulsed-field gel electrophoresis: criteria for bacterial strain typing. *J. Clin. Microbiol.* **33,** 2233–2239.

13. Akopyanz, N., Bukanov, N. O., Westblom, Y. U., Kresovich, S., and Berg, D. E. (1992) DNA diversity among clinical isolates of *Helicobacter pylori* detected by PCR-based RAPD fingerprinting. *Nucleic Acids Res.* **20,** 5137–5142.

14. Inagaki, Y., Myoga, F., Kawabata, H., Yamai, S., and Watanabe, H. (2000) Genomic differences in *Streptococcus pyogenes* serotype M3 between recent isolates associated with toxic shock-like syndrome and past clinical isolates. *J. Infect. Dis.* **181,** 975–983.

15. Williams, J. G. K., Kubelik, A. R., Livak, K. J., Rafalski, J. A., and Tingey, S. V. (1990) DNA polymorphisms amplified by arbitrary primers are useful as genetic markers. *Nucleic Acids Res.* **18,** 6531–6535.

16. Dijkshoorn, L., Aucken, H., Gerner-Smidt, P., Janssen, P., Kaufmann, M. E., Garaizar, J., et al. (1996) Comparison of outbreak and nonoutbreak *Acinetobacter baumannii* strains by genotypic and phenotypic methods. *J. Clin. Microbiol.* **34,** 1519–1525.

17. Nandi, S., Subundhi, P.K., Senadhira, D., Manigbas, N.L., Sen-Mandi, S., and Huang, N. (1997) Mapping QTLs for submergence tolerance in rice by AFLP analysis and selective genotyping. *Mol. Gen. Genet.* **255,** 1–8.

18. Iyoda, S., Wada, A., Weller, J., Flood, S.J., Schreiber, E., Tucker, B., et al. (1999) Evaluation of AFLP, a high-resolution DNA fingerprinting method, as a tool for molecular subtyping of enterohemorrhagic *Escherichia coli* O157:H7 isolates. *Microbiol. Immunol.* **43,** 803–806.

19. Madico, G., Akopyants, N. S., and Berg, D. G. (1995) Arbitrarily primed PCR DNA fingerprinting of *Escherichia coli* O157:H7 strains by using templates from boiled cultures. *J. Clin. Microbiol.* **33,** 1534–1536.
20. Izumiya, H., Masuda, T., Ahmed, R., Khakhria, R., Wada, A., Terajima, J., et al. (1998) Combined use of bacteriophage typing and pulsed-field gel electrophoresis in the epidemiological analysis of Japanese isolates of enterohemorrhagic *Escherichia coli* O157:H7. *Microbiol. Immun.* **42,** 515–519.

6

STEC in the Food Chain

Methods for Detection of STEC in Food Samples

Michael Bülte

1. Introduction

Since the first outbreak caused by Shiga toxin-producing *Escherichia coli* (STEC) of serovar O157:H7 in 1982, this agent has emerged as a food-borne pathogen leading to hemorrhagic colitis (HC), hemolytic uremic syndrome (HUS), and thrombotic thrombocytopenic purpura (TTP). In addition to the protopathotype O157:H7, other enterohemorrhagic *E. coli* (EHEC) serovars (e.g., O26:H11, O103:H2, O111:H⁻, O145:H⁻, and O157:H⁻, have caused severe infections in humans (*1*). Nearly all of STEC strains contain the *E. coli* attaching and effacing gene (*eae*) encoding the outer membrane protein intimin, which mediates the attachment to enterocytes leading to irreversible destruction of the microvilli (*2*).

Many outbreaks have been linked to foods containing raw or undercooked beef, raw milk, apple cider, salami sausages, sprouts, and potable water. Considering food safety management, rapid and reliable methods for the detection of STEC, especially O157 strains, in the food chain are urgently needed. The procedures described in this chapter for the detection of STEC in foods include (1) the immunomagnetic separation (IMS) of *E. coli* O157, (2) a polymerase chain reaction (PCR) screening method for Shiga toxin-producing *E. coli*, (3) a panel of different PCR amplification protocols for the detection of *stx* variants and the *eae* gene, and (4) a colony hybridization assay using DIG-labeled probes.

From: *Methods in Molecular Medicine, vol. 73: E. coli: Shiga Toxin Methods and Protocols*
Edited by: D. Philpott and F. Ebel © Humana Press Inc., Totowa, NJ

2. Materials

2.1. Media for Cultivation and Isolation of STEC Strains

1. Brilliant green lactose broth (BRILA [Merck]).
2. Modified Tryptone Soja broth with novobiocin (50 µg/L) (mTSB+N).
3. MacConkey agar.
4. Cefixime tellurite sorbitol MacConkey agar (CT-SMAC *[3]*).
5. Hemorrhagic colitis agar (HC *[4]*).

2.2. Immunomagnetic Separation

1. Immunomagnetic beads (Dynabeads® anti-*E. coli* O157, Dynal)
2. Stomacher 400 (Seward Ltd., London, UK)
3. Immunomagnetic separation equipment: magnetic particle concentrator (Dynal, cat. no. 120.09), sample mixer (Dynal).
4. Microbiological swabs.
5. O157 Agglutination test (Oxoid).

2.3. Polymerase Chain Reaction

1. Polymerase chain-reaction buffer (10X): 500 mM KCl, 100 mM Tris-HCl, 0.1% (w/v) gelatin, 15 mM MgCl$_2$.
2. Deoxynucleotide triphosphates: dATP, dCTP, dGTP, dTTP each at 10 mM in distilled water.
3. Digoxigenin 11 dUTP (Roche Molecular Biochemicals): 1 mM in distilled water.
4. Primer (*see* **Table 1**).
5. Ampli*Taq* DNA polymerase, at 5 U/µL (PE Biosystem).
6. Template DNA extracted from enrichment broth or bacterial colonies.
7. DNA size 100-bp ladder (Roche Molecular Biochemicals).
8. Agarose.
9. Ethidium bromide (stock solution): 10 mg/mL, stored at 4°C in a bottle wrapped in aluminum foil.
10. TBE buffer (10X): 890 mM Tris-borate, 20 mM EDTA, pH 8.3.
11. Gel loading buffer: 0.25% bromphenol blue, 0.25% xylene cyanol, 25% Ficoll (type 400) in distilled water.
12. Thermal cycler.
13. Equipment for agarose gel electrophoresis.

2.4. Colony Hybridization

1. Nylon membranes for colony hybridization (Roche Molecular Biochemicals).
2. Whatman 3 MM paper.
3. Denaturation solution: 0.5 M NaOH, 1.5 M NaCl.
4. Neutralization solution: 1 M Tris-HCl, 1.5 M NaCl, pH 7.4.
5. SSC (20X): 3 M NaCl, 0.3 M Na citrate, pH 7.0.
6. Proteinase K solution (final concentration: 2 mg/mL SSC (2X() (Roche Molecular Chemicals).

Table 1
Oligonucleotide Primers Used for *stx* Gene and the *eaeA* Gene

Primer acronym	Target gene	Primer sequence (5'–3')	Primer conc. (μM)	PCR conditions			Amplicon size (bp)	Ref.
				Denaturing	Annealing	Extension[a]		
SK1 SK1	*eaeA*[b]	CCC GAA TTC GGC ACA AGC ATA AGC CCC GGA TCC GTC TCG CCA GTA TTC G	30	94°C, 30 s	52°C, 60 s	72°C, 60 s	863	5
MK1 MK2	*stx* general A[c]	TTT ACG ATA GAC TTC TCG AC CAC ATA TAA ATT ATT TCG CTC	25	94°C, 60 s	43°C, 60 s	72°C, 90 s	~230	6
KS7 KS8	*stx₁* A[c]	CCC GGA TCC ATG AAA AAA ACA TTA TTA ATA GCCCC GAA TTC AGC TAT TCT GAG TCA ACG	15	94°C, 30 s	52°C, 60 s	72°C, 40 s	282	7
GK3 GK4	*stx₂* B[c] *stx₂c* B[d]	ATG AAG AAG ATG TTT ATG TCA GTC ATT ATT AAA CTG	10	94°C, 60 s	52°C, 60 s	72°C, 40 s	270 (128+142)	8
VT2d-AM-I VT2d-AM-II	*stx₂d* B[c]	AGG GCC CAC TCT TTA AAT ACA TCC CGT CAT TCC TGT TAA CTG TGC G	10	94°C, 30 s	54°C, 40 s	72°C, 35 s	242	Own data
FK1 FK2	*stx₂e* B[c]	CCC GGA TCC AAG AAG ATG TTT ATA G CCC GAA TTC TCA GTT AAA CTT CAC C	10	94°C, 30 s	53°C, 60 s	72°C, 40 s	280	9

[a]*See* **Note 7**.
[b]*eaeA*: *E. coli* attaching and effacing.
[c]*stx*: Shiga toxin, subunit A or B
[d]For differentiation of the *stx₂c* variant from the classic *stx₂* *Hae*III digestion is used.

7. DIG Easy Hyb (Roche Molecular Biochemicals, Catalog no. 1-603-558) serves as prehybridization and hybridization solution.
8. Washing solution 1: 1 g/L sodium dodecyl sulfate (SDS) in 2X SSC.
9. Washing solution 2: 1 g/L SDS in 0.5X SSC.
10. Maleic acid buffer: 100 mM maleic acid, pH 7.5, 150 mM NaCl, pH 7.5, 0.3% Tween-20.
11. Blocking solution: 1% blocking reagent in maleic acid buffer (*see* **ref. *14***).
12. Detection buffer: 0.1 M Tris-HCl, 0.1 M NaCl, pH 9.5.
13. Tris-EDTA buffer: 10 mM Tris-HCl, 1 mM EDTA, pH 8.1.
14. DIG–Nucleic Acid Detection Kit (Roche Molecular Biochemicals) with blocking reagent, anti-digoxigenin–AP conjugate and NBT/BCIP stock solution.
15. Ultraviolet crosslinker (Hoefer, UVC 500).
16. Hybridization oven/shaker with flasks.

3. Methods

3.1. Detection of E. coli *O157 Strains in Foods by IMS*

1. Homogenize 25 g or mL of food sample with 225 mL prewarmed enrichment broth (37°C) for 2 min in a stomacher, incubate the homogenized sample at 37°C for 6 h, and incubate with shaking for 16–18 h at 100 rpm (*see* **Note 1**).
2. After incubation times of 6 h (for sub culture on HC agar plates) and 16–18 h (for subculture on CT-SMAC–agar plates) dispense 20 µLdynabeads anti-*E. coli* 0157 in a 1.5-mL Eppendorf tube and add 1 mL of enrichment broth.
3. Invert the tube three times and incubate on the rotating sample mixer for 10 min at room temperature.
4. Separate the bead–bacteria complex with the magnetic strip (leave for 3 min and invert the tubes three times after each minute and discard the supernatant with a Pasteur pipet (*see* **Note 2**).
5. Remove the magnetic strip and add 1 mL washing buffer to each tube; invert three times.
6. Replace the magnetic strip and repeat **steps 4** and **5** three times.
7. Discard the supernatant, remove the magnetic strip, add 100 µL maleic acid buffer and resuspend by vortexing.
8. Add 50 µL of bead–bacteria solution on the agar plates and spread the inoculum on one-half of the plate using a sterile standard microbiological swab. With a sterile loop strike out 30–40 times in the third quadrant and with a sterile loop 20–30 times in the fourth quadrant of the plate (*see* **Note 3**).
9. Incubate HC agar for 16–18 h at 41°C and CT-SMAC-agar for 16–18 h at 37°C (*see* **Note 4**). Presumptive O157 strains appear on HC agar as flat, transparent colonies with a diameter of 1–2 mm on the blue background of the medium, on CT-SMAC agar as flat brownish colonies with a darker brown centrum, and a diameter of about 2 mm.
10. Presumptive O157 colonies are serologically confirmed using an O157-agglutination assay.

3.2. Detection of STEC in Foods by PCR (stx *Genes,* see *Table 2*)

1. Add 1 mL of enrichment broth to a 1.5-mL Eppendorf tube and centrifuge at 12,000g for 10 min.
2. Remove supernatant and resuspend in 1 mL distilled water; centrifuge at 12,000g for 10 min.
3. Remove the supernatant and resuspend in 500 µL distilled water. Boil at 100°C for 10 min and cool immediately on ice.
4. PCR mix (on ice): 17.3 µL distilled water, 2.5 µL PCR buffer, 1 µL of each primer (MK1 and MK2), 0.5 µL dNTP-mix, 0.2 µL GoldStar-DNA polymerase, (Eurogentec, Seraing, Belgium) (*see* **Notes 5** and **6**).
5. Perform the PCR according to the appropriate conditions (compare **Table 1** and *see* **Note 7**).
6. Add 2.5 µL gel loading buffer to each tube, mix vigorously, pipet 10 µL in the gel slot of an 1.4% agarose gel, and run the gel at 5 V/cm.
7. Visualize the PCR product by staining with ethidium bromide under ultraviolet (UV) light (256 nm).

3.3. Differentiation of stx2-Variants/Detection of eae *Gene*

1. Suspend three to five colonies into 1 mL of distilled water.
2. Boil for 100°C, 10 min, centrifuge at 12,000g for 10 min (*see* **Note 8**).
3. Add 1 µL of the supernatant to the respective PCR reaction mix (*see* **Table 1**).
4. Perform the specific PCR according to the conditions listed in **Tables 1** and **2** (*see* **Notes 5–7** and **9**).

3.5. Detection of STEC by Colony Hybridization

This method enables the user to detect *stx* positive colonies by hybridization and simultaneously to isolate the presumptive STEC strain. Inoculate the MacConkey agar plate with 0.1 mL enrichment broth of the food homogenate by means of, for example, the spatula technique. When spreading or streaking the inoculum onto the plate, leave a free zone of about 5 mm at the outward side to guarantee that the smaller-sized nylon membrane covers the complete inoculated area. Incubate for 16–18 h at 37°C.

3.5.1. DIG Labeled stx Probes

1. To 62 µL distilled water, add 10 µL PCR buffer (10X), 2 µL each dATP, dCTP, and dGTP (10 mM), 2 µL dTTP (7 mM), 7 µL DIG 11 dUTP (1 mM), 2 µL each primer MK1/MK2 (30 mM), 1 µL GoldStar-DNA polymerase, (5 U/µL), and 8 µL template DNA (from *stx1* or *stx2* reference strain, *see* **Note 10**).
2. Run the PCR according to **Table 1**.
3. Control the successful DIG labeling by gel electrophoresis. Because of the DIG molecule, the labeled oligonucleotide probe migrates slower in the gel than the unlabeled amplicon.

Table 2
PCR Constituents

	Constituent	Volume per tube (µL)	Final concentration
1	Water	17.3	—
2	PCR buffer (10X)	2.5	1X
3	dNTP (10 mM)	0.5	0.2 mM
4	Primer 1 (25 µM)^a	1.0	1 µM
5	Primer 2 (25 µM)	1.0	1 µM
6	Taq DNA polymerase (5 U/µL)	0.2	1 U/reaction
	Total	22.5	
7	Template DNA	2.5	

^{a}For primer MK1/MK2; for other primer concentration, *see* **Table 1**.

3.5.2. Colony Hybridization

1. Cool the incubated agar plate at 4°C for 30 min.
2. Lay and softly press the nylon membrane on the cultured agar plate (master plate) for 1 min (*see* **Note 11**).
3. Transfer the membrane with colonies upward on Whatman 3MM soaked with denaturation solution for 20 min and dry for 5 min on Whatman 3MM.
4. Transfer the membrane on Whatman 3 MM soaked with 2X SSC for 10 min.
5. Spot the control DNA on the edge of the membrane.
6. Fix the DNA to the membrane using an UV crosslinker apparatus (*see* **Note 12**).
7. Treat the membranes with proteinase K solution for 1 h at 37°C.
8. Preheat the prehybridization solution to 42°C and add 10 mL per membrane to the flask, incubate with gentle agitation for 1 h at 42°C.
9. Denature the DIG-labeled probes by boiling for 5 min and cool rapidly on ice.
10. Add the DIG-labeled probes to fresh DIG Easy Hyb solution (*see* **Note 13**).
11. Pour off the prehybridization solution and add the hybridization mixture to the flasks (2 mL/membrane) and incubate at least for 3 h at 42°C.
12. Pour off the hybridization solution and add 20 mL washing solution 1 per membrane and wash twice for 5 min at room temperature, repeat twice with washing solution 2 at 68°C for 15 min.
13. Visualize hybridization by means of immunochemical detection; that is, wash in maleic acid/Tween-20 buffer for 1 min, incubate the membranes in blocking

buffer for 60 min and then incubate the membrane with 15 mL anti-Digoxigenin–AP conjugate for 30 min.

14. Remove the unbound conjugate, wash twice with washing buffer for 15 min at room temperature, and transfer the membrane in 20 mL equilibration buffer for 5 min.
15. Transfer the membrane into a new chamber and add freshly prepared NBT/BCIP solution (200 µL stock solution to 10 mL detection buffer) and incubate in the dark for about 4 h (strong signals may already appear after 1–2 h).
16. Compare the hybridization signals on the membrane (dark blue spots) with the plate to identify the positive colonies on the master plate (*see* **Note 14**).
17. Isolate *stx*-positive colonies for further investigations

4. Notes

1. It is recommended to use the brilliant green lactose broth (BRILA) for meat and meat products, and the modified tryptone soya broth with novobiocin (freshly prepared stock solution of 20 mg/mL destilled water; 1 mL of this solution is added to 1 L (mTSB+N) for other foods.
2. Avoid vigorous washing that may remove the adhering red-brownish bead–bacteria complex.
3. It is strongly recommended to follow the plating technique described in order to obtain entirely separated colonies at least in the fourth quadrant of the agar plate. This is the basis for the detection of presumptive O157:H7 colonies.
4. With this method, only sorbitol-negative O157 strains will be detected. Nevertheless, nearly all strains of the STEC proto-pathotype serovar O157:H7 are sorbitol negative as well as approx 40% of O157:H⁻ strains *(10)*.
5. Always use a negative control that contains all components of the PCR reaction, except the template DNA. For positive controls use the relevant (*stx1*, *stx2*, *eae*) strains provided by type culture collection.
6. It is convenient to prepare master mixes in advance, which can be stored at –20°C for at least 3 mo.
7. The final extension is prolonged for 10 min to ensure that all of the amplifying DNA strands are fully extended.
8. We found that it is not necessary to extract the DNA by "classical" methods (i.e., chloroform isoamyl alcohol/phenol extraction). The ordinary boiling method works satisfactory.
9. For amplifying the *eae*-gene, Ampli*Taq*Gold polymerase should be used to avoid low specificity.
10. We successfully use the *stx* reference strains C600J1 (*stx1* gene) and C600W34 (*stx2* gene) described by Karch and Meyer *(6)*. Other *stx*-positive control strains may be used but should be tested according to sensitivity and specificity prior to use.
11. After laying and softly pressing the nylon membrane on the cultured agar plate, mark the membrane and agar by pricking with a needle. This simplifies the detection of the *stx*-positive colonies that correspond to the positive hybridization signals on the membrane.

12. As an alternative to an UV crosslinker apparatus an UV light source (254 nm) might be used. To fix the DNA, lay the membrane under UV light for 5 min at a distance of 15 cm from the light source.
13. The DIG Easy Hyb as well as the DIG detection sytems are strongly recommended because these kits alleviate the colony hybridization assay.
14. To facilitate the detection of the *stx*-positive colonies, lay the wet nylon membrane on the bottom of the corresponding Petri dish (note the pricked markings) and illuminate with a simple light source.

References

1. Schmidt, H., Geitz, C., Tarr, P. J., Frosch, M., and Karch, H. (1999) Non-O157:H7 pathogenic Shiga toxin-producing *Escherichia coli*: phenotypic and genetic profiling virulence traits and evidence for clonality. *J. Infect. Dis.* **179**, 115–123.
2. Paton, J. C. and Paton, A. W. (1998) Pathogenesis and diagnosis of Shiga toxin-producing *Escherichia coli* infections. *Clin. Microbiol. Rev.* **11**, 450–479.
3. Zadik, P. M., Chapman, P. A., and Siddons, C. A. (1993) Use of tellurite for the selection of verocytotoxigenic *Escherichia coli* O157. *J. Med. Microbiol.* **39**, 155–158.
4. Szabo, R. A., Todd, E. C. D., and Jean, A. (1986) Method to isolate Escherichia coli O157:H7 from food. *J. Food Protect.* **49**, 768–722.
5. Schmidt, H., Plaschke, B., and Franke, S. (1994) Differentiation in virulence patterns of *Escherichia coli* possessing *eae* genes. *Med. Microbiol. Immunol. Berl.* **183**, 23–31.
6. Karch, M. and Meyer, T. (1989) Single primer pair for amplifying segments of distinct Shiga-like-toxin genes by polymerase chain reaction. *J. Clin. Microbiol.* **27**, 2751–2757.
7. Schmidt, H., Rüssmann, H., Schwarzkopf, A., Aleksic, S., Heesemann, J., and Karch, H. (1994) Prevalence of attaching and effacing *Escherichia coli* in stool samples from patients and controls. *Int. J. Med. Microbiol. Virol. Parasitol. Infect. Dis.* 281, 201–213.
8. Gunzer, F., Böhm, H., Rüssmann, H., Bitzan, M., Aleksic, S., and Karch, H. (1992) Molecular detection of sorbitol-fermenting *Escherichia coli* O157 in patients with hemolytic-uremic syndrome. *J. Clin. Microbiol.* **30**, 1807–1810.
9. Rüssmann, H., Kothe, E., Schmidt, H., Franke, S., Harmsen, D., Caprioli, A., et al. (1995) Genotyping of Shiga-like toxin genes in non-O157 Escherichia coli strains associated with hemolytic uremic syndrome. *J. Med. Microbiol.* **42**, 404–410.
10. Gunzer, F., Böhm, H., Rüssmann, H., Bitzan, M., Aleksic, S., and Karch, H. (1992) Molecular detection of sorbitol-fermenting *Escherichia coli* O157 in patients with hemolytic-uremic syndrome. *J. Clin. Microbiol.* **30**, 1807–1810

7

STEC as a Veterinary Problem

Diagnostics and Prophylaxis in Animals

Lothar H. Wieler and Rolf Bauerfeind

1. Introduction

   Shiga toxin-producing *Escherichia coli* (STEC) are important human pathogens causing severe clinical syndromes in a high percentage of infected individuals (*see* Chapter 1). Naturally acquired STEC infections have also been
detected in a wide spectrum of animal species (cattle, sheep, goat, deer, moose,
swine, horse, dog, cat, pigeon, chicken, turkey, gull) sometimes even with considerable prevalences *(1–10)*. Several of these animal hosts, particularly ruminants, have been identified as major reservoires of STEC strains that are highly
virulent in the human host (e.g., EHEC O157:H7). However, in contrast to the
human host, most STEC infections of animals remain clinically inapparent.
Even in ruminant species, where shedding rates of up to 88% have been
reported *(2)*, the clinical significance of STEC infections seems to be rather
limited. A natural pathogenic role of STEC has been clearly identified in
weaned piglets (edema disease [ED] *E. coli* enterotoxemia) *(11)*, calves (watery
to bloody diarrhea) *(12,13)*, and greyhounds (cutaneous and renal glomerular
vasculopathy of greyhounds [CRVG], "Alabama rot") *(14,15)*. However, histological and in vitro studies revealed that certain epithelial and/or endothelial
cells are affected by shiga toxins even in animals appearing clinical healthy
(16–18). Thus, the true health importance of STEC infections for animal hosts
may be underestimated.
   Similar to the diagnostic approach in human STEC infection, a definitive
diagnosis of STEC infection in animals is based on the isolation of suspicious
bacteria from fecal specimen and subsequent confirmation by the demonstra-

From: *Methods in Molecular Medicine, vol. 73: E. coli: Shiga Toxin Methods and Protocols*
Edited by: D. Philpott and F. Ebel © Humana Press Inc., Totowa, NJ

tion of typical virulence factors or their genes. The invention of polymerase chain reaction (PCR) has made it possible to rapidly screen large numbers of isolates with great sensitivity and specificity for a variety of determinative virulence genes. The recent advent of multiplex PCR techniques offers further advantages because isolates can be assayed for several virulence genes in a single test tube, thus reducing the number of tests needed to detect pathogenic *E. coli* bacteria. Single-host species specific assays may be designed that allow the detection of several *E. coli* pathovars relevant in a particular host. In the present chapter, we would like to encourage the use of molecular tools for the diagnosis of important *E. coli* pathovars in animals. We will describe a protocol for isolation and subsequent confirmation of STEC (and ETEC) by two multiplex PCR assays and a single primer pair PCR assay. PCR assays had been established by Schütz et al. *(19)*, Bosworth et al. *(20)*, and Wieler et al. *(10)* and proved valuable in detecting "attaching and effacing" STEC in cattle and STEC and ETEC in pigs, respectively. In order to provide the interested reader with theoretical background information about these assays, some aspects of the pathogenesis of STEC-associated diseases in cattle, pigs, and greyhounds are summarized in **Subheadings 1.1.–1.3.**

1.1. Pathogenesis of STEC Infections in Pigs

Edema disease (ED, *syn. E. coli* enterotoxemia) is a common cause of illness and death loss in weaned piglets most generally during the first 2 wk after weaning. The disease represents an enterotoxemia that is caused by specific STEC strains able to colonize the porcine small intestine and to produce a particular variant of Stx, Stx2e (edema disease *E. coli*, EDEC). EDEC almost exclusively belong to O serogroups 138, 139, and 141. Bacterial colonization of the intestine is enabled by the capacity of the bacteria to adhere to villous epithelial cells via F18ab fimbriae (previously called F107 fimbriae). The expression of receptors for these fimbriae on the apical enterocyte surface is inherited as a dominant trait among pigs and determines susceptibility to edema disease. A second host factor in the pathogenesis of the disease is a sudden increase in nutrient concentration of the diet resulting from the change of feed during weaning. Clinical signs and lesions are largely the effect of Stx2e that has passed the epithelial barrier and has entered the circulatory system. Stx2e causes a systemic microangiopathy characterized by fibrinoid necrosis of endothelial and smooth-muscle cells in small arteries and arterioles. Subsequently, these lesions lead to perivascular edema and ischemic necrosis in several locations, notably in the subcutis of the forehead and the eye lids, the greater curvature of the stomach, and the brain. The perivascular damage of the brain is accompanied with progredient neurologic dysfunction (e.g., ataxia, paralysis, convulsions, and lateral recumbency). Infarction and malacia in the brainstem is the main cause of death in affected pigs *(11,21)*. Some EDEC strains also exhibit classical characteristics of ETEC, as they produce heat-labile *E. coli* Enterotoxin I (LT-I), or heat-stabile *E. coli* enterotoxins I or II (ST-I, ST-II) and/or F4 or F5 fimbriae in addition to Stx2e and F18ab fimbriae. In those cases diarrhea and subsequent dehydration may be dominant clinical findings in the herd *(22,23)*.

Detection of Stx2e and F18ab as the crucial virulence factors in the pathogenesis of edema disease has stimulated efforts to develop an effective immunization procedure. Genetically engineered and enzymatically generated Stx2e toxoids have proved successful in inducing protective immune responses, as well as oral immunization with F18 fimbriae *(24,25)*. In addition, reduction of protein contents of the diet is highly recommended.

The pathogenic role of other STEC strains in pigs is currently not clear. Suckling piglets develop a severe neurological disease resembling edema disease in several clinical and histopathological details when orally infected with a Stx2-producing strain of EHEC O157:H7 *(26)*. Pigs have not been identified as relevant reservoire of EHEC O157:H7, although, recently, a 1.4% carriage rate of this pathogen has been determined among healthy pigs in Japan *(27)*.

1.2. Pathogenesis of STEC Infections in Calves

Calves infected with STEC may suffer from watery or bloody diarrhea *(13)*. However, as the prevalence of subclinical STEC infections outnumbers clinical cases, the mechanisms of diarrhea are only partly understood. Up to 70.1% of bovine STEC have the ability to cause the attaching and effacing lesion (AE lesion) *(10)*, mediated by products encoded in the locus of enterocyte efface-ment (LEE) and it is this capacity that causes disease *(13)*. Bovine AE-positive STEC strains often belong to O serogroups 5, 26, 84, 111, 103, 118, 145, and O157, and the AE ability is highly associated with production of the EHEC hemolysin (*syn.* enterohemolysin) *(10,28)*. These strains colonize the ileum, cecum, colon, and rectum *(12)*, inducing the reorganization of the host cell cytoskeleton, thereby causing a dramatic loss of absorptive villi. In vitro studies discovered a decrease in monolayer resistance, indicating an increased intestinal permeability. Furthermore, netto efflux of chloride ions was detected in epithelial cells infected with AE-positive *E. coli*, by using Ussing chambers for analysis *(29)*. AE lesions are associated with epithelial degeneration, and sometimes with hemorrhage and pseudomembrane formation *(12)*. Structures confering adhesion to the intestine are currently unknown, but the outer membrane protein intimin is clearly involved in colonization. Calves get infected soon after birth through fecally contaminated milk and surroundings. Infection studies with STEC O157:H7 revealed, that such strains are only pathogenic for animals younger than 3 wk *(12)*, a finding pointing toward a possible age-dependent expression of STEC-specific intestinal receptors. The pathogenic significance of Shigatoxins has to be determined more thoroughly in the future. Translocations of STEC into mesenteric lymph nodes, the observation of a marked lymphodepletion in gut-associated lymphatic tissues, and the susceptibility of bovine lymphocytes against Stx led to the hypothesis, that Stx affects the mucosal immune response *(17)*.

In theory, vaccines based on intimin may confer protection, but, so far, they have not been tested thoroughly. Dams should be vaccinated, as they would provide protection to their offspring via the colostrum. Hygienic measures should be even more successful, as the primary goal is to avoid smear infection of newborn calves.

1.3. Pathogenesis of STEC infections in Greyhounds

The pathogenesis of CRVG is not fully understood. However, the clinical and pathophysiological findings in diseased racing greyhounds resemble those of the severe human disease known as hemolytic uremic syndrome (HUS; *see* Chapter 1). The dogs are orally infected by raw beef contaminated with STEC. As bacteriemia does not occur, presumably the toxins produced in the intestines are somehow translocated across the epithelial barrier and reach the kidney via the bloodstream. The animals express significant concentrations of globotriasylceramide, the receptor for Stx, particularly on the cells of the kidney cortex and the colon, giving evidence of the susceptibility of these tissues to the necrotic action of Stx. Cutaneous lesions and edema noted with CRVG are clearly secondary to vasculopathy, of which endothelial necrosis is a significant component *(15)*.

The prophylaxis of CRVG is rather simple in educating personal involved with racing greyhounds to not feed beef condemned for human food. Furthermore, hygienic meat feeding practices should be applied *(14)*.

2. Materials

1. Stool specimens collected from cattle or pigs to be tested.
2. *Escherichia coli* control strains. The following *E. coli* reference strains are used as controls (relevant properties are presented only): HUS-2/85 (Stx1, Intimin) *(30)*, E57 (Stx2e, ST-Ia, ST-II, F18) *(31)*, B41 (*syn.* ATCC 31619; ST-Ia, F5, F41) *(32)*, 987P (ST-Ia, F6) *(32)*, G7 (ST-II, LT-I, F4) *(32)*, and H10407 (*syn.* ATCC 35401; ST-Ia, ST-Ib, LT-Ih) *(33)*. Strains E57, B41, 987P, G7, and H10407 have been kindly provided by C. Wray, Central Veterinary Laboratory, Weybridge, UK. Strain HUS-2/85 is a kind gift of H. Karch, Institute for Hygiene, University of Münster, Germany. *E. coli* strain HB101 (K12-derived strain; *syn.* ATCC 33694) harbors none of the relevant virulence genes and is used as a negative control (Invitrogen).
3. Agar plates.

a. Sheep blood agar: Dissolve 40 g blood agar base (Merck KGaA) in 1 L of double distilled water (ddH$_2$O) and autoclave. Cool down to 45–50°C and add 5% (v/v) sterile defibrinated sheep blood.

b. Washed sheep blood agar: Wash sheep erythrocytes three times with phosphate-buffered saline (PBS) pH 7.4 (10.0 g NaCl, 0.25 g KCl, 0.25 g KH$_2$PO$_4$, 18.0 g Na$_2$HPO$_4$ · 2H$_2$O in 1 L of ddH$_2$O). Dissolve 30 g tryptose blood agar base (Oxoid Ltd.) and 1.11 g CaCl$_2$ in 1 L of ddH$_2$O and autoclave. Cool down to 45– 50°C and add 10% (v/v) washed sheep erythrocytes.

c. Gassner agar: Dissolve 77 g Gassner agar base (Oxoid Ltd.) in 1 L of ddH$_2$O and autoclave.

d. BPLS agar: Dissolve 51.5 g BPLS agar base (Merck KGaA) in 1 L of heated sterile ddH$_2$O. Do *not* autoclave.

e. Luria–Bertani (LB) agar: Dissolve 10 g Bacto-tryptone (Difco Laboratories GmbH), 5 g yeast extract (Merck KGaA), 10 g NaCl, and 16.0 g Bacto-agar (Difco Laboratories) in 1 L of ddH$_2$O. Add 4 mL of 1 M NaOH and autoclave. Pour the liquid agar into 9.0-mm plastic Petri dishes (15 mL/dish) and let dishes undisturbed for 1 h. Store dishes at 4°C.

4. LB broth: Dissolve 10 g Bacto-tryptone (Difco Laboratories), 5 g yeast extract (Merck), and 10 g NaCl in 1 L of ddH$_2$O. Add 4 mL of 1 M NaOH. Autoclave and store at 4°C.

5. Oligodesoxyribonucleotide primers (Roth GmbH & Co.) used for the identification of pathogenic *E. coli* isolates are listed in **Tables 1** and **2**. Primers have been designed [except MK1 and MK2 *(34)*] and evaluated for use in multiplex technology by Schütz et al. *(19)* (primers MK1, MK2, ST-I-1, ST-I-2, LT-I-1, LT-I-2) and Bosworth et al. *(20)* (primers ST-II-1, ST-II-2, ST-I-3, ST-I-4, F5-1, F5-2, LT-I-3, LT-I-4, F18-1, F18-2, F6-1, F6-2, F4-1, F4-2, F41-1, F41-2, Stx2e-1, Stx2e-2), respectively. Each primer is dissolved in sterile ddH$_2$O and adjusted to a stock concentration of 100 μM. Primer stock solutions are stored at –20°C (*see* **Notes 1** and **2**). Multiprimer mixes I and II are set up on ice. Multiprimer-Mix I contains 275 μL of sterile ddH$_2$O and 25 μL of each of the stock solutions of primers ST-II-1, F5-1, F5-2, LT-I-3, F18-1, F6-1, F6-2, F4-1, and F4-2. Multiprimer mix II is composed of 275 μL of sterile ddH$_2$O and 25 μL of each of the stock solutions of primers ST-II-2, ST-I-3, ST-I-4, LT-I-4, F18-2, F41-1, F41-2, Stx2e-1, and Stx2e-2. Mix solutions thoroughly and store solutions as 100-μL aliquots at –20°C.

6. Deoxyribonucleotide triphosphate (dNTP) stock solution (Hybaid GmbH, Heidelberg, Germany): Adjust aqueous solution of dATP, dCTP, dGTP, and dTTP to a concentration of 4 mM each, aliquotize, and store at –20°C.

7. *Taq* DNA Polymerase AmpliTaq™, Stoffel fragment (10 U/μL); Stoffel reaction buffer (10X) and 25 mM MgCl$_2$ stock solution are usually included (PE Applied Biosystems GmbH, Weiterstadt, Germany).

8. *Taq* DNA Polymerase PanScript™ (5 U/μL); NH$_4$ reaction buffer (10X) and 50 mM MgCl$_2$ stock solution are usually included (PAN Biotech).

9. *Tfl* DNA Polymerase (1 U/μL); *Tfl* Polymerase buffer (20X) is usually included (Biozym).

10. High-grade mineral oil (Sigma).

11. SeaKem® LE agarose (Biozym).

12. Ethidium bromide stock solution: 10 mg/mL (Sigma).
13. TAE stock solution (50X): 0.04 M Tris-acetate, 0.001 M EDTA. Dissolve 242 g Tris base, 57.1 mL of glacial acetic acid, and 100 mL of 0.5 M EDTA (pH 8.0) in ddH$_2$O to a final volume of 1000 mL and autoclave.
14. Gel-loading buffer (6X): Aqueous solution of 0.25% (w/v) xylene cyanol FF and 15% (w/v) Ficoll 400.
15. BioLadder™ 100 DNA fragment length standard (Hybaid): Dilute 10 µL of DNA with 40 µL of 6X gel-loading buffer and 150 µL of sterile ddH$_2$O. Store at 4°C. Load 8 µL per lane onto a minigel.
16 API 20E test system (Api-bioMérieux, Nürtingen, Germany).
17. Sterile ddH$_2$O: deionized, twice distilled water, autoclaved and stored at 4°C.
18. Disposable, ethidium bromide-resistant gloves (e.g., N-DEX).
19. Hinged 2.0-mL reaction tubes (Eppendorf) and 0.7-mL PCR tubes (Biozym).
20. Calibrated pipets and disposable pipet tips. Sealed pipets are used for all PCR procedures (Biozym). All other pipet tips are autoclaved prior to their use.
21. Parafilm®.
22. Shaking incubator and incubator both set at 37°C.
23. Boiling water bath.
24. Laboratory centrifuge and appropriate rotors for the centrifugation of 0.7 mL and 1.5-mL reaction tubes (0–16,000g) (e.g., model 5415 C [Eppendorf]).
25. Spectrophotometer (e.g., model DU®640 [Beckmann]) (*see* **Note 3**)
26. Electro-4 electrophoresis apparatus (Hybaid) and compatible power supply.
27. Thermal cycler, model TC-1 (PE Applied Biosystems).
28. Thermal cycler, model Mastercycler 5330 (Eppendorf).
29. Gel documentation system (e.g., model E.A.S.Y. Image Plus gel imager, Rev. 4.16, equipped with a CCD camera E.A.S.Y. 429 k, a video copy processor Mitsubishi P68E, and an ultraviolet (UV) transilluminator UVT-20 M/W [Herolab]).

3. Methods

3.1. Isolation of Putative E. coli from Fecal Samples

1. Inoculate fecal material from each stool sample on a set of agar plates consisting of a sheep blood agar, a Gassner agar, and a BPLS agar. Include a washed sheep blood agar additionally if specimens from cattle are examined. Streak for single colonies.
2. Incubate plates at 37°C for 16 h.
3. Select and transfer putative *E. coli* colonies onto LB agar plates (*see* **Note 4**).
4. Incubate plates at 37°C for 16–40 h (37°C, 16 h).
5. Proceed to **Subheading 3.2.** or seal LB agar Petri dishes with parafilm and store at 4°C until further use. Bacteria remain alive for at least 2 mo.

3.2. Identification of STEC by Polymerase Chain Reaction

3.2.1. Bovine STEC Multiplex PCR

1. Transfer bacteria from the putative *E. coli* subculture into a tube containing 1 mL of LB broth. Inoculate *E. coli* control strains HUS-2/85, H10407, and HB101 in the same manner (*see* **Note 5**).
2. Incubate cultures aerobically at 37°C for 16–18 h.

3. Set up master mix I on ice for $n + 1$ tests (*see* **Note 6**). Each test requires 13.0 µL of sterile ddH_2O, 2.5 µL of the Stoffel reaction buffer (10X), 0.25 µL of each of the stock solutions of primers MK1, MK2, ST-I-1, ST-I-2, LT-I-1, LT-I-2, 1.25 µL of the dNTP stock solution, 0.25 µL of the Stoffel fragment, and 4.0 µL of the 25 mM $MgCl_2$ stock solution.

4. Set up reaction mixes in PCR tubes on ice (one tube per culture and control, respectively). Each mix contains 22.5 µL of master mix I and 2.5 µL of the bacterial culture (or 2.5 µL of ddH_2O in case of reagent-negative control; *see* **Note 7**).

5. Overlay reaction mixes with 20 µL of mineral oil.

6. Spin tubes for 10 s at 10,000g and immediately place them into the thermal cycler TC-1.

7. Start DNA amplification with nine cycles of DNA denaturation (94°C, 40 s), primer annealing (55°C, 80 s), and primer extension (72°C, 60 s). These cycles are followed by 15 cycles with identical parameters except that the annealing temperature is decreased to 50°C. Finish amplification with an incubation step of 5 min at 72°C and hold the samples at 4°C.

8. Remove tubes from the cycler and proceed to **Subheading 3.2.4.** or store at 4°C until further use.

3.2.2. Bovine STEC eae-PCR

1. Transfer bacteria from the suspicious *E. coli* subculture into a tube containing 1 mL of LB broth. Inoculate *E. coli* control strains HUS-2/85 and HB101 in the same manner (*see* **Note 5**).

2. Incubate cultures aerobically at 37°C for 16–18 h.

3. Set up master mix II on ice for $n + 1$ tests (*see* **Note 6**). Each test requires 17.6 µL of sterile ddH_2O, 1.25 µL of the *Tfl* polymerase buffer (20X), 0.25 µL of each of the stock solutions of primers ECW-1 and ECW-2, 0.4 µL of the dNTP stock solution, and 0.25 µL of the *Tfl* DNA polymerase.

4. Set up reaction mixes in PCR tubes on ice (one tube per culture and control, respectively). Each mix contains 20.0 µL of master mix II and 5.0 µL of the bacterial culture (or 5.0 µL of ddH_2O in case of reagent-negative control; *see* **Note 7**).

5. Overlay reaction mixes with 20 µL of mineral oil.

6. Spin tubes for 10 s at 10,000g and immediately place them into the Mastercycler 5330.

7. Start DNA amplification with an initial incubation step at 94°C for 5 min. Proceed with 30 cycles of DNA denaturation (94°C, 60 s), primer annealing (64°C, 90 s), and primer extension (72°C, 90 s). Finish amplification with a final incubation step of 5 min at 72°C.

8. Remove tubes from the cycler and proceed to **Subheading 3.2.4.** or store at 4°C until further use.

3.2.3. Porcine STEC Multiplex PCR

1. Transfer bacteria from the suspicious *E. coli* subculture into a tube containing 1 mL of LB broth. Test at least 10 colonies per fecal sample. Inoculate *E. coli* control strains E57, G7, 987P, B41, and HB101 in the same manner (*see* **Note 5**).

2. Incubate cultures aerobically at 37°C for 16–18 h.
3. Set a reaction tube on ice and prepare master mix III for *n* + 1 tests (*see* **Note 6**). Each test requires 15.6 μL of sterile ddH$_2$O, 3 μL of the NH$_4$ reaction buffer, 1.2 μL of the 50 m*M* MgCl$_2$ stock solution, 3 μL of multiprimer mix I, 3 μL of multiprimer mix II, 1 μL of the dNTP stock solution, and 0.2 μL of the PanScript™ DNA polymerase.
4. Set PCR tubes on ice (one tube per culture and control, respectively) and add in the following order: 27.0 μL of master mix III, 3 μL of the bacterial culture (or 3 μL of ddH$_2$O in case of reagent negative control; *see* **Note 7**).
5. Overlay reaction mixes with 20 μL of mineral oil.
6. Spin tubes for 10 s at 10,000*g* and immediately place them into the Mastercycler 5330.
7. Start DNA amplification with an initial incubation step at 94°C for 5 min. Consecutively, perform 30 cycles of DNA denaturation (94°C, 1 min), primer annealing (55°C, 1 min), and primer extension (72°C, 1.5 min). Finish amplification with a final incubation step of 5 min at 72°C.
8. Remove tubes from the cycler and proceed to **Subheading 3.2.4.** or store at 4°C until further use.

3.2.4. Analysis of PCR Products by Horizontal Agarose Gel Electrophoresis

1. Prepare a 3.0% (w/v) agarose gel. Approximately 60 mL gel solution is needed for a gel of 80 × 120 × 6 mm. Heat the aqueous suspension of the agarose until it becomes completely clear. Let the solution cool down to 60°C. Add ethidium bromide to a final concentration of 0.5 μg/mL, mix thoroughly, and pour the solution into the prepared gel tray (*see* **Note 8**). Insert the comb and leave the gel undisturbed for at least 1 h at room temperature.
2. Place the gel into the electrophoresis tank, load tank with 1X TAE running buffer, and remove comb from the gel. Mix 5 μL of each PCR reaction with 2 μL of gel loading buffer and 5 μL of 1X TAE. Load samples and 8 μL/lane of the BioLadder™ 100 DNA fragment length standard into the wells (*see* **Note 9**).
3. Separate DNA molecules by electrophoresis at 100 V for 1–1.5 h.
4. Place gel on the UV transilluminator and visualize DNA fragments by UV illumination (*see* **Notes 10** and **11**).
5. Record the electropherogram as an image by the gel image system (*see* **Note 12**).
6. Calculate the sizes of PCR products by comparison of their migration distances with those of the standard DNA fragments.
7. Identify the genotype of the *E. coli* isolates by their PCR products according to the scheme presented in **Table 1** as well as the electropherograms depicted in **Figs. 1** and **2** (*see* **Notes 13** and **14**).

4. Notes

1. All reagents used are of standard molecular biology grade.
2. Several molecular biology companies maintain technology services, including the synthesis of oligodeoxyribonucleotides of desired sequences, modifications, purity, and amount. Highly purified primers (e.g., by high-performance liquid

Table 1
Primers Used for the Analysis of *E. coli* Isolated from Cattle

Virulence factor	Primer No.	Sequence (5' → 3')	MW	Length of product (bp)
LT-I	LT-I-1	TCTCTATGTGCATACGGAGC	6188	
	LT-I-2	CCATACTGATTGCCGCAAT	5828	322
Stx1, Stx2	MK1	TTTACGATAGACTTCTCGAC	6147	227[a]
	MK2	CACATATAAATTATTTCGCTC	6419	224[b]
ST-I	ST-I-1	CTTGACTCTTCAAAAGAGAAAATTAC	8026	
	ST-I-2	GATTACAACAAAGTTCACAGCAGT	7434	124
Intimin	ECW-1	TGCGGCACAACAGGCGGCGA	6257	
	ECW-2	CGGTCGCCGCACCAGGATTC	6159	629

[a]*stx-1* gene.
[b]*stx-2* gene.
Abbreviations: LT-I, heat-stabile *E. coli* enterotoxin type I; Stx, Shiga toxin; ST-I, heat-stabile *E. coli* enterotoxin type I (*syn.* STa); MW: molecular weight.

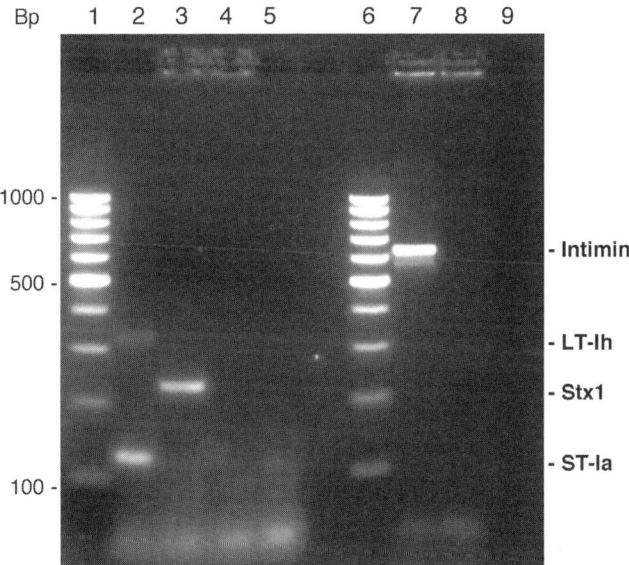

Fig. 1. Results of the Bovine STEC Multiplex PCR (lanes 1–5) and the Bovine STEC *eae*-PCR (lanes 6–9). Bacterial cultures of *E. coli* reference strains were used as the source of template DNA. Products were separated on a 3% (w/v) agarose gel. Bands corresponding to genes of intimin (*eae*), Stx1 (*stx1*), ST-Ia (*estA1*), and LT-Ih (*elt-Ih*) are indicated. Lane 1, BioLadder™ 100 length standard; lane 2, strain H10407; lane 3, strain HUS-2/85; lane 4, strain HB101 (negative control); lane 5, ddH₂O (reagent negative control); lane 6, BioLadder™ 100 length standard; lane 7, strain HUS-2/85; lane 8, strain HB101 (negative control); lane 9, ddH₂O (reagent negative control).

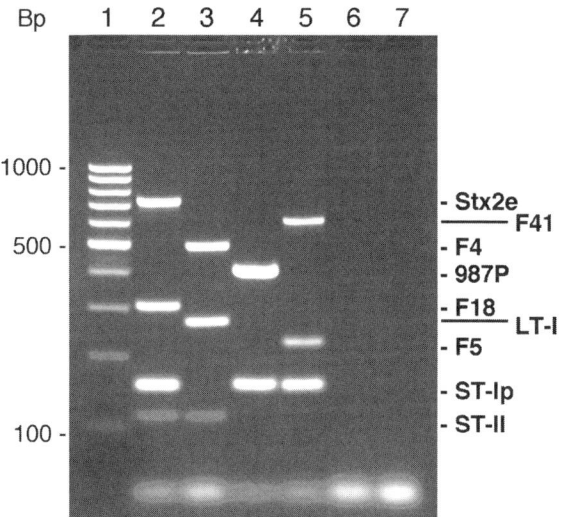

Fig. 2. Result of the Porcine STEC Multiplex PCR. Bacterial cultures of *E. coli* reference strains were used as source of template DNA. Products were separated on a 3% (w/v) agarose gel. Bands corresponding to genes of Stx2e (*stx2e*), F41 (*fimF41a*), F4 (*faeG*), F5 (*fanA*), F6 (*fasA*), F18 (*fedA*), LT-I (*eltB-I*), ST-Ia (*estA1*), and ST-II (*est-II*) are indicated. Lane 1, BioLadderTM 100 length standard; lane 2, strain E57; lane 3, strain G7; lane 4, strain 987P; lane 5, B41; lane 6, HB101 (negative control); lane 7, ddH2O (reagent negative control).

chromatography) are preferred but desaltet, nonpurified primers worked as well in our hands. However, quality differences may occur from lot to lot, independently of any purity. Therefore, check each lot of a primer pair with a number of positive and negative control strains separately before setting up the multiprimer mixes. It is also recommended to check the primer concentrations after setting up the primer stock solutions (*see also* **Note 3**).

3. The spectrophotometer is used to determine the DNA concentration of primer solutions.

4. Even early after the onset of clinical symptoms, the number of STEC or ETEC bacteria in a fecal sample can be very small in comparison to bacterial counts of commensal *E. coli* strains. No methods for the specific enrichment of bovine or porcine STEC can be recommended currently. In consequence, a valid diagnostic approach requires testing of a considerable number of individual colonies per fecal sample. A hemolytic phenotype can be used as additional screening marker for STEC colonies. Almost all EDEC strains produce α-hemolysin, whereas a significant percentage of bovine STEC strains produces enterohemolysin. The α-hemolytic phenotype is easily detected on the sheep agar plate. The detection of enterohemolysin activity requires an additional agar containing washed sheep erythrocytes *(10,28,35,36)*. Both agars are described in **Subheading 2.** How

Table 2
Primers Used for the Analysis of *E. coli* Isolated from Pigs

Virulence factor	Primer				Length of product (bp)
	No.	Sequence (5' → 3')		MW	
Stx2e	Stx2e-1	AATAGTATACGGACAGCGAT		6554	
	Stx2e-2	TCTGACATTCTGGTTGACTC		6154	733
F41 fimbriae	F41-1	AGTATCTGGTTCAGTGATGG		6283	
	F41-2	CCACTATAAGAGGTTGAAGC		6230	612
F4 fimbriae	F4-1	GAATCTGTCCGAGAATATCA		6205	
	F4-2	GTTGGTACAGGTCTTAATGG		6283	499
F6 fimbriae	F6-1	AAGTTACTGCCAGTCTATGC		6172	
	F6-2	GTAACTCCACCGTTTGTATC		6123	409
F18 fimbriae	F18-1	TGGTAACGTATCAGCAACTA		6205	
	F18-2	ACTTACAGTGCTATTCGACG		6172	313
LT-I	LT-I-3	GGCGTTACTATCCTCTCTAT		6114	
	LT-I-4	TGGTCTCGGTCAGATATGT		5930	272
F5 fimbriae	F5-1	AATACTTGTTCAGGGAGAAA		6269	
	F5-2	AACTTTGTGGTTAACTTCCT		6153	230
ST-I	ST-I-3	CAACTGAATCACTTGACTCTT		6420	
	ST-I-4	TTAATAACATCCAGCACAGG		6174	158
ST-II	ST-II-1	TGCCTATGCATCTACACAAT		6116	
	ST-II-2	CTCCAGCAGTACCATCTCTA		6077	113

Abbreviations: LT-I, heat-labile *E. coli* enterotoxin type I; Stx2e, edema disease variant of Shiga toxin 2; ST-I, heat-stabile *E. coli* enterotoxin type I (*syn.* STa); ST-II, heat-stabile *E. coli* enterotoxin type II (*syn.* STb); MW, molecular weight.

many colonies should be tested is still an open question. To our experience, 5–10 putative *E. coli* colonies per fecal sample can be sufficient if the sample has been collected from a sick animal or a herd diagnosis is required and several animals are tested per herd. However, if single apparently healthy animals are tested, more colonies should be tested (20–50) to increase the assay sensitivity. In those cases it is advantageous to perform two rounds of PCR analysis where pools of up to 10 LB broth cultures are tested in the first round.

5. To ensure the reliability of assay results, the use of positive and negative controls is mandatory. *E. coli* strains listed in the Materials section are well-characterized reference strains that work fine as positive and negative controls. However, other strains may be used as well. *E. coli* K12 HB101 harbors none of the virulence genes of interest and serves as a negative control. Usually, negative controls (PCR reaction mixes containing bacteria of strain HB101) make at least 5% of the test samples to be tested in the same PCR assay.

6. The routine performance of the same PCR over long periods increases the risk that PCR products may contaminate your PCR reaction mixtures or even stock solutions. Standard precautions to avoid contamination are as follows:

    a. Strictly separate all post-PCR work, equipment, and reagents from all pre-PCR work. Consider the thermal cycler as post-PCR equipment. Never allow any material (PCR reaction mixes, tubes, pipets and tips, gloves, etc.) that has been used in the post-PCR environment to get back into the pre-PCR laboratory.

    b. Use stock solutions only for PCR procedures. Set up stock solutions in a clean room separate from the laboratory where PCR reactions are set up routinely.

    c. In order to avoid carryover contamination, the last component (prior to the addition of mineral oil) are the bacterial cells to be tested.

    d. Use UV radiation to routinely decontaminate surfaces of the laboratory equipment.

7. At least one additional PCR reaction mix is supplemented with ddH$_2$O instead of bacterial material in each assay to control the PCR reagents used ("reagent negative control").

8. **Caution**: Ethidium bromide is a mutagen and should be handled with care. Wear gloves when working with ethidium bromide solutions or stained gels to avoid any skin contact. All solutions containing ethidium bromide should be decontaminated before disposal: Add activated charcoal or 0.1 N NaOH and stir the solution for 16–24 h.

9. The rest of the PCR reaction mixture may be stored for several days at 4°C. PCR products are stable for at least 6 mo when stored at –20°C.

10. **Caution**: Ultraviolet radiation is hazardous to your eyes and skin. Make sure that the UV light source is always adequately shielded. Always wear safety glasses or, even better, a full safety mask or use safety lids for your protection when working with UV radiation.

11. The fluorescence intensity of ethidium bromide-stained DNA decreases within minutes. Keep the time the gel is exposed to UV radiation to a minimum. Examine the electropherogram rapidly and decide whether the separation process has been completed or has to be continued. If necessary, take a gel image for preliminary documentation and continue electrophoresis.

12. A gel image system allows the banding patterns to be recorded by a video camera, transferred to a personal computer and stored as bitmap image files. Usually, the systems also generate photo image reprints which can be stored as hardcopies. Bitmap image files have the advantage that digitized images can be easily transferred into appropriate computer programs and be processed further. However, other techniques for recording and documentation of test results may also satisfy in the particular laboratory setting (e.g., a Polaroid system MP-4+ Instant Camera System equipped with a Wratten filter No.12 and loaded with Polaroid 667 films).

13. If a negative control yields a positive test result, repeat the whole test starting with the growth of bacterial isolates in LB broth. If negative controls become repeatedly positive, it is usually more effective and straightforward to discard all stock solutions that are currently used than to spend much energy in efforts to identify the particular contaminated component.

14. Note that a definitive genotypic identification of STEC or ETEC requires a biochemical confirmation as *E. coli* (e.g., utilizing the API 20E test system as recommended by the supplier).

References

1. Beutin, L., Geier, D., Steinrueck, H., Zimmermann, S., and Scheutz, F. (1993) Prevalence and some properties of verotoxin (Shiga-like toxin)-producing *Escherichia coli* in seven different species of healthy domestic animals. *J. Clin. Microbiol.* **31,** 2483–2488.
2. Beutin, L., Geier, D., Zimmermann, S., Aleksic, S., Gillespie, H. A., and Whittam, T. S. (1997). Epidemiological relatedness and clonal types of natural populations of *Escherichia coli* strains producing Shiga toxins in separate populations of cattle and sheep. *Appl. Environ. Microbiol.* **63,** 2175–2180.
3. Emery, D. A., Nagaraja, K. V., Shaw, D. P., Newman, J. A., and White, D. G. (1992) Virulence factors of *Escherichia coli* associated with colisepticemia in chickens and turkeys. *Avian Dis.* **36,** 504–511.
4. Chapman, P. A. and Ackroyd, H. J. (1997) Farmed deer as a potential source of verocytotoxin-producing *Escherichia coli* O157. *Vet. Rec.* **141,** 314.
5. Heuvelink, A. E., Zwartkruis-Nahuis, J. T., van den Biggelaar, F. L., van Leeuwen, W. J., and de Boer, E. (1999) Isolation and characterization of verocytotoxin-producing *Escherichia coli* O157 from slaughter pigs and poultry. *Int. J. Food Microbiol.* **52,** 67–75.
6. Holland, R. E., Schmidt, A., Sriranganathan, N., Grimes, S. D., Wilson, R. A., Brown, C. M., et al. (1996) Characterization of *Escherichia coli* isolated from foals. *Vet. Microbiol.* **48,** 243–255.
7. Schmidt, H., Scheef, J., Morabito, S., Caprioli, A., Wieler, L. H., and Karch, H. (2000) A new Shiga Toxin 2 variant (Stx2f) from *Escherichia coli* isolated from pigeons. *J. Appl. Environ. Microbiol.* **66,** 1205–1208.
8. Todd, E. C., Szabo, R. A., MacKenzie, J. M., Martin, A., Rahn, K., Gyles, C., et al. (1999) Application of a DNA hybridization-hydrophobic-grid membrane filter method for detection and isolation of verotoxigenic *Escherichia coli*. *Appl. Environ. Microbiol.* **65,** 4775–4780.
9. Wallace, J. S., Cheasty, T., and Jones, K. (1997) Isolation of vero cytotoxin-producing *Escherichia coli* from wild birds. *J. Appl. Microbiol.* **82,** 399–404
10. Wieler, L. H., Vieler, E., Erpenstein, C., Schlapp, T., Steinrueck, H., Bauerfeind, R., et al. (1996) Shiga toxin-producing *Escherichia coli* strains from bovines: association of adhesion with carriage of *eae* and other genes. *J. Clin. Microbiol.* **34,** 2980–2984
11. Moxley, R.A. (2000) Edema disease. *Vet. Clin. North Am. Food Anim. Pract.* **16,** 175–185.
12. Dean-Nystrom, E.A., Bosworth, B.T., Moon, H.W., and A.D. O´Brien (1998) Bovine infection with Shiga toxin-producing *Escherichia coli*, in Escherichia coli *O157:H7 and Other Shiga Toxin-Producing* E. coli *Strains.* (Kaper, J. B., O´Brien, A. D., eds.), American Society of Microbiology, Washington, DC, pp. 261–267.
13. Dean-Nystrom, E. A., Bosworth, B. T., and Moon, H. W. (1999) Pathogenesis of *Escherichia coli* O157:H7 in weaned calves. *Adv. Exp. Med. Biol.* **473,** 173–177.
14. Fenwick, B. W. and Cowan, L. A. (1998) Canine model of hemolytic-uremic syndrome, In Escherichia coli *O157:H7 and Other Shiga Toxin-Producing E. coli Strains.* (Kaper, J. B., O´Brien, A. D., eds.), American Society of Microbiology, Washington, DC, pp. 268–277.

15. Cowan, L. A., Hertzke, D. M., Fenwick, B. W., and Anderson, C. B. (1997) Clinical and clinicopathological abnormalities in greyhounds with cutaneous and renal glomerular vasculopathy: 18 cases (1992–1994). *J. Am. Vet. Med. Assoc.* **210,** 789–793.

16. Kausche, F. M., Dean, E. A., Arp, L. H., Samuel, J. E., and Moon, H. W. (1992) An experimental model for edema disease (*Escherichia coli* enterotoxemia) in pigs: subclinical disease manifest as vascular necrosis. *Am. J. Vet. Res.* **53,** 281–287.

17. Menge, Ch., Wieler, L. H., Schlapp, T., and Baljer, G. (1999) Shigatoxin 1 from *Escherichia coli* blocks activation and proliferation of bovine lymphocyte subpopulations in vitro. *Infect. Immun.* **67,** 2209–2217.

18. Wieler, L. H., Franke, S., Menge, Ch., Rose, M., Bauerfeind, R., Karch, H., et al. (1995) The immune response in edema disease of weaned piglets measured with a recombinant B subunit of shiga-like toxin II$_e$. *Deutsche Tierärztl. Wochenschr.* **102,** 40–43.

19. Schütz, B., Hack, B., Keiner, K., Zimmermann, K., and Rusch, V. (1993) Multiprimer-PCR als Screeningverfahren zum Nachweis der Toxingene LTI, STI VTI und VTII bei *E. coli*-Pathotypen. *Lab. Med.* 17, 496-501.

20. Bosworth, B. T. and Casey, T. A. (1997) Identification of toxin and pilus genes in porcine *Escherichia coli* using polymerase chain reaction (PCR) with multiple primer pairs. 97th General Meeting of the American Society for Microbiology.

21. Gyles, C. L. (1994) *Escherichia coli* verotoxins and other cytotoxins, in *Escherichia coli in Domestic Animals and Humans.* (Gyles, C. L., ed.), CAB International, Wallingford, UK, pp. 365–398.

22. Wray, C. and Woodward, M. J. (1997) *Escherichia coli* infections in farm animals, in Escherichia coli—*Mechanisms of Virulence* (Sussmann, M., ed.), Cambridge University Press, Cambridge, pp. 95–104.

23. Osek, J. (1999) Prevalence of virulence factors of *Escherichia coli* strains isolated from diarrheic and healthy piglets after weaning. *Vet. Microbiol.* **31,** 209–217.

24. Bosworth, B. T., Samuel, J. E., Moon, H. W., O'Brien, A. D., Gordon, V. M., and Whipp, S. C. (1996) Vaccination with genetically modified Shiga-like toxin IIe prevents edema disease in swine. *Infect. Immun.* **64,** 55–60.

25. Bertschinger, H. U., Nief, V., and Tschäpe, H. (2000) Active oral immunization of suckling piglets to prevent colonization after weaning by enterotoxigenic *Escherichia coli* with fimbriae F18. *Vet. Microbiol.* **71,** 255–267.

26. Dean-Nystrom, E. A., Pohlenz, J. F. L., Moon, H. W., and O´Brien, A. (2000) *Escherichia coli* O157:H7 causes more-severe systemic disease in suckling piglets than in colostrum-deprived neonatal piglets. *Infect. Immun.* **68,** 2356–2358.

27. Nakazawa, M., Akiba, M., and Sameshima, T. (1999) Swine as potential reservoire of Shiga Toxin-producing *Escherichia coli* O157:H7 in Japan. *Emerg. Infect. Dis.* **5,** 833.

28. Wieler, L. H.,Tigges, M., Schäferkordt, S., Ebel, F., Djafari, S., Schlapp, T., and Chakraborty, T. (1996) The enterohemolysin phenotype of bovine Shiga-like toxin producing *Escherichia coli* (SLTEC) is encoded by the EHEC-hemolysin gene. *Vet. Microbiol.* **52,** 153–164.

29. Nataro, J. P. and Kaper, J. B. (1998) Diarrheagenic *Escherichia coli. Clin. Microbiol. Rev.* **11,** 142–201.

30. Böhm, H. and H. Karch (1992) DNA fingerprinting of *Escherichia coli* O157:H7 strains by pulsed field gel electrophoresis. *J. Clin. Microbiol.* **30,** 2169–2172.
31. Woodward, M. J., Kearsley, R., Wray, C., and Roeder, P. L. (1989) DNA probes for the detection of toxin genes in *Escherichia coli* isolated from diarrhoeal disease in cattle and pigs. *Vet. Microbiol.* **22,** 277–290.
32. Woodward, M. J. and Wray, C. (1990) Nine DNA probes for detection of toxin and adhesin genes in *Escherichia coli* isolated from diarrhoeal disease in animals. *Vet. Microbiol.* **25,** 55–65.
33. Skerman, F. J., Formal, S. B., Falkow, S. (1972) Plasmid-associated enterotoxin production in a strain of *Escherichia coli* isolated from humans. *Infect. Immun.* **5,** 622–624.
34. Karch, H. and Meyer, T. (1989) Single primer pair for amplifying segments of distinct Shiga-like-toxin genes by polymerase chain reaction. *J. Clin. Microbiol.* **27,** 275–277.
35. Beutin, L., Prada, J., Zimmermann, S., Stephan, R., Orskov, I., and Orskov, F. (1988) Enterohemolysin, a new type of hemolysin produced by some strains of enteropathogenic *E. coli* (EPEC). *Int. J. Med. Microbiol. Virol. Parasitol. Infect. Dis.* **267,** 576–588.
36. Wieler, L. H., Bauerfeind, R. Weiss, R., Pirro, F., and Baljer, G. (1995) Association of enterohemolysin and non-fermenting of rhamnose and sucrose with shiga-like toxin genes in *Escherichia coli* from calves. *Int. J. Med. Microbiol. Virol. Parasitol. Infect. Dis.* **282,** 265–274.

8

Cellular Microbiology of STEC Infections

An Overview

Frank Ebel and Dana Philpott

Cellular microbiology defines an emerging discipline that brings together the study of pathogenic microbes with eurkaryotic cell biology in order to investigate in detail the complex interactions that occur between pathogen and host during the process of disease. Over the years, we have seen the study of "cellular microbiology" move from research that was largely observational to, more recently, where microbes have become powerful tools to probe the complex molecular workings of the eukaryotic host cell *(1)*. Examination of molecular mechanisms that characterize the interplay between bacteria and host cell has led to a new appreciation of microbial pathogenesis. A recurring theme that has emerged is that microorganisms have developed sophisticated mechanisms to subvert host cell signaling pathways in order to create a favorable environment for their own survival.

Shiga toxin-producing *Escherichia coli* (STEC) bacteria are an excellent example to study how the exchange of genetic DNA leads to specialized strains that can trigger certain diseases. A diversity of *E. coli* pathotypes that cause a variety of illnesses exists. However, the majority of *E. coli* strains are non-pathogenic and comprise a large proportion of the normal commensal flora. Indeed, the representative laboratory *E. coli* strain K12 is used extensively in all areas of research. In the background of K12, it is now possible to pinpoint those DNA sequences that are "pathogenic" and/or specific for STEC.

The first virulence factor identified for STEC was Shiga toxin (Stx), which led to the designation "Shiga toxin-producing *E. coli* (STEC)." Stx is a potent toxin with an AB architecture that completely blocks protein biosynthesis by

From: *Methods in Molecular Medicine, vol. 73: E. coli: Shiga Toxin Methods and Protocols*
Edited by: D. Philpott and F. Ebel © Humana Press Inc., Totowa, NJ

inactivating eukaryotic ribosomes in an irreversible manner. This contributes to the severity of STEC disease because of the action of the toxin on the kidney glomeruli and microvasculature of the intestine. The A-subunit contains the enzymatic active fragment that enters the host cell and cleaves a specific N-glycosidic bond in the 28S ribosomal RNA (2). The B-subunit is a lectin that targets the holotoxin to sensitive cells that present the receptor carbohydrate structure on their surface. Different chapters of this volume contain methods now used to isolate the toxin, to study its effects on host cell, or to make use of the unique features of the B-subunit to target or probe endogenous vesicular systems within the host cells (Chapters 15–22). The latter provides an excellent example of how pathogenic bacterial molecules are currently used to study and manipulate the cellular machinery of mammalian cells.

Apart from Stx, STEC possess additional "pathogenic tools" that are currently studied on genetic, protein, and cellular levels. The most prominent ones are the STEC hemolysin and a set of proteins encoded by the locus of enterocyte effacement (LEE). The STEC hemolysin, also called enterohemolysin, is expressed in some but not all STEC strains. Because highly virulent strains associated with the hemolytic uremic syndrome express this toxin with high frequency (3), it is likely that it plays a role in disease. STEC hemolysin is homologous to the α-hemolysin of uropathogenic E. coli strains and both toxins form pores by inserting into host membranes (4). Although this pore-forming activity has been studied in some detail (see Chapter 13), its precise role in pathogenicity remains to be determined. Other less well-studied virulence factors have been described. Most of them are encoded on the large STEC-specific plasmid that also harbors the hemolysin operon. These putative virulence factors comprise an exported serine protease (5,6), proteins with striking homologies to large clostridial toxins (7), a catalase (8), and a putative type II secretion system (9).

From the cellular microbiology viewpoint, probably the most exciting virulence mechanism found in pathogenic E. coli is the ability to induce a characteristic histopathological disorder, called attaching and effacing (A/E) lesions. This ability is associated with STEC, enteropathogenic E. coli (EPEC), rabbit pathogenic E. coli strains, and the related rodent pathogen Citrobacter rodentium and enables these pathogens to colonize the epithelial lining of the gut and, finally, to induce diarrhea (for review, see ref. 10). The genes responsible for the formation of attaching and effacing lesions are clustered in a large chromosomal pathogenicity island, LEE (11). The G + C content and the codon usage of this region differs from that of the chromosomal backbone of E. coli and it is thought that this element has been acquired by horizontal gene transfer as a distinct block of genetic material.

In the Gram-negative bacterial world, a pool of virulence genes exists that can be transferred between bacterial species by various, mostly unknown vectors. Shuttling these genes in a heterogeneous population by horizontal gene transfer is a powerful strategy to enable nonpathogenic or less pathogenic bacteria to acquire a broad variety of potentially useful virulence factors. Individual mosaics of acquired genes can then be selected for such combinations that provide the best pathogenic fitness. STEC have independently developed several times and similar pathogenic DNA elements have been inserted in different strains at different positions. Techniques to identify large stretches of inserted pathogenic DNA, designated "pathogenicity islands" (e.g., the LEE), within the *E. coli* chromosome have been developed and are described in detail in Chapter 9. To study the impact of potential virulence genes, it is necessary to inactivate such genes without effecting others. The generation of defined deletion mutants as it is described in Chapter 10 provides a unique possibility to delete a certain gene even in the context of an operon without any unwanted genetic side effects.

Whereas many pathogenic Gram-negative bacteria, including *Shigella* and *Salmonella*, have developed sophisticated strategies to enter and survive within host cells, STEC and the related EPEC interact with their host cell from an extracellular position. These Gram-negative bacteria mentioned here and many other Gram-negative pathogens of both animals and plants possess what is know as a type III secretion system. It is through this macromolecular apparatus that these bacteria are able to inject virulence factors directly into the host cell. These virulence proteins, also called effectors, are responsible for the complex actin rearrangements that are necessary for the uptake of *Shigella* or *Salmonella* into the host cell or, in the case of STEC and EPEC, the formation of the morphologically distinct attaching and effacing lesion *(12)*.

The A/E lesion induced by EPEC was first characterized in 1983 by Moon and colleagues *(13)*. This distinct morphological structure is now known to be induced by a number of pathogens, including STEC, rabbit EPEC (REPEC), and *Citrobacter rodentium*. The A/E lesion is characterized by the loss of microvilli in the area of bacterial attachment, recruitment of filamentous (F)-actin and other cytoskeletal components in the host cell cytoplasm beneath adherent bacteria and the formation of a pedestal that suspends the bacteria above the host cell surface. The role of pedestals and A/E lesion formation in EPEC- and STEC-mediated disease remains unclear, but the formation of A/E lesions may provide strong attachment of the bacteria to the cell surface in order to prevent dislodging during the subsequent diarrheal phase of the disease. Alternatively, it may represent an exaggerated response by the bacteria to remain extracellular, which is in keeping with the observation that EPEC and STEC resist phagocytosis by macrophages *(14,15)*, a process that was recently shown to depend on the inactivation of phosphatidylinositol (PI)-3-kinase *(16)*.

The LEE locus of EHEC and EPEC are seemingly identical in that both pathogenicity islands are responsible for the formation of similar A/E lesions.

The approx 40 open reading frames encoded by the LEE are organized into 5 polycistronic operons and this set of genes is found in an identical order and orientation in the LEE elements of STEC and EPEC. Although the cloned LEE island of EPEC confers on nonpathogenic *E. coli* K12 the ability to form A/E lesions on cultured epithelial cells, the LEE of STEC is unable to confer such an ability *(17)*. These findings suggest functional and or regulatory differences between the LEE of EPEC and STEC.

Regulatory proteins that control expression of LEE-encoded genes have been identified both in and outside of the island and LEE-encoded regulators were also shown to control expression of genes located outside of the island *(18)*. Another set of genes found within the LEE encodes for chaperones, which are proteins dedicated to stabilize effector proteins in the bacteria cytoplasm and to assist their delivery into the host cell via the type III secretion system. All structural components of the type III secretion apparatus itself are encoded in three LEE operons that also harbor the transcriptional regulator, *ler*. The function of a type III secretion apparatus is the translocation (injection) of effector proteins into the host cell cytosol, where they can directly interfere with host proteins and/or manipulate host cell functions. The LEE-encoded type III apparatus differs structurally from analogous translocation machines in *Yersinia*, *Salmonella*, and *Shigella* by an filamentous, extracellular extension that is formed by the type-III-dependently exported EspA protein *(14,19,20)*. High-resolution electron microscopic techniques, like the ones described in Chapter 12, have been essential for the identification and characterization of this highly specialized part of the translocation structure. EspD is another LEE-encoded and in vitro secreted protein that is essential for protein translocation. It has been localized in the membrane of infected cells in close proximity to the attached bacteria *(20,21)*, and in a recent study, its ability to form a translocation pore in the host cell membrane has been shown *(22)*. The third LEE-encoded protein that is secreted in high amounts in a type-III-dependent manner is EspB. Conflicting data suggest that EspB may be part of the translocation pore *(23)*, whereas other research suggests that EspB may be a translocated effector protein of yet unknown function *(24)*.

A fourth LEE operon contains genes whose protein products directly mediate the formation of the A/E lesion. These include intimin (encoded by the *eae* gene), Tir, and the chaperone for Tir (CesT). One of the most striking features of EPEC and STEC is that these bacteria insert their own receptor, Tir, into the host cell membrane that then allows binding of the outer membrane protein intimin to mediate intimate bacterial adherence *(25,26)*. In EPEC, Tir is phosphorylated on tyrosine residues, which is an event necessary for A/E formation. Recent data demonstrate that only the Tyr-phosphorylated Tir is able to interact directly with the host adaptor protein Nck *(27)*, which subsequently leads to the recruitment of N-WASP and the Arp 2/3 complex, which are proteins that are essential to start the actin polymerization process. The Tir protein of STEC diverges from that of EPEC, particularly at the C-terminus. Interestingly, phosphorylation of Tir is not required for A/E lesions produced by STEC strains *(28,29)* and this may reflect functional differences in the way translocated Tir interacts with the host actin system *(30,31)*. The ability of Tir to interact with several host cytoskeletal proteins *(26,30)* and additional data that suggest that intimin, via interaction with a yet unknown host membrane receptor, remodels the host cell surface *(32)* demonstrate that the A/E process represents an attractive and unique model to study the organization and regulation of the actin cytoskeleton and the subsequent consequences for the cell morphology.

Apart from Tir, three additional LEE-encoded effector proteins have been described: EspF, Orf19, and, most recently, EspG. Translocated EspF has been shown to be responsible for a loss of transepithelial electrical resistance, increased monolayer permeability, and a redistribution of the tight-junction-associated protein occludin *(33)*. Translocated Orf19 is specifically targeted to mitochondria, where it appears to interfere with the ability of these host organelles to maintain their membrane potential *(34)*. The molecular mechanism of this unique virulence mechanism is yet unknown. The most recently identified LEE-encoded effector molecule is EspG. This protein shows homology to the *Shigella* effector protein VirA and can complement a *Shigella virA* null mutant, but its function in LEE-based pathogenicity remains to be determined *(35)*.

Recent research clearly demonstrates that the virulence mechanisms triggered by LEE-encoded proteins are by far broader than expected. The formation of actin pedestals is a fascinating and instructive process, but induction of antiphagocytosis, modulation of the tight-junction barrier function, and disturbance of the functional integrity of mitochondria are additional virulence mechanisms that may well be just as important for the development of disease. The cellular microbiology of STEC pathogenesis now covers host–pathogen interactions that effect diverse cellular functions, such as cytoskeletal reorganization, ion homeostasis, vesicle trafficking, membrane organisation, mitochondrial function, and signal transduction.

References

1. Cossart, P., Boquet, P., Normark, S., and Rappuoli, R. (1996) Cellular microbiology emerging. *Science* **271,** 315–316.
2. Endo, Y., Tsurugi, K., Yutsudo, T., Takeda, Y., Ogasawara, T., and Igarashi, K. (1988) Site of action of a Vero toxin (VT2) from *Escherichia coli* O157:H7 and of Shiga toxin on eukaryotic ribosomes. RNA *N*-glycosidase activity of the toxins. *Eur. J. Biochem.* **171,** 45–50.

3. Schmidt, H. and Karch, H. (1996) Enterohemolytic phenotypes and genotypes of shiga toxin-producing *Escherichia coli* O111 strains from patients with diarrhea and hemolytic–uremic syndrome. *J. Clin. Microbiol.* **34,** 2364–2367.

4. Schmidt, H., Maier, E., Karch, H., and Benz, R. (1996) Pore-forming properties of the plasmid-encoded hemolysin of enterohemorrhagic *Escherichia coli* O157:H7. *Eur. J. Biochem.* **241,** 594–601.

5. Brunder, W., Schmidt, H., and Karch, H. (1997) EspP, a novel extracellular serine protease of enterohaemorrhagic *Escherichia coli* O157:H7 cleaves human coagulation factor V. *Mol. Microbiol.* **24,** 767–778.

6. Djafari, S., Ebel, F., Deibel, C., Krämer, S., Hudel, M., and Chakraborty, T. (1997) Characterization of an exported protease from Shiga toxin-producing *Escherichia coli. Mol. Microbiol.* **25,** 771–784.

7. Burland, V., Shao, Y., Perna, N.T., Plunkett, G., Sofia, H.J., Blattner, F.R. (1998) The complete DNA sequence and analysis of the large virulence plasmid of *Escherichia coli* O157:H7. *Nucleic Acids Res.* **26,** 4196–4204.

8. Brunder, W., Schmidt, H., and Karch, H. (1996) KatP, a novel catalase-peroxidase encoded by the large plasmid of enterohaemorrhagic *Escherichia coli* O157:H7. *Microbiology* **142,** 3305–3315.

9. Schmidt, H., Henkel, B., and Karch, H. (1997) A gene cluster closely related to type II secretion pathway operons of gram-negative bacteria is located on the large plasmid of enterohemorrhagic *Escherichia coli* O157 strains. *FEMS Microbiol. Lett.* **148,** 265–272.

10. Frankel, G., Phillips, A. D., Rosenshine, I., Dougan, G., Kaper, J. B., and Knutton, S. (1998) Enteropathogenic and enterohaemorrhagic *Escherichia coli*: more subversive elements. *Mol. Microbiol.* **30,** 911–921.

11. Perna, N. T., Mayhew, G. F., Posfai, G., Elliott, S., Donnenberg, M. S., Kaper, J.B., et al. (1998) Molecular evolution of a pathogenicity island from enterohemorrhagic *Escherichia coli* O157:H7. *Infect. Immun.* **66,** 3810–3817.

12. Cornelis, G. R. and Van Gijsegem, F. (2000) Assembly and function of type III secretory systems. *Annu. Rev. Microbiol.* **54,** 735–774.

13. Moon, H. W., Whipp, S. C., Argenzio, R. A., Levine, M. M., Giannella, R. A. (1983) Attaching and effacing activities of rabbit and human enteropathogenic *Escherichia coli* in pig and rabbit intestines. *Infect. Immun.* **41,** 1340–1351.

14. Ebel, F., Podzadel, T., Rohde, M., Kresse, A. U., Krämer, S., Deibel, C., et al. (1998) Initial binding of Shiga toxin-producing *Escherichia coli* to host cells and subsequent induction of actin rearrangements depend on filamentous EspA-containing surface appendages. *Mol. Microbiol.* **30,** 147–161.

15. Goosney, D. L., Celli, J., Kenny, B., and Finlay, B. B. (1999) Enteropathogenic *Escherichia coli* inhibits phagocytosis. *Infect. Immun.* **67,** 490–495.

16. Celli, J., Olivier, M., and Finlay, B. B. (2001) Enteropathogenic *Escherichia coli* mediates antiphagocytosis through the inhibition of PI 3-kinase-dependent pathways. *EMBO J.* **20,** 1245–1258.

17. Elliott, S. J., Yu, J., and Kaper, J. B. (1999) The cloned locus of enterocyte effacement from enterohemorrhagic *Escherichia coli* O157:H7 is unable to confer the attaching and effacing phenotype upon *E. coli* K-12. *Infect. Immun.* **67,** 4260–4263.

18. Elliott, S. J., Sperandio, V., Giron, J. A., Shin, S., Mellies, J. L., Wainwright, L., et al. (2000) The locus of enterocyte effacement (LEE)-encoded regulator controls expression of both LEE- and non-LEE-encoded virulence factors in enteropathogenic and enterohemorrhagic *Escherichia coli*. *Infect. Immun.* **68,** 6115–6126.

19. Knutton, S., Rosenshine, I., Pallen, M. J., Nisan, I., Neves, B. C., Bain, C., et al. (1998) A novel EspA-associated surface organelle of enteropathogenic *Escherichia coli* involved in protein translocation into epithelial cells. *EMBO J.* **17,** 2166–2176.

20. Sekiya, K., Ohishi, M., Ogino, T., Tamano, K., Sasakawa, C., and Abe, A. (2001) Supermolecular structure of the enteropathogenic *Escherichia coli* type III secretion system and its direct interaction with the EspA-sheath-like structure. *Proc. Natl. Acad. Sci. USA* **98,** 11,638–11,643.

21. Kresse, A. U., Rohde, M., and Guzman, C. A. (1999) The EspD protein of enterohemorrhagic *Escherichia coli* is required for the formation of bacterial surface appendages and is incorporated in the cytoplasmic membranes of target cells. *Infect. Immun.* **67,** 4834–4842.

22. Wachter, C., Beinke, C., Mattes, M., and Schmidt, M. A. (1999) Insertion of EspD into epithelial target cell membranes by infecting enteropathogenic *Escherichia coli*. *Mol. Microbiol.* **31,** 1695–1707.

23. Ide, T., Laarmann, S., Greune, L., Schillers, H., Oberleithner, H., and Schmidt, M.A. (2001) Characterization of translocation pores inserted into plasma membranes by type III-secreted Esp proteins of enteropathogenic *Escherichia coli*. *Cell. Microbiol.* **3,** 669–679.

24. Wolff, C., Nisan, I., Hanski, E., Frankel, G., and Rosenshine, I. (1998) Protein translocation into host epithelial cells by infecting enteropathogenic *Escherichia coli*. *Mol. Microbiol.* **28,** 143–155.

25. Rosenshine, I., Ruschkowski, S., Stein, M., Reinscheid, D. J., Mills, S. D., and Finlay, B. B. (1996) A pathogenic bacterium triggers epithelial signals to form a functional bacterial receptor that mediates actin pseudopod formation. *EMBO J.* **15,** 2613–2624.

26. Kenny, B., DeVinney, R., Stein, M., Reinscheid, D. J., Frey, E. A., and Finlay, B. B. (1997) Enteropathogenic *E. coli* (EPEC) transfers its receptor for intimate adherence into mammalian cells. *Cell* **91,**511–520.

27. Gruenheid, S., DeVinney, R., Bladt, F., Goosney, D., Gelkop, S., Gish, G.D., et al. (2001) Enteropathogenic *E. coli* Tir binds Nck to initiate actin pedestal formation in host cells. *Nat. Cell. Biol.* **3,** 856–859.

28. Ismaili, A., McWhirter, E., Handelsman, M. Y., Brunton, J. L., and Sherman, P. M. (1998) Divergent signal transduction responses to infection with attaching and effacing *Escherichia coli*. *Infect. Immun.* **66,** 1688–1696.

29. Deibel, C., Krämer, S., Chakraborty, T., and Ebel, F. (1998) EspE, a novel secreted protein of attaching and effacing bacteria, is directly translocated into infected host cells, where it appears as a tyrosine-phosphorylated 90 kDa protein. *Mol. Microbiol.* **28,** 463–474.

30. DeVinney, R., Puente, J. L., Gauthier, A., Goosney, D., and Finlay, B. B. (2001) Enterohaemorrhagic and enteropathogenic *Escherichia coli* use a different Tir-based mechanism for pedestal formation. *Mol. Microbiol.* **41,** 1445–1458.

31. Goosney, D. L., DeVinney, R., and Finlay, B. B. (2001) Recruitment of cytoskeletal and signaling proteins to enteropathogenic and enterohemorrhagic *Escherichia coli* pedestals. *Infect. Immun.* **69,** 3315–3322.

32. Phillips, A. D., Giron, J., Hicks, S., Dougan, G., and Frankel, G. (2000) Intimin from enteropathogenic *Escherichia coli* mediates remodelling of the eukaryotic cell surface. *Microbiology* **146,** 1333–1344.

33. McNamara, B. P., Koutsouris, A., O'Connell, C. B., Nougayrede, J. P., Donnenberg, M. S., and Hecht, G. (2001) Translocated EspF protein from enteropathogenic *Escherichia coli* disrupts host intestinal barrier function. *J. Clin. Invest.* **107,** 621–629.

34. Kenny, B. and Jepson, M. (2000) Targeting of an enteropathogenic *Escherichia coli* (EPEC) effector protein to host mitochondria. *Cell. Microbiol.* **2,** 579–590.

35. Elliott, S. J., Krejany, E. O., Mellies, J. L., Robins-Browne, R. M., Sasakawa, C., and Kaper, J. B. (2001) EspG, a novel type III system-secreted protein from enteropathogenic *Escherichia coli* with similarities to VirA of *Shigella flexneri*. *Infect. Immun.* **69,** 4027–4033.

9

Analysis of Pathogenicity Islands of STEC

Tobias A. Oelschlaeger, Ulrich Dobrindt, Britta Janke,
Barbara Middendorf, Helge Karch, and Jörg Hacker

1. Introduction

In general, virulent bacteria are distinguished from nonvirulent ones by the presence of genes-encoding virulence factors. Virulence genes might be detected by polymeraase chain reaction (PCR) or hybridization, which also allows the identification of attenuated pathogenic strains. These might have lost not only one virulence gene but a whole set thereof that is part of a mobile genetic element.

Shiga toxin-producing *Escherichia coli* (STEC) may harbor at least three different kinds of mobile genetic element: prophage(s), a plasmid and pathogenicity islands (PAIs) *(1)*. All three mobile genetic elements may be propagated via horizontal gene transfer. PAIs represent distinct pieces of DNA, which (1) encode one or more virulence factors, (2) are present in the genome of pathogenic bacteria but absent from that of nonpathogenic members of the same species, (3) occupy large genomic regions (10–200 kb), (4) differ from the rest of the genome in their G + C content and their codon usage, (5) are often flanked by small direct repeats, (6) are often associated with tRNA genes, and (7) often carry cryptic or functional genes-encoding mobility factors and might therefore be unstable *(2)*. Most likely, STEC evolved from a nonpathogenic ancestor by acquiring these genetic elements via horizontal gene transfer. However, evolution never stops and therefore, STEC might lose some of these elements and/or sample others in addition. To date, four different PAIs have been identified in STEC (*see* **Table 1**). Even so, a certain bacterial strain might carry several PAIs; so far, no STEC strains have been identified that harbor all four PAIs at the same time.

From: *Methods in Molecular Medicine, vol. 73: E. coli: Shiga Toxin Methods and Protocols*
Edited by: D. Philpott and F. Ebel © Humana Press Inc., Totowa, NJ

Table 1
Properties of the Identified Pathogenicity Islands of STEC

PAI	Chromosomal location	Size (kb)	Important features
LEE	Adjacent to *selC* (82') (e.g., O157:H7)	43.359 including 7.5 kb of prophage 933L	TTSS, Tir, Eae, Esp
	Adjacent to *pheU* (94') (e.g., O26:H⁻)	>43.359 (because of IS elements)	TTSS, Tir, Eae, Esp
HPI	Adjacent to asnT	~36	Encodes yersiniabactin iron uptake system
Efa-PAI	Unknown	Unknown	Encodes EHEC factor for adherence
TAI	Unknown	8.040	TlpA to D (Tellurite resistance proteins), Iha (IrgA homolog adhesin)

The first PAI discovered in STEC is the locus of enterocyte effacement (LEE). The core of this PAI is 35 kb in size, encoding a type III secretion system, the outer membrane protein intimin responsible for intimate attachment, the corresponding receptor protein Tir (translocated intimin receptor), and several Esp proteins (*E. coli* secreted proteins) secreted and/or injected into host cells via the type III secretion system. In contrast to the LEE of most enteropathologenic *E. coli* (EPEC), the LEE of certain STEC contains 13 additional open reading frames with high homology to type P4 prophages *(3)*. As is the case for many PAIs in different species, the LEE is also inserted in the genome adjacent to a tRNA gene. The insertion site is often the *selC* or the *pheU* tRNA gene. However, there are reports of other insertion sites as well *(4)*. The so-called high pathogenicity island (HPI), which is of similar size to the LEE, has been first reported to be absent in STEC. However, this is only true for certain STEC serotypes, like O157:H7/H⁻, O103:H2, O111:H⁻, O145:H⁻, whereas other serotypes like O26:H11/H⁻ and O128:H2/H⁻ were shown to harbor HPI *(1)*. Interestingly, usually either the LEE or the HPI is present in a certain strain, with the exception of strains of serotype O26:H11/H⁻, which contains both PAIs (LEE and HPI) *(1)*. Except for a minor modification, the HPI encodes in all HPI-positive strains a complete yersiniabactin iron sequestering system, consisting of 10 genes, 1 of which shows homology to phage P4 integrases. The latter gene is always located next to the integration site, which is the tRNA gene *asnT (1)*.

Recently, a large gene (9669 bp) encoding the EHEC factor for adherence (Efa1) has been identified, which seems to be part of another PAI of STEC *(5)*. In addition to Efa1, this PAI encodes proteins homologous to other potential virulence factors, like the *Shigella flexneri* enterotoxin ShET2. Furthermore, this island harbors several IS elements and transposase homologs *(5)*. These findings together with the low G+C content (40.9%) and the unusual codon usage argue for a PAI that is present in all *eae*-positive STEC strains tested and absent in all *eae*-negative STEC strains. Nevertheless, the Efa1 PAI and the LEE PAI are not colocated according to pulsed-field gel electrophoresis analysis (PFGE) *(5)*. The genomic insertion site is still unknown.

Another probable PAI encodes an adhesin and four tellurite-resistance proteins (Tlp) and is therefore termed TAI, for tellurite resistance- and adherence-conferring island *(6)*. Surprisingly, the adhesin Iha (IrgA homolgoue adhesin), encoded by TAI, shows high homology to the iron-regulated IrgA protein of *Vibrio cholerae* and to a variety of bacterial iron acquisition proteins but not to other known adhesins.

The TAI is present only in non-sorbitol-fermenting O157:H7 strains and not in O157:H⁻ strains, which are sorbitol fermenting and sensitive to tellurite. It seems to be also absent from certain STEC strains of serotype O103:H6. However, the TAI is present in all *eae*-positive STEC strains tested of serotype O26:NM, O85:NM, O126:H2, O103:H2, and O111:HN *(6)*. The presence of TAI even in some *stx*-negative EPEC strains might indicate the transmission of TAI via a mobile genetic element. Its G+C content of 45%, its distribution also among distantly related *E. coli*, and its low frequency in *E. coli* strains from nondiarrheal human stools are indications for TAI being a PAI. Nevertheless, other PAI characteristics like the presence of IS elements and/or transposons and the insertion into the genome adjacent to a tRNA gene have not (yet) been reported.

Obviously, the presence or absence of a certain PAI determines the virulence properties of STEC strains. The detection of PAIs might, therefore, represent a valuable diagnostic tool that helps to distinguish between different STEC serotypes that are characterized by the presence, absence, or combination of certain PAIs.

Transfer RNA genes frequently serve as target sites for the integration of bacteriophages into the bacterial chromosome and are therefore important for the acquisition of horizontally transferred genetic elements *(7,8)*. Additionally, many PAIs including the LEE of STEC are associated with the 3' end of tRNA-encoding genes, which implies that they may have been acquired via horizontal gene transfer *(9)*. Certain tRNA genes can serve as integration sites for different PAIs (*see* **Fig. 1**). Many different integration sites of "foreign" DNA within the *E. coli* core chromosome are already known from the characterization of the chromosomal sequence context of different PAIs and lysogenic bacteriophages in various pathogenic and nonpathogenic *E. coli* strains *(7,10)*.

In order to systematically analyze the genomes of STEC strains for "foreign" DNA elements acquired by horizontal gene transfer, the chromosomal

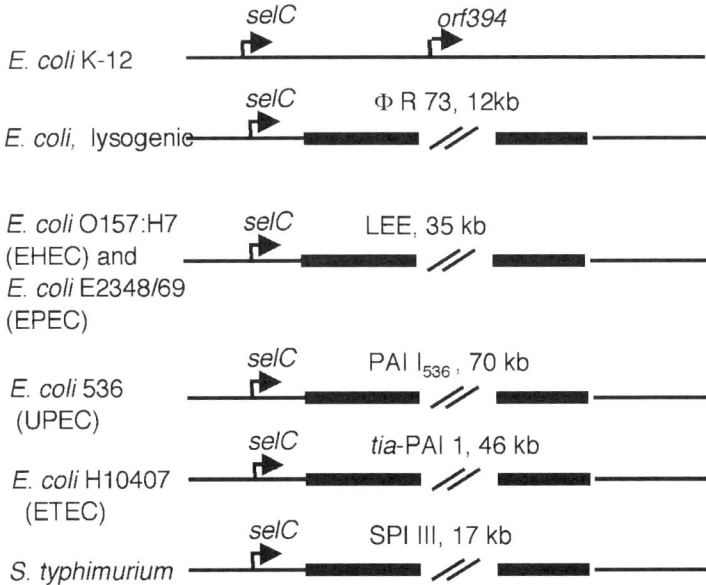

Fig. 1. The tRNA gene *selC* as a paradigm for tRNA-encoding genes as chromosomal integration sites for foreign DNA. In contrast to *E. coli* K-12 strains, different foreign DNA elements such as a bacteriophage as well as different pathogenicity islands can be associated with the 3'-end of tRNA genes in different *E. coli* pathotypes and other pathogenic enterobacteria.

context downstream of tRNA genes from *E. coli* K-12 strain MG1655 and different STEC strains can be compared. In this chapter, we describe a method to detect mobile genetic elements or their derivatives integrated into tRNA loci. For this purpose, tRNA genes, including their flanking regions, are amplified with PCR primers that are specific for the corresponding chromosomal regions of the *E. coli* strain MG1655. Sequence alterations, DNA rearrangements, and the integration of mobile genetic elements will either result in PCR products of different lengths or in the absence of a PCR product. Chromosomal loci and DNA fragments of interest can then be further characterized by Southern hybridization and DNA sequence analysis.

In the light of the vast amount of sequence information, recently developed molecular genetic techniques such as suppression subtractive hybridization (SSH) *(11)* have become more and more important for the analysis of pathogenic bacteria. The SSH approach allows one to identify genomic differences between related bacterial strains *(12,13)*. This method or the similar approach of representational difference analysis (RDA) was successfully used in modified forms to analyze genomic differences between strains of two closely related *Neisseria* species *(12)*.

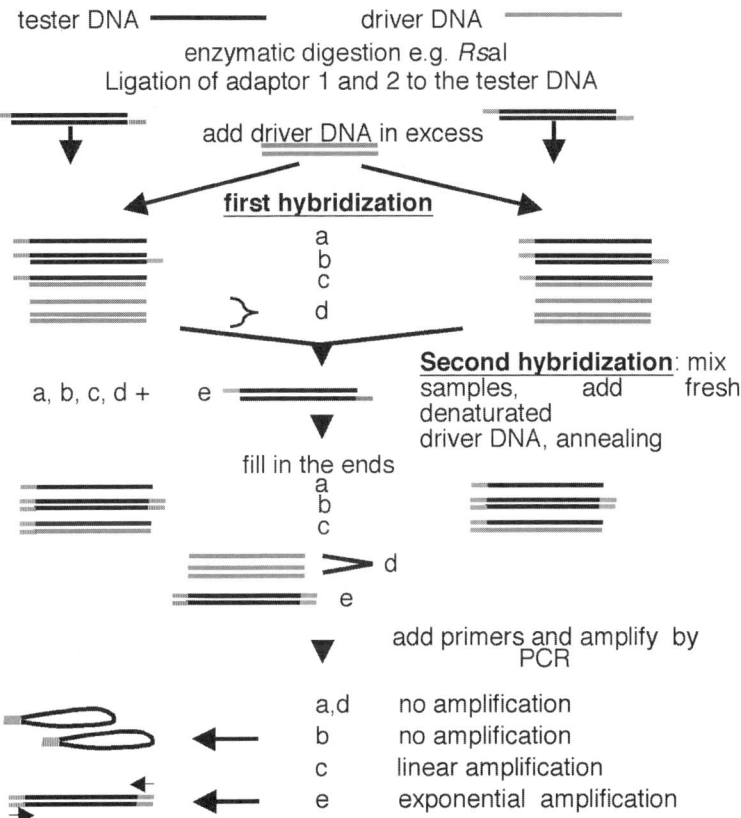

Fig. 2. Schematic overview of the subtractive hybridization method. Black lines represent the digested tester DNA and grey lines represent the digested driver DNA. The two different adaptor sequences are given in dotted and hatched thin lines. (Modified from **ref. *13*.**)

Here, we describe a method that is based on a publication of Diatchenko et al. *(11)* and helps to distinguish between pathogenic and nonpathogenic bacteria. A suppression subtractive hybridization technique is used to identify sequences that are present in one strain (so-called tester strain) and absent in another strain (driver strain). A schematic overview of this approach is given in **Fig. 2**. The genomic DNA from bacterial strains of interest is digested with a suitable four-base cutting restriction enzyme. The tester DNA is subdivided in two portions and ligated with two different adaptors. Then, two hybridization steps are performed, followed by two PCR reactions. Subsequently, a pattern of different PCR bands is obtained and can be further analyzed.

The PAIs of certain bacteria have the tendency to be lost with high frequencies *(14)*. The deleted parts of the chromosome are often flanked by either

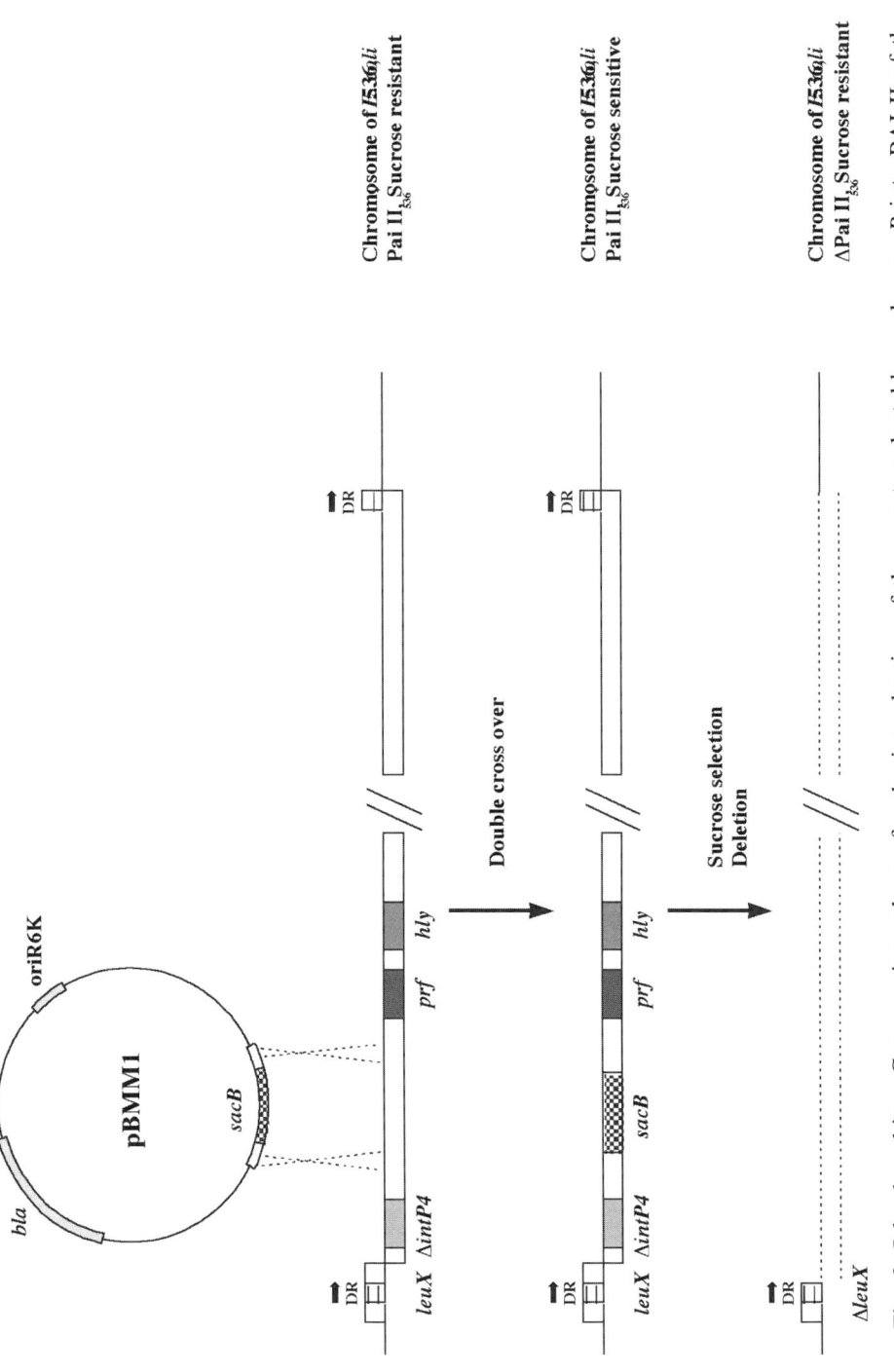

Fig. 3. Island probing. Construction scheme for the introduction of the counterselectable marker *sacB* into PAI II of the uropathogenic *E. coli* strain 536. *bla*, ampicillin resistance gene; *sacB*, counterselectable marker; *leuX*, leucine-specific tRNA; *ΔintP4*, cryptic integrase gene; *prf*, P-related fimbrial gene cluster; *hly*, hemolysin gene cluster; DR, direct repeat

perfect or nonperfect repeats, which may play a role in site-specific recombination events. Recently, the so-called method of "island probing" has been established to analyze the instability of PAIs of *Shigella flexneri* and the uropathogenic *E. coli* strain 536 *(15,16)* (*see* **Fig. 3**). This approach is based on the insertion of counterselectable markers in the islands by homologous recombination *(17)*. They inhibit growth of bacteria on selective media until the "host" PAI is lost by spontaneous deletion. Therefore, this method provides a useful tool for studying the stability of pathogenicity islands in general.

2. Material
2.1. tRNA Screening

1. STEC strains.
2. Growth medium: LB agar plates.
3. Disposable plastics: 1.5-mL and 0.5-mL Eppendorf tubes.
4. Sterile water.
5. Sterile tooth picks.
6. Thermocycler.
7. Oligonucleotide primer pairs specific for each chromosomal locus of interest (100 pmol/μL).
8. *Taq* DNA polymerase.
9. dNTPs (10 m*M*).
10. 50X TAE (tris–acetate–EDTA) buffer stock solution: 242 g Tris-HCl, 57.1 mL glacial acetic acid, 100 mL of 0.5 *M* EDTA, pH 8.0; water added to 1 L.

2.2. Subtractive Hybridization

1. 2 μg Chromosomal DNA of each tester and the driver strain.
2. Restriction enzymes (e.g., *Rsa*I or *Alu*I and restriction buffers).
3. 80% Ethanol and 95% Ethanol.
4. Phenol:chloroform:isoamylalcohol (25:24:1).
5. Chloroform:isoamylalcohol (24:1).
6. 0.2 *M* EDTA.
7. 10 m*M* Tris-HCl, pH 7.5.
8. T4 DNA ligase (400 U/μL) and ligation buffer.
9. Adaptor oligonucleotides (10 μ*M* each);
   Adaptor: 1 5'-CTAATACGACTCACTATAGGGCTCGAGCGGCCGCCCGGG CAGGT-3' and 3'-GGCCCGTCCA-5'
   Adaptor 2: 5'-CTAATACGACTCACTATAGGGCAGCGTGGTCGCGGC CGAGGT-3'and 3'-GCCGGCTCCA-5'
10. 5X Hybridization buffer: 2.5 *M* NaCl, 250 m*M* HEPES, pH 8.3, 1 m*M* EDTA.
11. Dilution buffer: 50 m*M* NaCl, 5 m*M* HEPES, pH 8.3, 0.2 m*M* EDTA.
12. 0.5-mL PCR reaction tubes.
13. DNA polymerase, PCR buffer, dNTP mix for PCR (10 m*M* each dNTP).
14. PCR primer P1 (10 μ*M*). The sequence of this primer matches the long strands of adaptors 1 and 2 at their 5'ends: 5'-CTAATACGACTCACTATAGGGC-3'.

15. Nested primer NP1 (10 μ*M*) 5'-TCGAGCGGCCGCCCGGGCAGGT-3'.
    Nested primer NP2 (10 μ*M*) 5'-AGCGTGGTCGCGGCCGAGGT-3'.
16. T/A cloning vector (e.g., pGEM® T Easy vector).
17. *E. coli* strain DH5α.

2.3. Island Probing

1. Disposable plastics: 1.5-mL reaction tubes; sterile pipet tips; 90-mm Petri dishes triple vent.
2. Bacterial strains and plasmids: *E. coli* strains to be analyzed; *E. coli* SY327 and SM10λ*pir* for cloning experiments and conjugational transfer of plasmid DNA; pGP704 as cloning vector and pCVD442 as source for the counterselectable marker *sacB* and its cis-acting regulatory locus *sacR* *(18,19)*.
3. Growth medium, liquid; LB medium: 10 g Tryptone, 5 g yeast extract, 5 g NaCl per liter; if required add ampicillin and streptomycin at a concentration of 100 μg/mL.
4. Growth medium, solid: Prepare agar plates by the addition of 1.5% agar to LB medium; if necessary, supplement the medium with 7% sucrose.
5. *Pfu Turbo*® DNA polymerase (Stratagene) in addition to a PAI-specific primer pair.
6. Restriction enzymes, alkaline phosphatase (calf intestine), T4 polynucleotide kinase, Klenow fragment, and T4 DNA ligase together with their corresponding buffer systems (Amersham/Pharmacia).
7. Ca^{2+}-competent cells of *E. coli* SY327 and SM10λ*pir* (store at –80°C).
8. 0.9% NaCl solution (sterile).
9. Qiagen Plasmid Miniprep Kit, Qiagen Gel Extraction Kit, and Qiagen PCR Purification Kit (Qiagen).
10. 37°C Orbital shaker; 37°C and 20°C incubator.
11. Ultraviolet (UV) spectrophotometer (Amersham/Pharmacia).

3. Methods

3.1. tRNA Screening

1. Grow the reference strain MG1655 and the STEC strains to be investigated overnight on LB agar plates at 37°C.
2. Prepare template DNA by resuspending a bacterial colony from the agar plate in 100 μL water and subsequent boiling for 5 min in order to lyse the cells.
3. Preparation of the PCR reactions: For every individual PCR, transfer the following into a separate Eppendorf tube: 5 μL 10X reaction buffer, 1 μL dNTP stock solution, 1 μL of each primer of the required primer pair (*see* **Notes 1** and **2**), 1 μL of boiled STEC cells, and 1–4 U of *Taq* DNA polymerase. $MgCl_2$ stock solution should be added to the reaction mix depending on the size of the expected PCR product and as recommended by the supplier of the *Taq* DNA polymerase. Add water to the final volume of 50 μL.
4. PCR control reactions: Prepare the same PCR reactions as described in **step 3** with boiled cells of the *E. coli* K-12 strain MG1655 as template.
5. PCR: After 2 min denaturation at 95°C, run 30 PCR cycles consisting of the following steps: step 1 (denaturation; 45 s, 95°C), step 2 (annealing: 45 s, tem-

perature depends on the primer pair used), step 3 (elongation; time depends on the size of the expected PCR product (1 min/1 kb), 72°C).

6. Analysis of the PCR products: Separate 10 μL of the PCR samples by agarose gel electrophoresis (1% agarose in 1X TAE buffer). Stain with ethidium bromide (0.1% solution in water) and compare the PCR products obtained with template DNA from STEC strains and *E. coli* K-12 strain MG1655 (*see* **Notes 4** and **5**).

3.2. Suppressive Subtractive Hybridization (13)

3.2.1. Restriction of Driver (Reference Strain) and Tester (Strain of Interest) DNA and Adaptor Ligation

1. 2 μg of chromosomal tester and driver DNA were each digested with 20 U of restriction enzyme overnight.
2. Extract both restriction samples with phenol and precipitated with ethanol. Resuspend the samples in 10 m*M* Tris-HCl, pH 7.5, at a final concentration of 200 ng/μL.
3. Ligate two aliquots of tester DNA with 120 ng DNA separately to 2 μL of the two adaptors, each in a total volume of 10 μL and a final concentration of 2 μ*M*. Perform the ligation overnight at 16°C using 1 μL T4 DNA ligase (*see* **Note 3**).
4. For inactivation of the ligase, add 1 μL of 0.2 *M* EDTA, heat samples 5 min at 70°C, and store them at –20°C.

3.2.2. Subtractive Hybridization

1. Add 200 ng (3 μL) of the restricted driver DNA to 12 ng (1 μL) of each of the adaptor-ligated tester DNAs (ratio 50:1) (*see* **Note 12**).
2. Add 1μL of 5X hybridization buffer to each sample and overlay with mineral oil.
3. Denaturate the DNAs 1.5 min at 98°C and subsequently anneal for 1.5 h at 65°C.
4. Combine the two samples with adaptor 1 and adaptor 2 and add 300 ng of fresh heat-denatured driver DNA in 3 μL of 1X hybridization buffer. Perform the second hybridization step for 14 h at 65°C (*see* **Note 12.**)
5. Dilute the reaction to 200 μL with dilution buffer and heat for 10 min at 65°C. The samples can be stored at –20°C until use in PCR.

3.2.3. PCR Amplification (see **Note 14**)

1. Mix 1 μL of genomic DNA (obtained in **Subheading 3.2.2.**, **step 5**) and 2 μL of PCR primer P1 with 47 μL of a PCR master mix containing DNA polymerase, PCR buffer, and dNTP mix (10 m*M* each dNTP).
2. Perform PCR reaction using the following conditions: 2 min at 72°C followed by 25 cycles at 95°C for 30 s, 66°C for 30 s, and 72°C for 1.5 min.
3. Dilute the amplified products 20-fold in 10 m*M* Tris-HCl, pH 7.5.
4. Use 1 μL of the diluted sample in the nested PCR with 2 μL of nested PCR primer NP1 and NP2 and 47 μL of a PCR master mix (described in **Subheading 3.2.3.**, **step 1**) followed by 10 cycles at 95°C for 30 s, 68°C for 30 s, and 72°C for 1.5 min (*see* **Note 13**).

5. Excise the resulting enriched tester-specific DNA fragments from the gel, purify, and ligate then to the pGEM T Easy vector. After ligation and transformation into *E. coli* DH5α, the DNA fragments can be further analyzed.

3.3. Island Probing with the Counterselectable Marker sacB

3.3.1. Basic Methods for DNA Manipulation

1. PCR: The reaction should be carried out in a total volume of 100 μL containing 1X PCR buffer with 2 mM MgCl$_2$, 0.2 mM dNTPs, 1 μM of each primer, 1 μL cell lysate or chromosomal DNA as template, and 5 U *Pfu Turbo* DNA polymerase.
2. Restriction of plasmids: Incubate the DNA with 1X reaction buffer and the suitable restriction enzyme (1 U/1μg DNA) in a total volume of 20 μL for 1 h at the recommended temperature.
3. Dephosphorylation of plasmid DNA: Incubate 1–20 pmol of linearized plasmid DNA with alkaline phosphatase (0.1 U/pmol DNA) and the corresponding 1X reaction buffer for 30 min at 37°C in a total volume of 50 μL.
4. Phosphorylation of DNA: Incubate the DNA with 1X reaction buffer, 1 μL 100 mM ATP, and 10 U of T4 polynucleotid kinase for 30 min at 37°C.
5. Generate blunt ends: Incubate 0.1–4 μg of DNA with 1X reaction buffer, 1 μL 20 mM dNTPs, and 1–5 U of Klenow fragment at 37°C for 30 min in a total volume of 20 μL.
6. Ligation: Mix insert and vector DNA in a 3:1 ratio (i.e., 180 fmol:60 fmol). Incubate it with 1X ligation buffer and 1 U of T4 DNA ligase in a total volume of 20 μL for 24 h at 16°C.
7. Transformation: Mix 10 μL ligation reaction with 100 μL Ca^{2+}-competent cells (4°C). Perform a heat shock for 2 min at 42°C, mix the cells with 1 mL LB medium, and incubate them for 1 h at 37°C.

3.3.2. Construction of a Mutagenesis Plasmid for Allelic Exchange

1. Grow the strain to be analyzed on a LB agar plate at 37°C.
2. Harvest a small amount of bacteria with an inoculation loop and resuspend in 50 μL sterile water. Boil 5 min at 100°C to lyse the cells, put them on ice for 5 min and pellet the debris by short centrifugation (>10,000g, 4°C, 2 min). The clear supernatant can be used as a template in PCR experiments.
3. Amplify a PAI-specific fragment by PCR with a suitable primer pair (*see* **Notes 6** and **7**). Purify the PCR product with the Qiagen PCR Purification kit.
4. Phosphorylate the PAI-specific fragment and repurify it using the Qiagen PCR Purification kit. Determine the concentration with a UV spectrophotometer.
5. Linearize the vector pGP704 with *Eco*RV and heat inactivate the restriction enzyme subsequently in a total volume of 50 μL (65°C, 20 min). Dephosphorylate the DNA and purify the vector with the Qiagen PCR Purification kit. Determine its concentration photometrically.
6. Ligate 180 fmol of the PAI-specific fragment with 60 fmol of the linearized plasmid pGP704. Transfer half of the ligation reaction into competent SY327 by heat

shock. Plate 100 µL aliquots on LB agar supplemented with 100 µg/mL ampicillin and grow the bacteria for approx 16 h at 37°C.

7. Choose colonies from **step 6** and use them to inoculate overnight cultures in LB medium supplemented with 100 µg/mL ampicillin. Use 3 mL of these cultures to isolate plasmid DNA using the Qiagen Plasmid Miniprep kit and confirm the insertion of the PAI-specific fragment into pGP704 by restriction analysis. Select one of the positive clones for further experiments and linearize the plasmid by using a suitable unique restriction site in the middle of the insert. Heat inactivate the enzyme in a total volume of 50 µL, generate blunt ends if necessary, and dephosphorylate the 5' ends to avoid religation. Finally, purify the vector with the Qiagen PCR Purification kit and determine its concentration in a UV spectrophotometer.

8. Cut the *sacB–sacR* cartridge out of the vector pCVD442 with *Pst*I as a fragment of approximately 2.6 kb (*see* **Note 8**). Separate it by submarine gel electrophoresis on a 1% (w/v) agarose gel from the plasmid DNA and purify it with the Qiagen Gel Extraction Kit. Generate blunt ends and purify the DNA with the Qiagen PCR Purification kit. Determine its concentration photometrically.

9. Ligate 60 fmol plasmid DNA from **step 7** and 180 fmol insert DNA from **step 8**. Transfer 10 µL of the ligation into SY327. Verify the correct insertion of the *sacB–sacR* cartridge by restriction analysis of the resulting constructs.

10. Transfer one of the mutagenesis plasmids by heat shock into *E. coli* SM10λ*pir* and use the resulting recombinant strain as donor in the following conjugational transfer.

3.3.3. Insertion of the sacB–sacR Cartridge into the PAI

1. Prepare overnight cultures of SM10λ*pir* carrying the mutagenesis plasmid (donor) and the *E. coli* strain to be analyzed (recipient).

2. Dilute the overnight cultures 1:100 into fresh medium and grow the bacteria to OD_{600} of 0.7 and 0.8 for the recipient and the donor strain, respectively. Mix 100 µL of each culture, plate it on LB agar, and incubate at 37°C. Harvest the cells after 5–12 h with 2 mL 0.9% NaCl and plate serial dilutions on LB agar containing 100 µg/mL ampicillin and 100 µg/mL streptomycin to isolate single crossover mutants (*see* **Note 9**).

3. Choose one of the single crossover mutants, subculture it in LB medium without antibiotics, and plate serial dilutions on LB agar plates containing either ampicillin or sucrose. Select a double-crossover mutant by searching for a colony with an ampicillin- and sucrose-sensitive phenotype.

4. Confirm the correct integration of the counterselectable marker by Southern blot experiments with PAI- and *sacB*-specific probes or PCR experiments.

3.3.4. Determination of Deletion Frequencies (16)

1. Inoculate 30 mL LB medium 1:100 with a fresh overnight culture of the *sacB*-marked *E. coli* strain and incubate it in an orbital shaker at 37°C. If necessary, modifiy the experimental conditions to simulate different environmental condi-

tions (i.e., osmotic stress, elevated or lower temperatures, and subinhibitory antibiotic concentrations).

2. Take 1 mL samples during the late lag and mid-log phases as well as during the early and late stationary phases. This ensures an optimal ratio of donor and acceptor cells. Plate serial dilutions on LB agar and LB agar supplemented with 7% sucrose to determine the total colony-forming units (CFU) and the number of PAI-negative (sucrose-resistent) colonies, respectively. Incubate the bacteria for 48 h at 20°C to reduce the growth rate and establish the toxic effect of the levansucrase (*see* **Notes 10** and **11**).

3. Calculate the frequency of PAI-negative cells as the quotient of sucrose-resistant cells per total CFU. Resulting data should be mean values of at least three independent experiments.

4. Notes

1. The binding of the primer pairs should only result in PCR products in the absence of integrated foreign DNA or if the size of the integrated fragment is suitable for amplification. Therefore, the primers should hybridize within the upstream and downstream regions of the tRNA genes of interest. The insertion of foreign DNA into the 3'-end of a tRNA gene will consequently result in no or in larger PCR products in comparison to the positive control.

2. To increase the specificity, primers that hybridize to repetitive sequences should be avoided.

3. The SSH experiments could result in false-positive fragments. To minimize the production of these fragments, it could be helpful to use high-performance liquid chromatographic purified oligonucleotides/primer.

4. This approach allows only the limited comparison of flanking regions of tRNA genes between the *E. coli* K-12 strain MG1655 and any other *E. coli* strain. A negative PCR result in the strain studied in comparison to the positive control *E. coli* MG1655 may be indicative of differences within the flanking regions of this particular tRNA gene as a result of the integration of foreign DNA elements. However, a negative PCR result can also be caused by point mutations, deletions, or smaller DNA rearrangements.

5. To facilitate the analysis of the different PCR products as well as their comparison between the positive control and the STEC strain of interest, PCR samples with identical primer pairs should be applied pairwise (*E. coli* MG1655, STEC strain) to the agarose gel.

6. The counterselectable marker should be inserted into a noncoding region of the pathogenicity island to avoid knockout mutations of virulence factors that could be detrimental during further experiments.

7. The PCR conditions during amplification depend on the choosen primer pair, but in any case an elongation time of 2 min/kb is necessary if using *Pfu Turbo* DNA polymerase. Ideally, the PAI-specific fragment should contain an internal unique restriction site to enable the insertion of the counterselectable marker.

8. Instead of the *sacB–sacR* cartridge, other counterselectable markers can be used for island-probing experiments as well *(17)*. However, the wild-type *E. coli* strain

should be able to grow on the corresponding selective medium without any disadvantage.

9. The suicide vector pGP704 cannot be replicated in the recipient strain because of its origin of replication (oriR6K). Therefore, an ampicillin/streptomycin-resistant phenotype is only established if the mutagenesis plasmid is inserted into the chromosome by homologous recombination. The complete allelic exchange by a second recombination event will result in loss of the vector sequence and lead to an ampicillin/sucrose-sensitive phenotype.

10. A reduction in copy number of the counterselectable marker could lead to diminished toxic effects. Therefore, it is sometimes necessary to decrease the growth rate of marked *E. coli* strains by lower incubation temperatures or a reduced nutrient composition of the medium to re-establish this effect.

11. To ensure that the deletion of the pathogenicity island and no mutation has led to the sucrose-resistant phenotype, prepare PCR control experiments with PAI- or *sacB*-specific primer pairs.

12. To avoid unspecific binding, it could be useful to change the ratio between driver and tester DNA from 50:1 to 80:1 and increase the hybridization temperature additionally.

13. If you do not see any products after 10 cycles, perform 10 more cycles and/or use 1 μL of the diluted sample of the nested PCR with 2 μL of nested PCR primer NP1 and NP2 and 47 μL of a PCR master mix (*see* **Subheading 3.2.3., step 1**) and perform a second nested PCR under the following conditions: 25 cycles at 95°C for 30 s, 68°C for 30 s, and 72°C for 1.5 min.

14. To avoid problems with buffers or oligonukleotides, one might, for example, use the CLONTECH PCR-Select™ bacterial genome subtraction kit, which contains a positive control for all of your reactions.

References

1. Karch, H., Schubert, S., Zhang, D., Zhang, W., Schmidt, H., Oelschlaeger, T. A., et al. (1999) A genomic island, termed high-pathogenicity island, is present in certain non-O157 shiga toxin-producing *Escherichia coli* of clonal lineages. *Infect. Immun.* **67,** 5994–6001.

2. Hacker, J. and Kaper, J. B. (1999) The concept of pathogenicity islands, in *Pathogenicity Islands and Other Mobile Virulence Elements* (Hacker, J. and Kaper, J. B., eds.), ASM Washington, DC, pp. 1–11.

3. Perna, N. T., Mayhew, G. F., Posfai, G., Elliot, S., Donnenberg, M. S., Kaper, J. B., et al. (1998) Molecular evolution of a pathogenicity island from enterohemorrhagic *Escherichia coli* O157:H7. *Infect. Immun.* **66,** 3810–3817.

4. Sperandio, V., Kaper, J. B., Bortolini, M. R., Neves, B. C., Keller, R., and Trabulsi, L. R. (1998) Characterization of the locus of enterocyte effacement (LEE) in different enteropathogenic *Escherichia coli* (EPEC) and shiga-toxin producing *Escherichia coli* (STEC) serotypes. *FEMS Microbiol. Lett.* **164,** 133–139.

5. Nicholls, L., Grant, T. H., and Robinson-Browne, R. M. (2000) Identification of a novel genetic locus that is required for in vitro adhesion of a clinical isolate of enterohemorrhagic *Escherichia coli* to epithelial cells. *Mol. Microbiol.* **35,** 275–288.

6. Tarr, P. I., Bilge, S. S., Vary, J. C., Jelacic, S., Habeeb, R. L., Ward, T. R., et al. (2000) Iha: a novel *Escherichia coli* O157:H7 adherence-conferring molecule encoded on a recently acquired chromosomal island of conserved structure. *Infect. Immun.* **68,** 1400–1407.

7. Cheetham, B. F. and Katz, M. E. (1995) A role for bacteriophages in the evolution and transfer of bacterial virulence determinants. *Mol. Microbiol.* **18,** 201–208.

8. Reiter, W., Palm, D. P., and Yeats, S. (1989) Transfer RNA genes frequently serve as integration sites for prokaryotic genetic elements. *Nucleic Acids Res.* **17,** 1907–1914.

9. Elliott, S. J., Wainwright, L. A., McDaniel, T. K., Jarvis, K. G., Deng, Y. K., Lai, L. C., et al. (1998) The complete sequence of the locus of enterocyte effacement (LEE) from enteropathogenic *Escherichia coli* E2348/69. *Mol. Microbiol.* **28,** 1–4.

10. Hacker, J. and Kaper, J. B. (2000) Pathogenicity islands and the evolution of microbes. *Annu. Rev. Microbiol.* **54,** 641–679.

11. Diatchenko, L., Lau, Y.-F. C., Campell, A. P., Chenik, A., Moqadam, F., Huang, B., et al. (1996) Suppression subtractive hybridization: a method for generating differentially regulated or tissue-specific cDNA probes and libraries. *Proc. Natl. Acad. Sci. USA* **93,** 6025–6030.

12. Tinsley, C. R. and Nassif, X. (1996) Analysis of the genetic differences between *Neisseria meningitidis* and *Neisseria gonorrhoeae*: two closely related bacteria expressing two different pathogenicities. *Proc. Natl. Acad. Sci. USA* **93,** 11,109–11,114.

13. Akopyants, N. S., Fradkov, A., Diatchenko, L., Hill, J. E., Sibert, P. D., Lukyanov, S. A., et al. (1998) PCR-based subtractive hybridization and differences in gene content among strains of *Helicobacter pylori*. *Proc. Natl. Acad. Sci. USA* **95,** 13,108–13,113.

14. Hacker, J., Blum-Oehler, G., Mühldorfer, I., and Tschäpe, H. (1997) Pathogenicity islands of virulent bacteria: structure, function and impact on microbial evolution. *Mol. Microbiol.* **23,** 1089–1097.

15. Rajakumar, K., Sasakawa, C., and Adler, B. (1997) Use of a novel approach, termed island probing, identifies the *Shigella flexneri she* pathogenicity island which encodes a homolog of the immunoglobulin A protease-like family of proteins. *Infect. Immun.* **65,** 4606–4614.

16. Middendorf, B., Blum-Oehler, G., Dobrindt, U., Mühldorfer, I., Salge, S., and Hacker, J. (2001) The pathogenicity islands of the uropathogenic *Escherichia coli* strain 536: Island probing of PAI II$_{536}$. *J. Infect. Dis.* **183(Suppl 1),** S17–S20.

17. Reyrat, J. M., Pelicic, V., Gicquel, B., and Rappuoli, R. (1998) Counterselectable markers: untapped tools for bacterial genetics and pathogenesis. *Infect. Immun.* **66,** 4011–4017.

18. Ried, J. L. and Collmer, A. (1987) An *nptI-sacB-sacR* cartridge for constructing directed, unmarked mutations in Gram-negative bacteria by marker exchange-eviction mutagenesis. *Gene* **57,** 239–246.

19. Miller, V. L. and Mekalanos, J. J. (1988) A novel suicide vector and its use in construction of insertion mutations: osmoregulation of outer membrane proteins and virulence determinants in *Vibrio cholerae* requires *tox*R. *J. Bacteriol.* **170,** 2575–2583.

10

Generation of Isogenic Deletion Mutants of STEC

Soudabeh Djafari, Nadja D. Hauf, and Judith F. Tyczka

1. Introduction

Genetic tools are necessary to unravel complex phenotypes, like the formation of attaching and effacing lesions by Shiga toxin-producing *Escherichia coli* (STEC). To inactivate a particular gene, we use a precise and convenient method that is based on the generation of an in-frame deletion within this gene. The advantages of this technique in comparison with others (e.g., transposon mutagenesis or site-directed insertional mutagenesis) are as follows:

- No polar effects on the transcription of downstream located genes by disrupting polycistronic RNA transcripts.
- Genetic stability of the generated deletion mutants, because a reversion to the wild-type is impossible.
- No continuous selective pressure is needed to maintain the mutant.

In order to create isogenic deletion mutants in STEC, we routinely use the low-copy-number suicide vector pMAK700*ori*T *(1,2)*. The features of this plasmid are the temperature sensitive (ts) replicon of plasmid pSC101 (*ori*101/rep101ts), the RP4/RK2 origin of transfer (*ori*T), and a chloramphenicol resistance gene (*cat*) as a selective marker (*see* **Subheading 2.**, **Fig. 1**).

In this chapter, we provide a detailed protocol for the generation of isogenic deletion mutants employing this vector. The underlying strategy is depicted in **Fig. 2**. It is based on a site-specific replacement of a particular wild-type allele with a cloned allele carried on pMAK700*ori*T harboring an internal in-frame deletion. The allelic exchange occurs in two steps via successive homologous recombination events:

1. Following introduction of pMAK700*ori*T containing the shortened version of the gene of interest (pMAK::Δ*orf*) into the wild-type STEC strain, it is possible

From: *Methods in Molecular Medicine, vol. 73: E. coli: Shiga Toxin Methods and Protocols*
Edited by: D. Philpott and F. Ebel © Humana Press Inc., Totowa, NJ

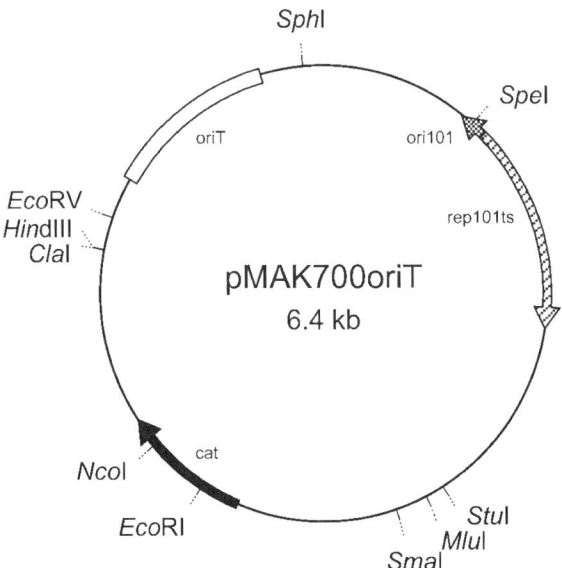

Fig. 1. Restriction map of the temperature-sensitive suicide vector pMAK700*ori*T. The vector pMAK700*ori*T is a derivate of plasmid pMAK700 *(2)* carrying a chloramphenicol resistance gene *cat*. Other features of the plasmid are the origin of replication *ori*101 from plasmid pSC101, the gene for the temperature-sensitive replication–initiation protein rep101ts, and the RP4/RK2 origin of transfer *ori*T. (From **ref. *1*,** with permission of the publisher.)

   to select for the integration of the recombinant plasmid into the target gene. This is carried out at the restrictive temperature of 42–44°C and simultaneous selection for chloramphenicol resistance.
2. Subsequent growth of the obtained cointegrates at the permissive temperature of 28–30°C without selection leads to the second homologous recombination event. This results in the excision of the suicide vector DNA from the chromosome.

To cure the excised plasmid from the bacterial cell, an additional temperature shift to 42–44°C without selective pressure can be performed. Because replication of the episomal plasmid is inhibited at the elevated growth temperature, descendants will subsequently be free of plasmids.

Depending on the site of the second recombination event, the corresponding wild-type gene is either replaced by the truncated allele as desired or the intact copy of the wild-type gene is restored. Therefore, chloramphenicol sensitive clones have to be distinguished by polymerase chain reaction (PCR) and nucleotide sequence analysis into the desired deletion mutants and unwanted revertants carrying the wild-type gene.

A   **in vitro steps**

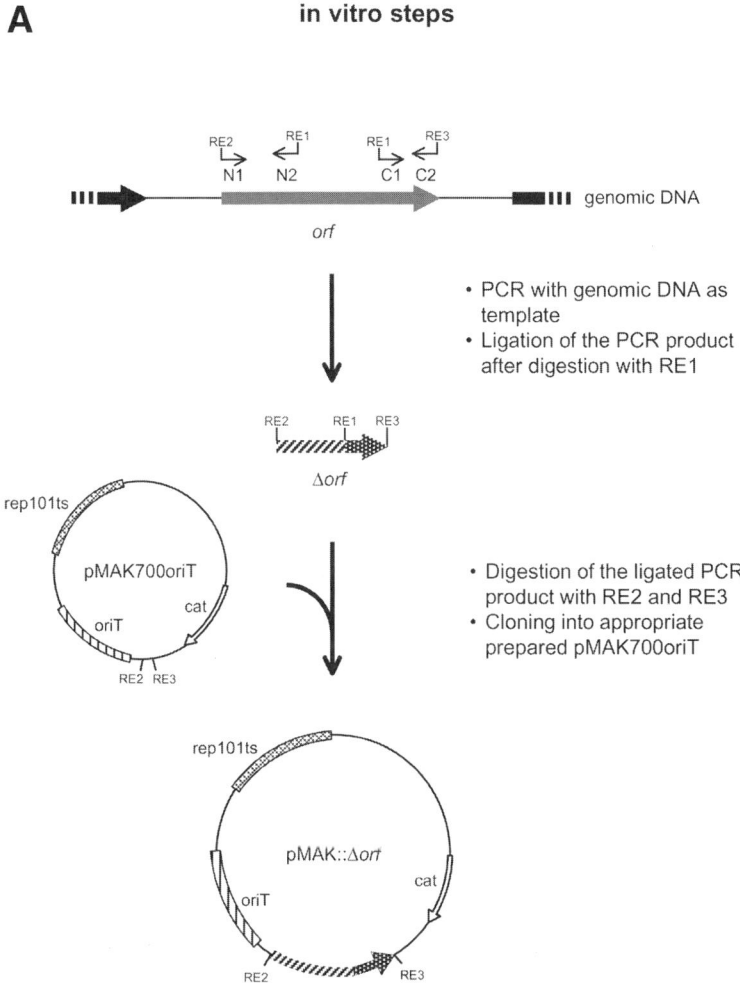

Fig. 2. Schematic representation of the generation of deletion mutants in STEC. (**A**) Following polymerase chain reaction-based amplification of short N- and C-terminal regions of the gene of interest (*orf*) using specific primers (N1/N2; C1/C2) with incorporated restriction sites (RE), the both resulting fragments are digested with RE1 and ligated. Next, the obtained ligation product (harboring the desired in-frame deletion) is digested with RE2 and RE3 and cloned into vector pMAK700*ori*T, giving rise to pMAK::Δ*orf*. For further details, see the text. The figure is not drawn in scale. (**B**) (*see* p. 116) After introduction of the knockout vector pMAK::Δ*orf* into the STEC wild-type strain, the desired isogenic deletion mutant is obtained via two homologous recombination events between the STEC DNA sequence on the plasmid and their corresponding wild-type allele. For further details, see the text. The figure is not drawn in scale.

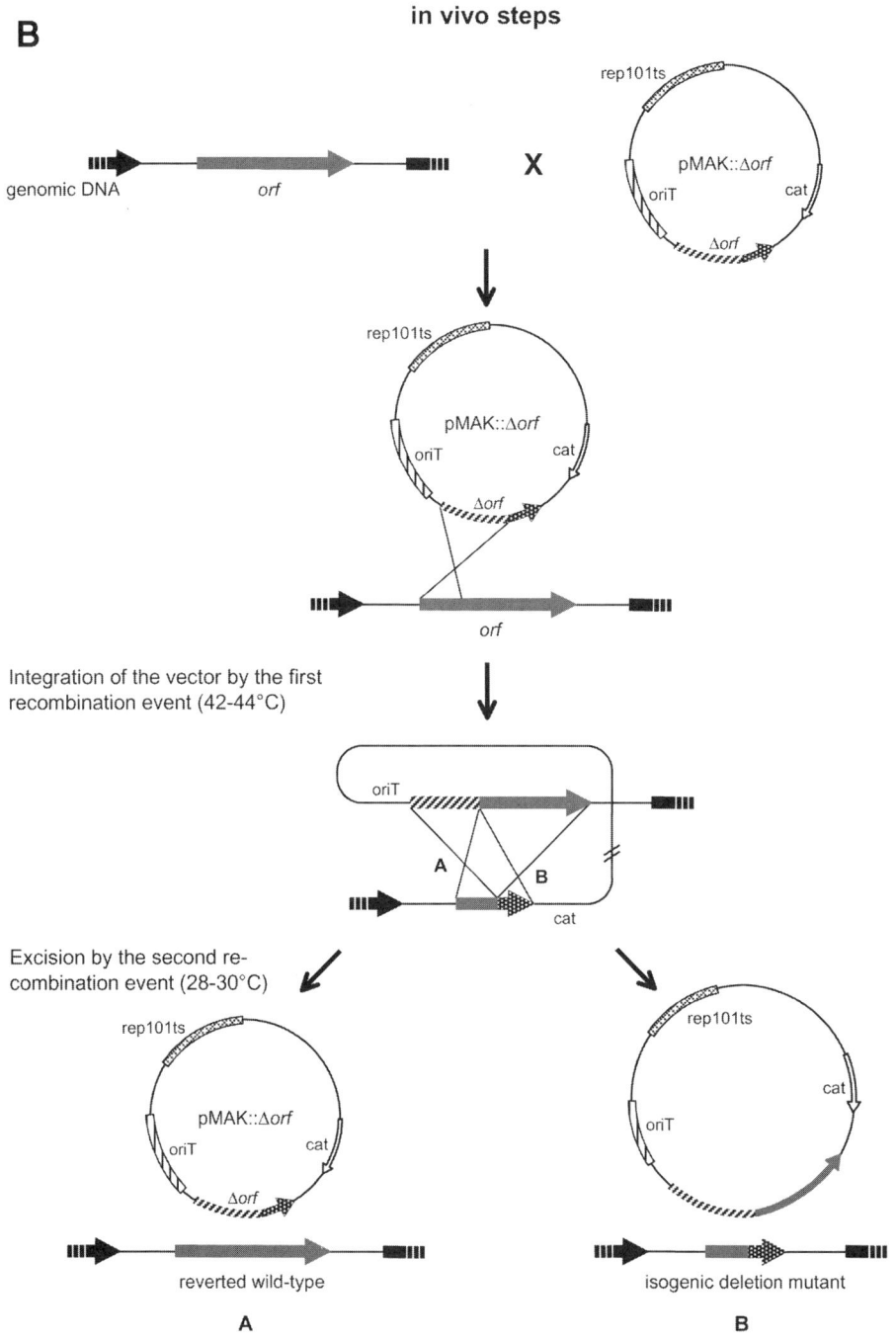

Fig. 2B.

Similar strategies for the generation of deletion mutants are based on the use of other suicide vector systems (e.g., containing the *pir*-dependent replicon of plasmid R6K combined with the *sacB* gene of *Bacillus subtilis*) *(3)*. In our experience, the use of a R6K vector system is limited, probably because of the presence of an endogenous plasmid harboring this replication origin in many wild-type isolates of *E. coli*. Additional vectors bearing the temperature-sensitive pSC101 origin of replication together with *sacB* have been developed *(4)*. The *sacB* gene encodes the enzyme levan sucrase, which is toxic for Gram-negative bacteria in the presence of sucrose. The combination of a suicide vector with a conditional lethal gene allows a positive selection for the loss of vector sequences after a second homologous recombination event.

Using the protocol given in this chapter isogenic deletion mutants can be obtained at frequencies up to 90%. The method is not restricted to chromosomal genes but is also suitable for the deletion of plasmid genes. In this case, isolated plasmid DNA instead of chromosomal DNA is used as a template for the subsequent manipulations (compare **Subheading 3.1.1.**).

2. Materials

1. STEC wild-type strain.
2. *Escherichia coli* K-12 strains; for example, S17-1 {*thi*, *pro*, *rec*A, *hsd* [r⁻ m⁺], RP4-2-Tc::Mu-Km::Tn7, Tp^R, Sm^R *(5)*}, or INVαF' (*end*A1, *rec*A1, *hsd*R17 (r^K,m^{+K}), *sup*E44, λ–, *thi*-1, *gyr*A, *rel*A1, φ80, *lac*ZαΔ, M15Δ (*lac*ZYA-*arg*F), deoR+, F'; Invitrogen).
3. The low-copy-number suicide vector pMAK700*ori*T (**Fig. 1**) isolated using a Genomed Plasmid Midiprep kit (*see* **Note 1**) according to the manufacturer's instructions.
4. Luria–Bertani (LB) broth: 10 g casein hydrolysate (Peptone No. 140) (Gibco-BRL), 5 g Bacto yeast extract (Gibco-BRL), 10 g NaCl, and 1 L distilled water. Adjust the pH to 7.5 and autoclave.
5. LB agar plates: Add 1% (w/v) agar (Gibco-BRL) to LB broth before autoclaving.
6. LB agar plates supplemented with 20 µg/mL chloramphenicol.
7. SOB medium without Mg²⁺: dissolve 20 g Bacto tryptone (Difco), 5 g Bacto yeast extract (Difco), and 0.58 g NaCl in 950 mL distilled water; add 10 mL of 250 m*M* KCl in distilled water; adjust the pH to 7.0; distilled water at 1 L; autoclave.
8. Chloramphenicol stock solution: 20 mg/mL (w/v) in 70% ethanol. Store at 4°C up to 2–3 mo.
9. SOC medium *(6)*: 2% (w/v) Bacto tryptone (Gibco-BRL), 0.5% (w/v) Select yeast extract (Gibco-BRL), 10 m*M* NaCl, 2.5 m*M* KCl, 10 m*M* MgCl₂, 10 m*M* MgSO₄, 20 m*M* glucose; distilled water at 1 L; autoclave.
10. CCMB80 buffer: 80 m*M* CaCl₂, 20 m*M* MnCl₂, 10 m*M* MgCl₂, 10% glycerol (98%), and 10 m*M* potassium acetate, pH 7.0: Adjust to pH 6.4 with 0.1 *N* HCl; do not use base to adjust the pH because the solution might become brownish. Sterilize the solution through a 0.2-µm-filter (Nalgene) and store at 4°C.
11. 0.85% NaCl in distilled water.
12. 1X TE buffer *(6)*: 10 m*M* Tris-HCl, 1 m*M* EDTA in double distilled water, pH 8.0.

13. DNase-free RNase (Roche Molecular Biochemicals; stock solution: 500 µg/mL).
14. GES reagent: 5 *M* guanidium thiocyanate (Sigma), 100 m*M* EDTA, pH 8.0, and 0.5% (v/v) sarkosyl (sodium-*N*-lauroyl sarcosinate; Fluka). GES reagent is prepared as follows: guanidium thiocyanate (60 g), 0.5 *M* EDTA, pH 8.0 (20 mL), and deionized water (20 mL) are heated at 65°C and stirred until dissolved. Cool down and add 5 mL of 10% (v/v) sarkosyl. Fill up with deionized water to 100 mL and sterilize through a 0.45-µm-pore filter (Nalgene). Store in the dark and at room temperature.
15. 7.5 *M* Ammonium acetate.
16. Phenol/chloroform/isoamylalcohol mixture (25 : 24 : 1).
17. Isopropanol (100%).
18. Ethanol (70%).
19. Washing buffer: 10% (v/v) glycerol (ultrapure) in deionized water.
20. Special disposable plastics: Cut pipet tips (1 mL), 1/2 microcuvets and cuvets for electroporation (0.1 cm; Invitrogen).
21. Additional PCR equipment: Four primers with suitable incorporated restriction sites (**Fig. 2A**).
22. Filter: type HA, 0.45-µm-pore size (Millipore, Germany); autoclave and airdry.
23. Liquid nitrogen.
24. Shaker for culturing bacteria.
25. Microcentrifuge for Eppendorf tubes (1.5 mL and 2.0 mL).
26. Refrigerated centrifuge for 15-mL and 50-mL tubes.
27. Speed-Vac.
28. Photometer (Amersham Pharmacia Biotech; Ultraspec 3000).
29. Gene Pulser II System (Bio-Rad).

Materials and solutions that come in contact with bacteria or DNA have to be sterile or DNase-free, respectively.

3. Methods

3.1. Construction of the Knockout Vector pMAK::Δorf

3.1.1. Isolation of Chromosomal DNA from STEC (7, Modified)

1. Inoculate STEC into 20 mL LB medium and incubate overnight at 37°C.
2. Harvest 1.5 mL of the 20-mL overnight culture in an Eppendorf tube and centrifuge at 1400*g* for 3–5 min at room temperature. Carefully decant the supernatant.
3. Resuspend the cell pellet in 0.5 mL of 0.85% NaCl. Mix gently by pipetting or vortexing.
4. Centrifuge again using the same conditions and decant the supernatant carefully. Resuspend the cells in 100 µL of 1X TE buffer.
5. Add 1 µL of RNase to cell suspension and mix gently. Incubate the suspension at room temperature for 5 min.
6. Lyse the cells with 0.5 mL of GES reagent and mix gently by inversion up to five times. Incubate at room temperature for 5 min.
7. Keep the lysate on ice for 2 min, then add 0.25 mL of cold (4°C) 7.5 *M* ammonium acetate, and mix gently by inverting four times. Keep the sample on ice for

a further 10 min, then add 0.5 mL of the phenol/chloroform/isoamylalcohol mixture. Avoid separation of the phases by gentle inversion for at least 15 min.

8. Centrifuge at $17,300g$, 4°C for 10–15 min.
9. Transfer the aqueous phase into a fresh 1.5-mL Eppendorf tube using one of the cut pipet tips (*see* **Subheading 2.**).
10. Add 0.5 mL of ice cold (–20°C) isopropanol. Mix the solution carefully by rotating the tube in a slanted position until it becomes homogenous. The threadlike DNA is obtained by careful inversion of the tube. If several samples are prepared in parallel, each tube must be handled separately.
11. Centrifuge the sample at $17,300g$ at 4°C for 5 min. Carefully remove the supernatant.
12. Wash the DNA pellet five times with 0.5 mL of cold (–20°C) 70% ethanol. Remove the ethanol extremely carefully. At the end of the washing procedure, the precipitated DNA is hardly visible.
13. Dry the DNA pellet in a Speed-Vac under vacuum for about 10 min. Keep in mind that after drying, the DNA pellet sticks only weakly to the wall of the tube.
14. Add 100 µL of 1X TE buffer to the DNA pellet. Leave the sample without shaking overnight at room temperature to dissolve the chromosomal DNA. After resuspension the chromosomal DNA has to be stored at 4°C (*see* **Note 2**).

3.1.2. Cloning of the Deletion Allele (Δorf) into pMAK700oriT

1. The 5'- and 3'-regions of the gene of interest (*orf*) are amplified by PCR using the primer combinations N1/N2 and C1/C2 and isolated chromosomal DNA (*see* **Subheading 3.1.1.**) as a template. Each PCR primer must contain a appropriate restriction site (*see* **Fig. 2**). The resulting PCR products should be of different size (*see* **Fig. 2** and **Note 3**). If necessary, short sequences flanking the gene might be part of the amplified region.
2. Digest the resulting amplification products with restriction enzyme 1 (RE1) and ligate both fragments.
3. Digest both the ligated product and pMAK700*ori*T with RE2 and RE3 and ligate the DNA fragment into the linearized vector.
4. Introduce the resulting plasmid pMAK::Δ*orf* into an *E. coli* K-12 strain (*see* **Note 4**) by chemical transformation (*see* **Subheading 3.2.**). Incubate the bacterial cells at the permissive temperature of 28–30°C, selecting for chloramphenicol resistance.
5. Identify bacteria harboring the recombinant plasmid pMAK::Δ*orf* by PCR. An appropriate primer combination would be N1/C2.
6. Isolate plasmid pMAK::Δ*orf* (*see* **Note 1**).

3.2. Introduction of pMAK::Δorf into E. coli K-12 Strains by Chemical Transformation (8)

3.2.1. Preparation of Chemical Competent E. coli K-12 Cells

1. Inoculate a single colony of an *E. coli* K-12 strain (INVαF' or S17-1; *see* **Note 4**) into 5 mL SOB medium and incubate at 30°C overnight with shaking.
2. Dilute the overnight culture 1:50 in 60 mL SOB medium and incubate at 30°C with shaking until the optical density reaches $OD_{600} = 0.3$.

3. Transfer 50 mL of the culture into a centrifugation tube and leave it on ice for 10 min. Harvest the bacteria by centrifugation at 700g at 4°C for 10–15 min. Completely remove the supernatant.

4. Resuspend the pellet in 17 mL of cold (4°C) CCMB80 buffer and incubate for 20 min on ice.

5. Centrifuge the bacterial suspension at 700g at 4°C for 10 min and decant the supernatant carefully.

6. Resuspend the cells in 4.2 mL of cold (4°C) CCMB80 buffer and keep them on ice.

7. Transfer aliquots of 200 µL to Eppendorf tubes. Tubes not required immediately should be shock frozen in liquid nitrogen and stored at –80°C. These cells can be used for 2–3 mo.

3.2.2. Transformation

1. Add 100–200 ng (approx 5 µL) of plasmid pMAK::Δ*orf* to 0.2 mL of competent *E. coli* K-12 cells (*see* **Subheading 3.2.1.**) and mix gently.

2. Keep the tube on ice for 30 min, then incubate at 42°C for 90 s, and, again, leave the sample on ice for 2 min.

3. Immediately add 0.8 mL of SOC medium to the prepared suspension. Incubate at 28–30°C for 1 h with shaking.

4. Plate 100–200 µL aliquots onto LB agar plates supplemented with chloramphenicol. Incubate the plates at 28–30°C for 1–2 d until colonies are easily visible.

3.3. Introduction of pMAK::Δorf *into STEC*
by Conjugation or Electroporation

3.3.1. Conjugation

1. Inoculate a single colony of both the donor (S17-1 with pMAK::Δ*orf*) and the STEC recipient into 10 mL LB medium each and incubate the cultures overnight at 28–30°C with shaking. In the case of the donor strain, add 20 µg/mL chloramphenicol.

2. Determine the optical density (OD$_{600}$) of the overnight cultures photometrically. Transfer equal volumes of the donor and the recipient culture in two 1.5-mL Eppendorf tubes.

3. Harvest both strains by centrifugation at 2000g for 4 min at room temperature. Carefully remove the supernatants.

4. Resuspend the cell pellets in 1 mL LB medium each and centrifuge one more time using the same conditions. Again, remove the supernatants carefully.

5. Repeat **step 4** once.

6. Under sterile conditions, place three filters (type HA, Millipore) on a nonselective LB agar plate.

7. Mix three times 100 µL of both the donor and recipient cells in 1.5-mL Eppendorf tubes. Carefully pipet the resulting 200 µL of each cell suspension on one of the filters and incubate the plate at 28–30°C for about 6–7 h.

8. Transfer every filter separately into a 15-mL tube containing 2 mL LB medium and mix vigorously by vortexing for about 2 min.

9. Transfer each of the three cell suspensions into a fresh 2.0-mL Eppendorf tube.
10. Harvest the bacteria by centrifugation at $17,300g$ for 2 min at room temperature and remove approx 1.5 mL of the supernatants. Resuspend the cell pellets in the residual broth volumes.
11. Plate aliquots of 100–200 µL of the bacterial suspensions onto LB agar plates containing 20 µg/mL chloramphenicol. Incubate the plates at 28–30°C for 1–2 d.

3.3.2. Transformation of pMAK::Δorf into STEC by Electroporation

3.3.2.1. PREPARATION OF ELECTROCOMPETENT CELLS

1. Inoculate a single colony of your STEC strain into 10 mL LB medium and incubate at 37°C overnight with shaking.
2. Dilute the overnight culture 1:100 in 60 mL LB medium and incubate at 37°C with shaking until the optical density reaches $OD_{600} = 0.8–1.0$.
3. Transfer 50 mL of the culture into a centrifugation tube. Harvest the culture by centrifugation at $6200g$ at 4°C for 10 min. Completely remove the supernatant.
4. Resuspend the cell pellet in 50 mL of cold (4°C) washing buffer. Again, centrifuge the bacterial suspension and decant the supernatant carefully.
5. Repeat **step 4** once.
6. Resuspend the pellet in 200 µL of cold (4°C) washing buffer and keep the suspension on ice.
7. Aliquot the cells in 40 µL per Eppendorf tube. Tubes not required immediately should be shock-frozen in liquid nitrogen and stored at –80°C.

3.3.2.2. ELECTROPORATION

1. Prepare the following on ice:
   • Electrocompetent STEC cells (prepared freshly or thawed on ice; *see* **Subheading 3.3.2.1.**).
   • Plasmid DNA (pMAK::Δ*orf*).
   • A sufficient amount of SOC medium.
   • A sterile cuvet for electroporation.
   If possible, keep the materials to be used in close vicinity to the Gene Pulser II System.
2. Pipet 100–500 ng (max. 5 µL) of plasmid DNA to 40 µL of the electrocompetent STEC cells and mix gently.
3. Transfer the mixture (avoid air bubbles!) into the chilled electroporation cuvet. Keep on ice for at least 1 min.
4. Set the Gene Pulser for 25 µF, 1.8 kV/cm, 200 Ω. Place the cuvet between the electrodes and pulse. The pulse should last approx 4 ms. We would like to stress that the conditions given here may not be suitable for all STEC strains and might have to be determined empirically.
5. Replace the cuvet on ice. Immediately add 1 mL of cold (4°C) SOC medium and transfer the suspension to a fresh 1.5-mL Eppendorf tube (*see* **Note 5**). Incubate at 28–30°C for about 1 h with shaking.
6. Plate the entire volume in 100–200-µL aliquots onto LB plates with chloramphenicol. Incubate at 28–30°C for 1–2 d.

3.4. Allelic Exchange by Homologous Recombination

3.4.1. Integration of pMAK::Δorf into the Target Gene

1. Pick individual chloramphenicol-resistant exconjugants/transformants and grow them overnight in LB medium with 20 μg/mL chloramphenicol at the permissive temperature of 28–30°C.
2. For the integration step, prepare the following:
   - A sufficient number of LB agar plates containing chloramphenicol should be prewarmed for at least 1 h at 42–44°C.
   - For the dilutions, a sufficient number of Eppendorf tubes with an appropriate volume of LB broth should be prewarmed at 42–44°C.
3. To select for integration of the plasmid pMAK::Δ*orf* into the chromosomal target gene, plate two times 100 μL of the appropriate dilutions (e.g., 10^{-2}–10^{-4}) of the overnight cultures onto prewarmed agar plates. Incubate at the restrictive temperature of 42–44°C for 1–2 d.
4. Pick single bacterial colonies onto prewarmed LB agar plates with chloramphenicol and incubate them once again at 42–44°C overnight. In our experience it is sufficient to pick about 50 colonies.
5. Screen the obtained colonies by PCR for the desired integration of pMAK::Δ*orf* into the chromosomal gene using different primer combinations. Useful primer combinations are as follows (*see* **Note 6**):
   - One primer located upstream of the amplified 5'-region of the gene of interest together with the primer C2 (*see* **Fig. 2A**).
   - A primer flanking the amplified 3'-region of the gene in combination with the oligonucleotide N1.
6. Choose an insertional mutant, in which the recombination occurred via the smaller homologous fragment (*see* **Note 3**).

3.4.2. Excision of the Integrated Plasmid

1. Propagate a pure colony of the selected insertional mutant (*see* **Subheading 3.4.1.**) in 20 mL LB broth at the permissive temperature of 28–30°C overnight with shaking. Do not add chloramphenicol!
2. Dilute the overnight culture 1:1000 in fresh LB medium without chloramphenicol and again grow the bacteria to stationary phase. Repeat this step up to six times.
3. In order to eliminate the excised plasmids from the bacterial cells, plate appropriate dilutions (e.g., 10^{-5}–10^{-7}) of the latest culture onto prewarmed (42–44°C) LB plates without chloramphenicol. Incubate at the restrictive temperature of 42–44°C for 1–2 d.
4. Transfer single bacterial colonies via replica plating to LB agar with and without chloramphenicol. Recover chloramphenicol-sensitive clones and screen them for the presence of the deletion by PCR with specific primers flanking the respective gene (e.g., 5'- and 3'-oligonucleotides of **Subheading 3.4.1., step 5**).

Confirm the deletion in the gene of interest by other suitable techniques (e.g., Southern hybridization or nucleotide sequencing). Detailed protocols for the phenotypic analysis of the deletion mutants are given in other chapters of this volume.

4. Notes

1. For the isolation of plasmid DNA, we routinely use the Genomed Plasmid Midiprep kit. If alternative kits are used, one should keep in mind that pMAK700*ori*T has a low copy number and that the purity of the isolated plasmid DNA should be as high as possible. The plasmid DNA for electroporation has to be resolved in deionized distilled water. Do not use 1X TE buffer, because the salt concentration might cause arcing.
2. Do not freeze chromosomal DNA, which otherwise might be nicked. In our hands, chromosomal DNA can be used for more than 1 yr, if appropriately stored.
3. The desired mutant will be obtained, if both recombination events occur on the opposite sides of the intended deletion; otherwise, the wild-type gene is restored. A major factor, which influences the frequency of recombination is the length of the homologous regions of the altered and wild-type allele. Therefore, we recommend amplifying two PCR fragments of different lengths. After the first recombination event, one should choose an insertional mutant harboring pMAK::Δ*orf* in the smaller fragment. Because of the larger size of the other (homologous) fragment, it is more likely that the intended spontaneous excision of the plasmid will take place, yielding the desired deletion mutant.
4. If pMAK::Δ*orf* has to be introduced into an STEC strain by conjugation, the plasmid is transferred first to a conjugation competent *E. coli* K-12 host (e.g., S17-1). Alternatively it may be possible to use electroporation with some STEC wild-type strains (this has to be tested empirically). In this case, the plasmid can be introduced into any *E. coli* K-12 strain (e.g., INVαF').
5. To transfer the bacterial suspension into the tube, we normally use a Pasteur pipet. Likewise, one may employ a disposable syringe with needle.
6. In case you are not able to detect a PCR product using both primer combinations, it is possible that the conjugation/transformation took place, but that the plasmid was not integrated into the target gene. In our experience, the rep101ts of pMAK700*ori*T reverts to temperature independence at a significant frequency.

References

1. Virct, J.-F., Cryz, S. J., Jr., and Favre, D. (1996) Expression of *Shigella sonnei* lipopolysaccharide in *Vibrio cholerae. Mol. Microbiol.* **19,** 949–963.
2. Hamilton, C. M., Aldea, M., Washburn, B. K., Babitzke, P., and Kushner, S. R. (1989) New method for generating deletions and gene replacements in *Escherichia coli. J. Bacteriol.* **171,** 4617–4622.
3. Donnenberg, M. S. and Kaper, J. B. (1991) Construction of an *eae* deletion mutant of enteropathogenic *Escherichia coli* by using a positive-selection suicide vector. *Infect. Immun.* **59,** 4310–4317.
4. Blomfield, I. C., Vaughn, V., Rest, R. F., and Eisenstein, B. I. (1991) Allelic

exchange in *Escherichia coli* using the *Bacillus subtilis sacB* gene and a temperature-sensitive pSC101 replicon. *Mol. Microbiol.* **5,** 1447–1457.

5. Simon, R., Priefer, U., and Pühler, A. (1983) A broad host range mobilization system for in vivo genetic engineering: transposon mutagenesis in gram negative bacteria. *Bio/Technology* **1,** 784–791.

6. Sambrook, J., Fritsch, E. F., and Maniatis, T. (1989) *Molecular Cloning.* Cold Spring Harbor Laboratory, Cold Spring Harbor, NY.

7. Pitcher, D. G., Saunders, N. A., and Owen, R. J. (1989) Rapid extraction of bacterial genomic DNA with guanidinium thiocyanate. *Lett. Appl. Microbiol.* **8,** 151–156.

8. Hanahan, D., Jessee, J., and Bloom, F. (1991) Plasmid transformation of *Escherichia coli* and other bacteria. *Methods Enzymol.* **204,** 63–113.

11

Generation of Monoclonal Antibodies Against Secreted Proteins of STEC

Kirsten Niebuhr and Frank Ebel

1. Introduction

Bacterial pathogens have evolved a variety of strategies to hijack the host cell signaling systems, membrane trafficking events, as well as the cytoskeletal machinery, and the discovery of these strategies has contributed significantly to recent advances in cell biology. The relatively new scientific discipline "cellular microbiology" focuses on the moleular crosstalk between pathogens and their host cells, as the elucidation of these interactions not only helps to understand the molecular basis of disease but has also given rise to new model systems for probing host cell functions.

Pathogenic micro-organisms are commonly equipped with a whole set of virulence factors whose activities are orchestrated to attack and subvert host target cells. For detailed studies, specific tools are required to detect and localize such proteins during infection. One possibility is to tag bacterial proteins with an epitope for which a suitable antibody is already available (1). An alternative is to purify such proteins to raise specific antibodies in either rabbits or mice.

The advantage of polyclonal antibodies in rabbits is that they are easy to make and that they recognize various epitopes on a polypeptide chain and are, therefore, suitable to detect the protein even in denatured form (e.g., after a harsh fixation procedure, as it is often necessary for electron microscopic studies). Additionally, polyclonal antibodies are often crossreactive with homologous proteins from related organisms and it can be advantageous to identify such proteins. However, the major disadvantage of polyclonal antibodies is the presence of large quantities of antibodies with other specificities, a problem that is especially severe if *Escherichia coli* proteins are studied.

An alternative approach is to raise monoclonal antibodies in mice. This method requires smaller amounts of antigen but deserves some additional tis-

From: *Methods in Molecular Medicine, vol. 73: E. coli: Shiga Toxin Methods and Protocols*
Edited by: D. Philpott and F. Ebel © Humana Press Inc., Totowa, NJ

sue culture work. Monoclonal antibodies are tools of high specificity that can theoretically be produced in unlimited quantities. A major advantage of this approach is that a complex mixture of proteins can be used as an antigen, which can, after a screening and subcloning procedure, finally result in a set of hybridomas that secrete highly specific antibodies against various proteins and are derived from a single experiment. If all hybridomas of a fusion are stored under proper conditions, this will result in a real hybridoma bank that can be rescreened again and again to isolate antibodies against various antigens step by step.

The protocol described in this chapter is a very rapid and effective method to generate monoclonal antibodies. The immunization procedure takes only 17 d and has originally been developed to raise antibodies against poor antigens *(2–4)*. We applied this method, which elicits a local immune response and uses the lymph nodes as the donor organ, to generate monoclonal antibodies against various virulence-associated proteins of Shiga toxin-producing *E. coli* (STEC). These antibodies were successfully used in enzyme-linked immunosorbent assay (ELISA), immunoblot, immunofluorescence, and electron microscopic studies. Readers who are interested in using a similar technique to raise monoclonal antibodies in rats are referred to **ref. 5**.

2. Materials

2.1. Antigen Preparation

1. Luria–Bertani (LB) medium.
2. HEPES-buffered tissue culture medium: minimal essential medium (MEM) (e.g., Life Technologies), 25 m*M* *N*-2-hydroxethylpiperazine-*N*'-2-ethansulfonic acid (HEPES, Sigma), pH 7.4, sterilize by filtration using a 0.2 µm bottle top filter (e.g., Nalgene).
3. Erlenmeyer flasks (500 mL, 100 mL).
4. Water bath (37°C) equipped with a shaking device.
5. Standard photometer to measure the optical density at 600 nm.
6. Centrifuge.
7. 100% Trichloroacetic acid (TCA): 500 g TCA in 500 mL distilled water.
8. 1.5 *M* Tris-HCl, pH 8.8.
9. Equipment to run and stain sodium dodecyl sulfate–polyacrylamide gel electrophoresis (SDS-PAGE) gels (e.g., Biometra-Whatman, Bio-Rad).

2.2. Isolation of Feeder Cells

1. BALB/c mice and access to an animal facility equipped for housing of mice.
2. CO_2 gas supply.
3. Glass beakers of different sizes.
4. Preparation plate covered with aluminum foil.
5. Surgical instruments (fine scissors and tweezers), sterilized.
6. 70% Ethanol.

7. Tissue culture work bench.
8. Ice bucket.
9. Sterile Pasteur pipets.
10. Centrifuge.
11. Tissue culture incubator (37°C, 5% CO_2).
12. 24- and 96-Well plates and Petri dishes (9 cm and 4 cm) (sterile, tissue culture grade; e.g., NUNC)
13. Freezing medium: 8% dimethyl sulfoxide (DMSO, high-performance liquid chromatography [HPLC] grade, Merck) in fetal calf serum (e.g., Life Technologies).
14. Optional: growth supplement that may be used instead of feeder cells (e.g., Hybdidoma Cloning Factor, Origen, IGEN International Inc., Gaithersburg, MD).

2.3. Immunization

1. BALB/c mice (female, 8–10 wk old) and access to an animal facility equipped for housing of mice.
2. Ether or Metofane (Janssen Pharmaceutica, Beerse, Belgium).
3. 500-mL Glass beaker covered with aluminium foil.
4. 1-mL Syringe with 0.45 × 13-mm needles (e.g., Braun, Melsungen, Germany)
5. Freund's complete/incomplete adjuvant (Sigma) or alternative adjuvants like Alu-GelS (Serva, Heidelberg, Germany) or ImmunEasy (Qiagen).
6. Ether or Metofane.
7. Optional: Metal pipe with a diameter of approx 4 cm and a length of approx 15 cm.

2.4. Culture of Myeloma Cells

1. Myeloma medium: RPMI 1640 medium, 5% fetal calf serum (FCS), 4 m*M* glutamine, 50 IU penicillin, 50 μg streptomycine (e.g. Life Technologies).
2. P3-X63-Ag8 myeloma cells (ATCC, catalog no. CRL 1580).
3. Petri dishes (tissue culture grade).
4. Sterile 50-mL tubes.

2.5. Fusion

1. CO_2 gas supply and a 500-mL glass beaker covered with aluminum foil.
2. 100-mL Glass beaker with 70% ethanol.
3. Preparation plate covered with aluminum foil.
4. Surgical instruments (fine scissors and tweezers), sterilized.
5. Plastic ware (tissue culte grade): Petri dishes, 15- and 50-mL tubes, 5- and 10-mL pipets.
6. Sterile phospate-buffered saline (PBS) or Hank's buffered salt solution (e.g., Life Technologies).
7. Sterile microscopic glass slides with frosted ends.
8. Hybridoma medium: OPTI-MEM 1, 5% FCS, 4 m*M* glutamine, 50 IU penicillin, 50 μg streptomycine (Life Technologies).
9. HEPES medium: minimal essential medium supplemented with 25 m*M* HEPES, pH 7.4.
10. Fusion medium: Dissolve 5 g polyethylenglycol (PEG) 4000 (Sigma) in 5 mL HEPES medium using a water bath prewarmed to 60°C. Adjust the pH to 7.4

using 1 *M* NaOH. Sterilize by filtration through a 0.2-µm filter. Freeze in 1-mL aliquots and store at 4°C.

11. Selective medium: Hybridoma medium supplemented with 20 mg/L hypoxan-thine and 1.5 mg/L azaserine (Sigma).
12. 500-mL Beaker with approx 400 mL water, prewarmed to 37°C.
13. Tissue culture incubator and an inverse microscope.

2.6. Freezing and Storage of Hybridoma Cells

1. Centrifuge and sterile 50-mL tubes.
2. Freezing medium (*see* **Subheading 2.2.**).
3. Cryo tubes for tissue culture cells (e.g., NUNC).
4. Freezing device (e.g., 5100 Cryo 1°C [Nalgene] containing 2-propanol).
5. Freezer (–80°C).
6. Container with liquid nitrogen and equipment to host cryo tubes.

2.7. Subcloning

1. 96-Well plates (tissue culture grade, e.g., NUNC).
2. Neubauer chamber (e.g., Brand, Wertheim, Germany) for counting cells and an inverse microscope.
3. Adjustable multichannel pipet.
4. Freezing equipment (*see* **Subheading 2.6.**)

3. Methods

3.1. Antigen Preparation

Apart from the surface protein intimin, all known or putative STEC virulence factors are released into the culture supernatant, at least under some conditions. Different culture conditions can result in very different profiles of proteins present in the supernatant. The determination of conditions optimal for secretion is, therefore, a first and important step. Proteins encoded by the LEE pathogenic-ity island are produced and released in large amounts if STEC bacteria are grown at 37°C in a HEPES-buffered minimal cell culture medium (e.g., minimal essen-tial medium [MEM] or Dulbecco's minimal eagle medium [DMEM]).

1. Inoculate 200 mL of HEPES-buffered MEM or DMEM (pH 7.4) with 4 mL of an overnight culture grown in LB at 37°C. Incubate for 3–4 h at 37°C in a shaking device, which will result in a final OD_{600} of 0.8–1.2.
2. Centrifuge for 15 min at 3500*g*, transfer the supernatant into a clean centrifuge tube, and add 10% (v/v) of 100% TCA (*see* **Note 1**).
3. Store overnight at 4°C.
4. Centrifuge at 4000*g* for 20 min to pellet denatured proteins (*see* **Note 2**). Care-fully discard the supernatant.
5. Place tubes upside down on a thick layer of paper. Make sure that they dry com-pletely and resuspend the pellet in 200 µL of 1.5 *M* Tris-HCl, pH 8.8. Mix a small

sample with an appropriate volume of SDS sample buffer (*see* **Note 3**).

6. Separate proteins by SDS-PAGE and stain the gels with Coomassie brillant blue. To estimate the protein concentration, it is advisable to run defined amounts of a standard protein (usually bovine serum albumin) in parallel.

7. For immunization of three mice, 200 µL containing 10–30 µg of protein are usually sufficient (*see* **Note 4**).

3.2. Isolation of Feeder Cells

After the fusion, the vast majority of cells will die within the first 4–6 d as a result of the presence of hypoxanthine and azaserine. This is a necessary step for selecting sensible myeloma cells from resistant, successfully fused hybridoma cells. The unfused lymphocytes, which are also resistent, will die unless they are further stimulated. To improve growing conditions for the fragile hybridoma cells, it is recommended to use either feeder cells or commercially available growth supplements (*see* **Note 5**).

1. Sacrifice the mice (e.g., with CO_2 [*see* **Note 6**]), and dip them into a beaker with 70% ethanol for 30 s.

2. Fix the body with pins to a plate covered with aluminum foil and cleaned with 70% ethanol.

3. Carefully remove the skin from the ventral side of the abdomen (*see* **Note 7**).

4. Use forceps to lift a small piece of the peritoneal membrane and inject 4–5 mL of ice-cold sterile PBS. Massage the abdomen using a forceps. Remove the medium, either with a syringe or apply a small cut and use a sterile Pasteur pipet. Transfer the cell suspension to a precooled 50-mL tube placed on ice (*see* **Note 8**).

5. Centrifuge at 300g and 4°C for 5 min. Discard the supernatant and either freeze the cells in cold freezing medium or resuspend them in selective medium (containing hypoxanthine and azaserine) and plate immediately. In our hands, feeder cells from one mouse are sufficient for six multiwell tissue culture plates or ten 9-cm Petri dishes. Apply 1 mL of feeder cells per well of a 24-well plate and 200 µL per well of a 96-well plate.

6. Incubate at 37°C in a CO_2 incubator. Feeder cells should be seeded 3–6 d before the scheduled fusion to allow conditioning of the medium.

3.3. Immunization (see Note 9)

As a matter of routine we immunize two to three mice in parallel using the same antigen.

1. Anesthesize mice in a 500-mL glass beaker covered with aluminum foil using ether or Metofane (*see* **Note 6**).

2. Inject the antigen subcutaneously into both footpads of the hindlegs (approx 30 µL per pad) (*see* **Note 10**). Immunize on d 17, 14, 11, 7, 3, and 1 before the fusion. For the first three immunizations, the antigen may be mixed with incomplete Freund's adjuvant in a ratio of 2:1 (v/v, antigen:adjuvants). All subse-

quent immunizations are performed without adjuvants (*see* **Note 11**). Emulgate the antigen and the adjuvants using a 1-mL syringe or two syringes connected by an adaptor. Inject this solution immediately to prevent swelling of the plunger (*see* **Notes 12–14**).

3.4. Cell Culture of Myeloma Cells

We use the myeloma cell line P3-X63-Ag8 as a fusion partner for lymphocytes isolated from BALB/c mice.

1. Grow P3-X63-Ag8 cells in RPMI1640 culture medium. Myeloma cells from eight Petri dishes (9 cm) containing a dense population of growing cells in the log phase are sufficient for a fusion with lymphocytes isolated from the popethial lymph nodes of three mice (*see* **Note 15**).
2. Harvest the cells by centrifugation (5 min at 300g) shortly before the fusion.
3. Resuspend cells in 20 mL fresh RPMI1640 culture medium and transfer them to a fresh 50-mL tube.
4. Incubate at 37°C in a CO_2 incubator until use.

3.5. Fusion

1. First make sure that the myeloma (*see* **Subheading 3.3.**) and the feeder cells (*see* **Subheading 3.2.**) are not contaminated and in good condition.
2. Kill the mice in a covered beaker using carbon dioxide or cervical dislocation (*see* **Note 6**).
3. Dip the body into a beaker with 70% ethanol for 30 s. Cover a preparation plate with aluminum foil, clean the surface using 70% ethanol, and place it under a tissue culture hood.
4. Transfer the mouse to the plate and fix the body with pins. All subsequent steps are done under sterile conditions!
5. Make a cut around the thigh, pull down the skin like a stocking, and fix it with pins to the plate in a way that the leg is straightened. The popliteal lymph nodes are located in the hollow of the knee and should be enlarged because of the local immune response. They can easily be localized by pushing up the hollow of the knee from the back (*see* **Fig. 1** and **Note 16**).
6. Clean the leg briefly with 70% ethanol (e.g., by using a plastic bottle with a pump-spray head). Isolate the lymph node with a generous cut (*see* **Fig. 1**). Make sure that the node is not damaged, as this will lead to loss of most lymphocytes. Transfer the explant to a small Petri dish with sterile PBS. Proceed in the same manner with lymph nodes from the other legs.
7. Collect the explants in one Petri dish and remove all attached tissue with a tweezer (*see* **Note 17**). Transfer lymph nodes to a Petri dish with fresh sterile PBS.
8. Place lymph nodes on the frosted end of a sterile glass slide, cut them in half, and mince them between the frosted ends of two glass slides to obtain a single cell suspension (*see* **Note 18**).
9. Collect the solution in a 15-mL tube, place it in a rack, and wait for 2 min to allow larger debris to sediment.Transfer the supernatant (single-cell solution) to a fresh tube.

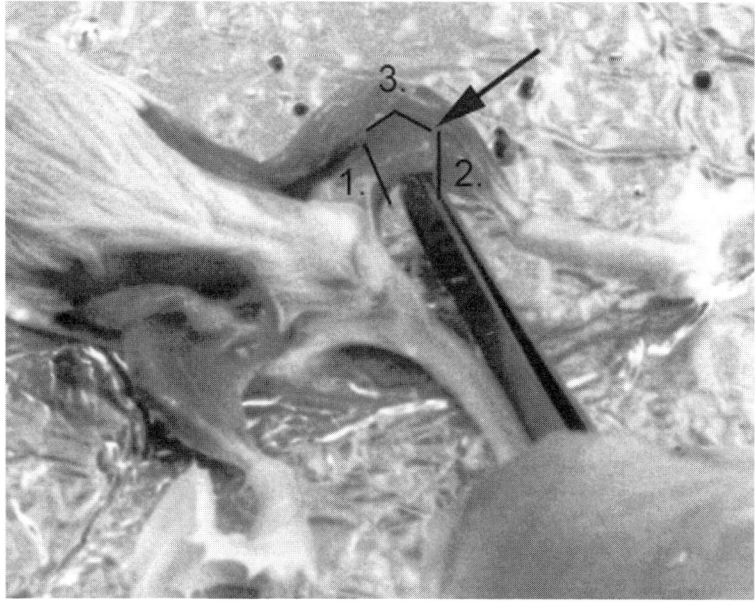

Fig. 1. Isolation of lymph nodes. After removal of the skin, popliteal lymph nodes (arrow) can be identified by pushing up the hollow of the knee ßhe back. Three cuts should be set as indicated (dark lines) to isolate the lymph nodes and its surrounding tissue.

10. Remove myeloma cells from the incubator and centrifuge both tubes in parallel (5 min, 300g, room temperature [RT]).
11. Resuspend both pellets in 30 mL of HEPES medium, pool them, and centrifuge again (5 min, 300g, RT).
12. Aspirate the supernatant as completely as possible. Tap the tube to disperse the pellet, transfer it to a 37°C water bath (e.g., a 500-mL beaker containing 400 mL prewarmed water placed under the tissue culture working bench), and incubate for 5 min.
13. Add dropwise 1 mL of prewarmed fusion medium within 1 min while gently shaking the tube in the water bath. Continue this gentle shaking for another minute.
14. While shaking, slowly add 20 mL of prewarmed HEPES medium according to the following scheme: 1 mL within 30 s, 3 mL in 30 s, 16 mL in 1 min.
15. Centrifuge at 220g for 5 min. Gently resuspend the pellet in an appropriate volume of selective medium using a 10-mL pipet, (e.g., suspend cells derived from three mice into 50 mL). Transfer approx 140 µL (2–3 drops from a 10 mL pipet) to each well of a 24-well tissue culture plate. Fifteen plates will be sufficient to plate all cells of the fusion.
16. Incubate at 37°C in a tissue culture incubator. After 2–3 d, the myeloma cells that make up the majority of cells will die. After 5–7 d, small colonies of hybridoma cells should be visible under the microscope. The supernatants should be screened

when the diameter of the colonies reaches about 1–2 mm. For the screening, remove 600 µL (*see* **Note 19**) and replace it by fresh selective medium. To prevent growth of myeloma cells selective conditions should be maintained for at least 2 wk. Use hybridoma medium for all subsequent tissue culture work.

3.6. Generation of a Hybridoma Bank

Under normal conditions immunization with a complex antigen will result in growth of hybridoma clones in more than 90% of the wells with 10 colonies per well in average. Rapid freezing of these cells is required to generate a bank of hybridoma cells.

1. Grow hybridoma cells in 24-well plates until the largest colonies are about 3–4 mm in diameter.
2. Resuspend cells, transfer them to fresh tubes, and centrifuge for 5 min at 300g and 4°C.
3. Store the individual supernatants (*see* **Note 20**).
4. Resuspend cells derived from one well in 0.5-mL ice-cold freezing medium.
5. Transfer this suspension to at least two cryo tubes (250 µL each) (*see* **Note 21**).
6. Store overnight in a freezing device.
7. Store tubes in a cryo container under liquid nitrogen (*see* **Note 22**).

3.7. Screening

If not all hybridoma cells are stored, a strategy for a rapid screening has to be developed to identify those wells that contain cells that secrete antibodies of interest. The screening is highly dependent on the antigen that has been used and the protein(s) of interest. If the antigen is a purified protein, a first screening in ELISA is recommendable. Such an ELISA screen is not helpful in the case of a complex antigen mixture because all supernatants are then likely to be positive. In this case, the fusion should be tested directly by immunoblot and/or immunofluorescence (*see* **Note 20**).

3.8. Maintentance of Hybridomas

1. Expand cells of interest on six-well plates and subsequently on 10-cm Petri dishes.
2. Harvest cells from a 10-cm Petri dish containing a dense but still growing culture, centrifuge (5 min, 300g) and resuspend the pellet in 2.5-mL ice-cold freezing medium. Transfer the suspension to five 1.5-mL freezing tubes (0.5 mL/tube) and place them into a freezing device overnight at –80°C.
3. Store tubes under liquid nitrogen.

3.9. Subcloning

1. Prepare feeder cells in 96-well plates (*see* **Subheading 3.2.**).
2. Take hybridoma cells from a growing culture and count them using a Neubauer chamber.

3. Plate two plates with (theoretically) 0.1 cell/well, two plates with 1 cell/well, and 1 plate with 5 cells/well.
4. Grow cells until colonies are visible.
5. Inspect plates and mark wells that contain only one colony.
6. Remove supernatants and replace them by fresh medium.
7. Test supernatants using an appropriate screening procedure.
8. Expand positive cultures.
9. Freeze cells and repeat the subcloning at least one more time.

4. Notes

1. Trichloracetic acid is a very aggressive agent that has to be handled with care (use gloves!). Make sure that the tubes are resistent to TCA!
2. After growth in minimal medium, the precipitate is usually very small and often hardly detectable. It is therefore advisable to mark the tube in a way that will enable you to identify the position of the pellet after centrifugation.
3. The remaining liquid is acidic. The 1.5 M Tris-HCl buffer used to resuspend the precipitate is sufficient to buffer traces of TCA. After mixing small samples with SDS–sample buffer (that contains bromphenol blue), the solution will be either neutral (blue) or acidic (yellow). In the latter case, it is necessary to add more buffer or small amounts of 1 M NaOH until it is blue.
4. The amount of protein necessary to achieve a good immune response can vary depending on the antigen and its preparation, but we found that that even very small amounts of a protein present in a complex protein mixture can be sufficient.
5. Make sure that macrophages used as feeder cells are derived from mice that have the same major histocompatibility complex as the myeloma cells and the mice used for immunization. P3-X63-Ag8 cells are compatible with BALB/c mice.
6. Killing and anesthetization of animals requires proper skill and education. The reader should refer to requirements specified in their local animal protection law!
7. It is critical to avoid any damage of the peritoneal membrane! Lift the skin with a forceps and set a small cut in the middle of the lower abdomen that is just large enough to introduce one arm of a fine scissor into the opening. Cut vertically upward and pull the abdominal skin to both sides of the body. Fix the skin to the plate using pins or needles.
8. If the solution appears turbid, a small amount of the fluid should be checked microscopically for the presence of bacteria to avoid contaminations that occur if the gut has been damaged.
9. Immunization of animals requires proper education and the reader should refer to requirements specified in their local animal protection law!
10. In countries where injection into the footpad is discouraged, the antigen can be applied subcutanously into the hind leg. For this purpose, it is recommended to shave the areas where the immunizations are applied.
11. Complete Freund's adjuvant may be used for the first immunization instead of incomplete Freund's adjuvant; however, for most bacterial antigens, this is not required.

12. Use a 1-mL syringe without a sealing around the plunger to prevent swelling problems.

13. For footpad immunizations, it is convenient to use a short metal or plastic pipe with a diameter sufficiently large to take up the body of the mouse. The foot can then be positioned on the rim of the pipe.

14. In principle, it is possible to check the antibody titer before the fusion by ELISA or immunoblot. However, one has to keep in mind that the immune response triggered by the immunization scheme described is only local and may, therefore, not result in a strong antibody titer in the blood. Because bacterial antigens are generally highly immunogenic, it is not mandatory to test serum samples before the fusion.

15. To ensure that the myeloma cells are in optimal condition, it is important to split the cells frequently and to replace the culture fluid by fresh medium the night before the fusion. This supply with nutrients ensures that the cells are growing and in good condition.

16. In case that you have problems in identifying the lymph nodes, you may coinject amido black with the last immunization. This dye is enriched in the lymph nodes, which then can be easily distinguished from the surrounding tissue.

17. Lymph nodes are hard bulbs with a characteristic pale and slightly yellowish color. Hold one end of the explant with a forceps and move a second fine forceps over the tissue. This should enable you to distinguish the rigid bulb of the lymph node from the surrounding soft tissue.

18. Only small pieces of connecting tissue should remain after mincing, as most of the cells are released in a now turbid single-cell suspension. The frosted ends of the slides may be additionally rinsed with 1 mL of sterile PBS to release all cells that stick to the glass.

19. The handling of supernatant samples is facilated by storage in plastic tubes and corresponding holders that ressemble in their formats 24- or 96-well plates, respectively (*see* **Note 20**).

20. The handling of a large number of supernatants is a problem. Sample tubes (1 and 6 mL) and corresponding tube holders (8×12 and 4×6 tubes) (Linbro Tubes-96 and -24, ICN) are helpful and can also be used to store supernatants derived from a hybridoma bank in a freezer for later rescreening. For the screening of large numbers of supernatants by immunoblot, multichamber incubation devices are recommended (e.g., Miniblotter, Biometra-Whatman), whereas multiwell slides are helpful for screening supernatants by immunofluorescence.

21. It is dangerous to make only one tube per well, as this might be lost as soon as there are problems with contamination after defrosting.

22. Liquid nitrogen is a dangerous agent and has to be handled carefully. Use protective equipment!

Acknowledgments

This work was supported by grants (to F. E.) from the Deutsche Forschungsgemeinschaft and the Fondation pour la Recherche Médicale.

References

1. Fritze, C. E. and Anderson, T. R. (2000) Epitope tagging: general method for tracking recombinant proteins. *Methods Enzymol.* **327,** 3–16.
2. Kubagawa, H., Mayumi, M., Kearney, J. F., and Cooper, M. D. (1982) Immunoglobulin VH determinants defined by monoclonal antibodies. *J. Exp. Med.* **156,** 1010–1024.
3. Holmdahl, R., Moran, T., and Andersson, M. (1985) A rapid and efficient immunization protocol for production of monoclonal antibodies reactive with autoantigens. *J. Immunol. Methods.* **83,** 379–384.
4. Brodsky, F. M. (1985) Clathrin structure characterized with monoclonal antibodies. II. Identification of in vivo forms of clathrin. *J. Cell Biol.* **101,** 2055–2062.
5. Sado, Y. and Okigaki, T. (1996) A novel method for production of monoclonal antibodies. Evaluation and expectation of the rat lymph node method in cell and molecular biology. *Cell Biol. Int.* **20,** 7–14.

12

Microscopic Methods to Study STEC

Analysis of the Attaching and Effacing Process

Stuart Knutton

1. Introduction

Attaching and effacing (A/E) is the name given to the striking and characteristic mechanism of intestinal colonization in which bacteria destroy brush-border microvilli, adhere very intimately to the intestinal epithelial cell surface, often on raised pedestal-like structures, and induce actin-rich cytoskeletal accumulation beneath intimately attached bacteria. First described in entero-pathogenic *Escherichia coli* (EPEC), A/E adhesion is also a feature of intestinal colonization by enterohemorrhagic *E. coli* (EHEC), a subset of Shiga toxin-producing *E. coli* (STEC), which includes the most common STEC sero-type, O157:H7 *(1)*. All of the genes necessary to produce the A/E histopathology reside in a chromosomal pathogenicity island designated the locus of enterocyte effacement (LEE). A/E lesion formation is a complex process involving a number of distinct stages: (1) initial bacterial attachment; (2) assembly of a type III protein translocation apparatus and translocation of viru-lence proteins into the host cell; (3) intimate bacterial attachment that involves the bacterial surface adhesin, intimin binding to a translocated intimin receptor Tir/EspE; (4) cytoskeletal accumulation beneath intimately attached bacteria *(2)*. Microscopical techniques have contributed significantly to our current understanding of this complex multi-stage bacterial adhesion process in terms of characterization of the lesion itself, the different stages involved, and the role of specific bacterial virulence proteins. The initial description of the A/E lesion originally came from conventional transmission electron microscopy (TEM) of thin sections and from scanning electron microscopy (SEM) of

From: *Methods in Molecular Medicine, vol. 73: E. coli: Shiga Toxin Methods and Protocols*
Edited by: D. Philpott and F. Ebel © Humana Press Inc., Totowa, NJ

infected gut tissue, but major advances came with the development of in vitro tissue culture cell models of A/E adhesion *(3)* and fluorescence actin staining as a simple diagnostic test for the A/E lesion *(4)*. This led to identification of the genes encoding A/E lesion formation, characterization of the proteins involved, and the production of protein-specific antibodies for use in immunolabeling studies. The subsequent application of microscopical techniques to both bacteria and bacteria–host cell interaction have significantly advanced our knowledge of the A/E lesion formation process.

Fluorescence microscopy has been widely used to examine aspects of A/E lesion formation, including actin accretion *(4)*, intimin expression during A/E lesion formation *(5)*, components of the type III translocon, in particular the large filamentous organelles composed of the secreted EspA protein (EspA filaments) *(6,7)*, the translocated intimin receptor and other host cell cytoskeletal proteins. Multiple fluorescence labeling studies have also demonstrated co-localization of key components of the A/E lesion *(8)*.

The higher resolution of the electron microscope has also been exploited in studies of A/E lesion formation. Immunonegative staining, which allows the visualization of specific bacterial protein surface antigens, has successfully been applied to examination of intimin expression *(5)* and EspA filament production *(6,7)* during bacterial growth. Immuno-TEM of thin sections allows the interaction between bacteria and host cells to be examined at high resolution and visualization of specific bacteria and host cell antigens associated with A/E lesion formation. Pre-embed immune staining has been employed to examine surface antigens involved in A/E lesion formation such as intimin *(9)* and EspA filaments *(7)*, whereas post-embed staining is required to localize intracellular antigens or epitopes such as cellular cytoskeletal proteins and translocated bacterial proteins. SEM is an ideal technique to examine bacterial–host cell interactions and the high resolution of modern SEMs has also made immunolabeling feasible. Compared to TEM, SEM has the advantage that large areas of the cell surface can be rapidly examined and screened for bacterial association and A/E lesion formation and immuno-SEM has recently been employed to characterize early stages of A/E lesion formation when EspA filaments promote interaction of STEC to host cells *(6)*. This chapter will focus on the use of tissue culture cell models of STEC A/E adhesion and on these most commonly used light and electron microscopic methods that have been employed to dissect the A/E process.

2. Materials
2.1. Infection of Cell Monolayers

1. Commonly used epithelial cell lines are HeLa and Hep-2 cells. Caco-2 and T84 intestinal epithelial cell lines have also been used.

2. Cell culture medium: HEPES-buffered modified Eagle's medium plus 2% fetal calf serum (MEM/FCS).
3. Glass coverslips (9 mm in diameter for SEM; 13 mm in diameter for immunofluorescence).
4. Cell culture dishes: 24-well tissue culture plates and 3- and 9-cm diameter cell culture dishes (Costar).
5. Luria broth (Oxoid).
6. STEC strains for analysis grown overnight in Luria broth.

2.2. Immunofluorescence Microscopy

1. Fixative: 4% Formaldehyde in phosphate-buffered saline (PBS) (commercially available).
2. Blocking buffer: PBS containing 0.2% bovine serum albumin (Sigma) (PBS/0.2% BSA) (*see* **Note 1**).
3. Permeabilization buffer: PBS containing 0.1% (v/v) Triton X-100.
4. Primary antibody: Specific monoclonal or polyclonal antibodies (*see* **Note 2**).
5. Secondary antibodies: Commercial fluorochrome conjugated antibodies specific for the primary antibody (*see* **Note 3**).
6. Fluorochrome conjugated phalloidin (*see* **Note 4**).
7. Mountant: Citifluor (PBS/glycerol antifade) mountant (Agar Scientific Ltd.).
8. Clear nail varnish.
9. Plate shaker.
10. Incident light fluorescence microscope with appropriate fluorescence filters.

2.3. Immuno-negative Staining Electron Microscopy

1. Carbon or carbon/formvar-coated 400-mesh copper grids (*see* **Note 5**).
2. Buffers: PBS and PBS/0.2% BSA.
3. Primary antibody: Specific monoclonal or polyclonal antibodies.
4. Secondary antibodies: Gold conjugated antibodies specific for primary antibody (British BioCell International) (*see* **Note 6**).
5. Negative stain: 1% Ammonium molybdate, pH 7.
6. Transmission electron microscope.

2.4. Immuno-TEM

2.4.1. Immuno-TEM (for Pre-Embed Staining of Antigens)

1. Buffers: PBS; PBS/0.2% BSA; and 0.1 M phosphate buffer, pH 7.3.
2. Prefixation: 0.1% glutaraldehyde in PBS (*see* **Note 7**).
3. Primary antibody: specific monoclonal or polyclonal antibodies.
4. Secondary antibodies: gold particle conjugated antibodies specific for primary antibody.
5. Primary fixative: 3% glutaraldehyde (EM Grade) in 0.1 M phosphate buffer, pH 7.3.
6. Secondary fixative: 1% osmium tetroxide in 0.1 M phosphate buffer, pH 7.3.
7. Dehydration reagents: 70% (v/v) ethanol; 90% (v/v) ethanol; 100% (v/v) ethanol; propylene oxide.

8. Epoxy/araldite embedding resin. We use Agar 100 resin (Agar Scientific Ltd).
9. Section staining: 1% toluidine blue; 25% uranyl acetate in methanol.
10. Rocking table.
11. Transmission electron microscope.

2.4.2. Immuno-TEM (for Post-Embed Staining of Antigens)

1. Buffers: PBS and PBS/0.2% BSA.
2. Prefixation: 0.1% glutaraldehyde in PBS.
3. Dehydration reagents: 70%, 90%, and 100% ethanol.
4. Hydrophilic acrylic embedding resin. We use Unicryl (British BioCell International).
5. Primary antibody: specific monoclonal or polyclonal antibodies.
6. Secondary antibodies: gold particle conjugated antibodies specific for primary antibody (British BioCell) (*see* **Note 6**).
7. Section staining: 25% uranyl acetate in methanol.
8. Transmission electron microscope.

2.5. Immuno-SEM

1. Buffers: PBS, PBS/0.2% BSA, 0.1 M phosphate buffer, pH 7.3.
2. Primary antibody: specific monoclonal or polyclonal antibodies.
3. Secondary antibodies: 30-nm gold particle conjugated antibodies specific for primary antibody (*see* **Note 6**).
4. Primary fixative: 3% glutaraldehyde (EM Grade) in 0.1 M phosphate buffer, pH 7.3.
5. Secondary fixative: 1% osmium tetroxide in 0.1 M phosphate buffer, pH 7.3.
6. Dehydration reagents: 50%, 70%, 90%, and 100% (v/v) acetone solutions.
7. Critical Point Drying apparatus.
8. Sputter coater with platinum target.
9. Scanning electron microscope.

3. Methods
3.1. Infection of Cell Monolayers

1. Wash glass coverslips in ethanol, dry, place in 9-cm-diameter cell culture dish, and sterilize under a UV lamp.
2. Seed dish with epithelial cells and grow to subconfluency. Use 13-mm-diameter coverslips for immunofluorescence and 9-mm-diameter glass coverslips for SEM; for TEM, grow cells to confluency on 3-cm-diameter cell culture dishes.
3. When ready to use, wash cells with warm PBS (3X 1 min with gentle shaking).
4. Place coverslips in the wells of a 24-well plate containing 1 mL MEM/FCS and add 10 µL of the STEC culture; for cells grown in 3-cm dishes, cover cells with 2 mL MEM/FCS and add 20 µL STEC culture.
5. Incubate cells for up to 6 h at 37°C (for incubations over 3 h, replace the medium with fresh medium at 3 h).
6. Remove nonadhering bacteria by washing with PBS (3X 1 min with gentle shaking).
7. Fix cells using appropriate fixation protocol for microscopical method being used (*see* **Subheading 3.2.–3.6.**).

3.2. Immunofluorescence Microscopy

1. In wells of a 24-well plate, fix infected cell monolayers with 1 mL of 4% formaldehyde for 20 min.
2. Wash 3X with 1 mL PBS (all washing performed on a plate shaker with gentle shaking).
3. Permeabilize cells in 1 mL of 0.1% (v/v) Triton X-100 for 4 min (if surface antigens only are being stained, this permeabilization step and subsequent washing can be omitted).
4. Wash 3X with 1 mL PBS and place in 1 mL PBS/0.2% BSA.
5. Transfer coverslips to a nonpolar surface (e.g., parafilm, silicone rubber) and incubate cells with 20 µL of the primary antibody diluted in PBS/0.2% BSA for 45 min at room temperature. Cover the coverslips with a lid to prevent drying.
6. Transfer coverslips back to the 24-well plate and wash 3X with 1 mL PBS/0.2% BSA.
7. Incubate coverslips with 20 µL secondary antibody (fluorochrome conjugated antibody specific for the primary antibody used in **step 5**) diluted in PBS/0.2% BSA for 45 min.
8. Wash 3X with PBS.
9. Mount coverslips cells side down on microscope slides using 2 µL of Citifluor. Make sure upper surface of coverslip is free of PBS by drying with filter paper or blowing gently with a gas "Duster".
10. If an oil immersion lens is to be used, seal the edge of the coverslip with clear nail varnish.
11. Examine cells by incident light fluorescence using an appropriate fluorescence filter and objective lens. **Figure 1** shows an example of immunofluorescence staining of STEC. The specificity of the staining should always be checked by omitting the primary antibody or by using a heterologous antibody. Double labeling of different antigens can be achieved using the same protocol if two noncrossreacting antisera raised in different animals and appropriate secondary antibodies are available. A multipass fluorescence filter block will be required to image more than one fluorochrome simultaneously.
12. Fluorescence actin staining (FAS) is an important variation of the above method because FAS can be used as a diagnostic test for STEC with the ability to produce A/E lesions *(4)*. FAS involves a single-stage staining of actin not using an anti-actin antibody but using a commercially available fluorochrome conjugated phalloidin, which binds specifically to polymerized actin. Thus, for FAS, **step 6** involves staining with a 5-µg/mL solution of phalloidin for 30 min and the procedure then continues from **steps 9 11**. FAS preparations are examined by both fluorescence and phase contrast using a ×40 objective. A positive FAS test is indicated by strong localized actin fluorescence coincident with the position of attached bacteria seen by phase contrast. **Figure 1B,C** is an example of a typical FAS test.

3.3. Immunonegative Staining Electron Microscopy

1. Wash and prepare a concentrated suspension of bacteria in PBS (*see* **Note 8**).
2. Apply 10 µL of the suspension to coated 400-mesh electron microscope grids held in fine tweezers for 2 min.

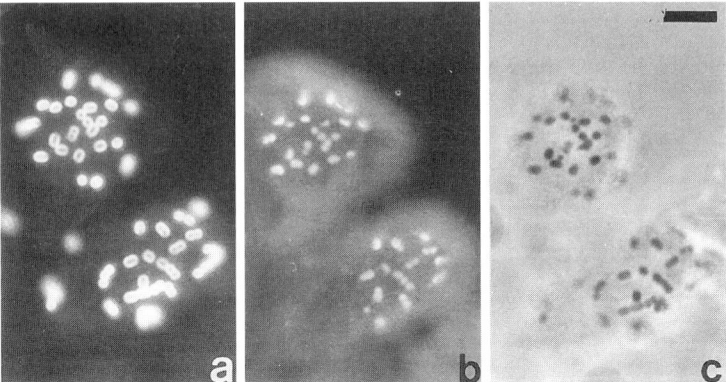

Fig. 1. Fluorescence micrographs showing STEC adhering to Hep-2 cells and double stained for intimin (**A**) and actin (**B**); a phase-contrast micrograph of the same field is shown in (**C**). Micrographs (**B**) and (**C**) constitute a positive fluorescence actin staining test and reveal intense concentrations of cytoskeletal actin at the site of each adherent bacterium which is diagnostic for A/E lesion formation. All of the A/E bacteria express surface intimin (**A**). Scale bar, 10 μm.

3. Remove excess liquid with filter paper and immediately place the grid face down on a 20-μL drop of primary antibody diluted with PBS/BSA for 30 min. Antibody staining and washing is performed on a hydrophobic surface (e.g., Parafilm or a sheet of dental wax).

4. Wash the grid by transferring sequentially onto five 25- to 30-μL drops of PBS/ 0.2% BSA for 1 min each. Grid transfer is best achieved using a loop of thin platinum wire slightly larger than the grid.

5. Transfer grid to a 20-μL drop of an appropriate gold-labeled secondary antibody diluted in PBS/0.2% BSA for 30 min (*see* **Note 6**).

6. Wash grids by transferring sequentially onto 3 drops of PBS/0.2% BSA followed by 3 drops of distilled water.

7. Negative stain bacteria by placing grid on to a drop of 1% ammonium molybdate. Pick grid up with fine tweezers, invert, stand for 1 min, then remove excess negative stain with filter paper and allow grid to air-dry.

8. Examine grids by TEM.

9. **Figure 2** shows an example of immunonegative staining. As for immunofluorescence, the specificity of the staining should always be checked by omitting the primary antibody or by using a heterologous primary antibody. Double labeling of different antigens can be achieved using this protocol if two noncrossreacting antisera raised in different animals and appropriate secondary antibodies with different sized gold particles are available. Negative staining is commonly used without immunolabeling to visualize bacterial surface structures and can be achieved essentially by eliminating the immunolabeling stages. Simply mix 10 μL

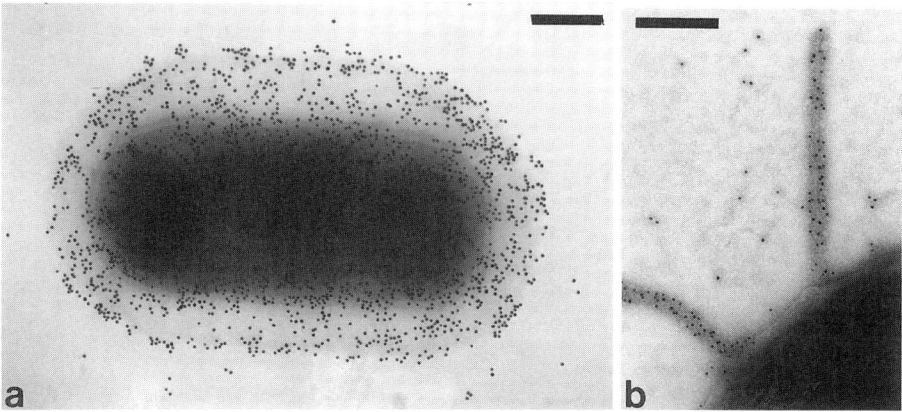

Fig. 2. Electron micrographs illustrating the immunonegative staining technique. When grown in tissue culture medium STEC express intimin (γ) uniformly over the cell surface (**A**), EspA filaments are also expressed at the bacterial surface (**B**). Scale bar, 0.2 μm.

of the bacterial suspension with 10 μL of negative stain, apply 10 μL to a grid for 2 min, remove excess liquid thoroughly, and allow the grid to air-dry.

3.4. Immuno-TEM (for Pre-Embed Staining of Surface Antigens)

1. Prefix cells grown and infected with STEC in 3-cm plastic dishes with 2 mL of 0.1% glutaraldehyde in PBS for 15 min.
2. Wash 3X PBS, 1X PBS/0.2% BSA.
3. Incubate cells with 0.5 mL primary antibody diluted in PBS/BSA for 2–4 h on a rocking table (using a rocking table keeps the amount of antibody required to a minimum).
4. Wash 3X PBS/0.2% BSA.
5. Incubate with gold-labeled secondary antibody diluted in PBS/0.2% BSA for 2–4 h.
6. Wash 3X PBS.
7. Fix cells with 1 mL 3% glutaraldehyde in 0.1 M phosphate buffer, pH 7.2, for 1 h (perform **steps 7–14** in a fume hood).
8. Postfix in 1 mL of 1% buffered osmium tetroxide for 1 h.
9. Wash 3X with distilled water.
10. Dehydrate samples through a graded series of alcohol solutions as follows: 50%, 70%, 90%, 2X 100%, 10 min/solution.
11. At this stage the cells are removed from the culture dish. Score the cell monolayer with a pointed needle to form small approx 3 mm squares of monolayer. Add 1–2 mL propylene oxide and agitate gently. Immediately, the cells come away from the plastic; transfer them carefully to a conical glass tube using a wide-bore Pasteur pipet. These pieces of cell monolayer can be quite delicate depending on the cell type and so must be handled carefully for the rest of the procedure.
12. Wash a further 2X 10 min with propylene oxide.

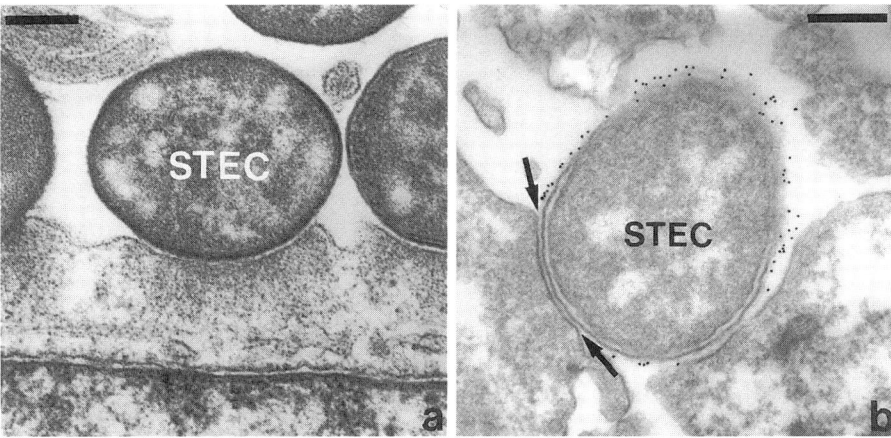

Fig. 3. Attaching and effacing lesion formation on Caco-2 intestinal cells. Bacteria are seen intimately attached to the apical cell surface devoid of microvilli on a raised pedestallike structure with accumulation of cytoskeletal elements beneath intimately attached bacteria (**A**). Intimin is seen uniformly expressed over the STEC surface, but note that intimin cannot be stained in the region of intimate adhesion to the host cell using this methodology (**B**, arrows). Scale bar, 0.2 μm.

13. Make up embedding resin according to manufacturer's specification and infiltrate cells with embedding resin as follows: 3:1; 1:1; 1:3 (v/v) propylene oxide/resin (1 h/solution); 100% resin overnight. For infiltration the pieces of cell monolayer can be transferred by pouring gently from the glass tube to a small widemouth glass vial.
14. Place fresh resin into a flat silicone rubber embedding mold and gently transfer pieces of the cell monolayer into each well using a fine metal spatula, orient as necessary, and polymerize resin in an embedding oven at 60°C for 24 h. In order to section more than a single monolayer at once, several pieces of cell monolayer can be placed on top of each other and embedded together as a stack.
15. Select areas of the sample for ultrathin sectioning by light microscopical examination of 1-μm-thick sections stained with 1% toluidine blue.
16. Using an ultramicrotome cut 60- to 100-nm-thick (gray-gold interference color) sections and collect on formvar-coated electron microscope grids.
17. Stain grids with uranyl acetate for 10 min, wash 3X 5 min in methanol and allow to dry.
18. Examine grids in a transmission electron microscope.
19. **Figure 3** shows examples of conventional and immuno-TEM of the A/E lesion.
20. Conventional TEM (without immunolabeling) can be performed by simply eliminating **steps 1–6**.

3.5. Immuno-TEM (for Post-Embed Staining of Antigens)

1. Fix cell monolayers grown on 3-cm plastic dishes in 0.1% glutaraldehyde in PBS for 15 min.

2. Wash 3X PBS.
3. Scrape cells from the plastic dish using a teflon scraper, transfer to a microfuge tube and spin to form a cell pellet; gently loosen pellet from side of tube with a fine spatula if necessary.
4. Dehydrate cell pellet in increasing alcohol concentrations (30%, 50%, 70%, 95%, 100%) (10 min each) at 4°C (perform **steps 4** and **5** in a fume hood).
5. Infiltrate with 100% Unicryl resin (2X 1 h; 1X 8 h or overnight) preferably with gentle agitation.
6. Transfer sample to a 1-mL Eppendorf vial and polymerize resin using ultraviolet (UV) light at 4°C or lower temperature according to the manufacturer's specifications (*see* **Note 9**).
7. Using an ultramicrotome, cut sections ranging from gray to green interference colors (<0.1–0.2 μm) and collect on formvar-coated nickel grids.
8. Immunostaining is performed on sheets of dental wax by placing grids sequentially onto 25-μL drops of appropriate reagent and wash solutions; a platinum loop is used to transfer grids from drop to drop. Initially, grids are washed with PBS/0.2% BSA (3X 1 min).
9. Incubate with primary antibody diluted in PBS/0.2% BSA (4°C overnight); to prevent drying out, place grids inside a small plastic box with a saturated atmosphere.
10. Wash with PBS/0.2% BSA (3X 1 min).
11. Incubate with secondary gold-labeled antibody (2 h at room temperature).
12. Wash 3X PBS/0.2% BSA; 3X distilled water.
13. Stain sections with 25% uranyl acetate in methanol (10 min); wash 3X in methanol and allow to dry.
14. Examine grids in a transmission electron microscope.
15. **Figure 4** shows examples of the post-embed staining of antigens. The advantage of this technique is that sections from the same specimen can be stained for as many different antigens as one has antibodies available.

3.6. Immuno-SEM

1. Prefix cells with 0.1% glutaraldehyde in PBS for 15 min. (For this method, we use 9-mm-diameter coverslips because they are a convenient size to fit into the Polaron Critical Point Drying apparatus we use.)
2. Wash 3X PBS.
3. Incubate cells with primary antibody diluted in PBS/BSA for 2–4 h.
4. Wash 3X PBS/0.2% BSA.
5. Incubate with 30-nm gold-labeled antibody diluted in PBS/BSA for 2–4 h (*see* **Note 6**).
6. Wash 3X PBS.
7. Fix cells with 3% glutaraldehyde in 0.1 *M* phosphate buffer, pH 7.2, for 1 h (perform **steps 7–9** in a fume hood)
8. Postfix in 1% buffered osmium tetroxide for 1 h.
9. Dehydrate samples through the following graded series of alcohol solutions (50%, 70%, 90% [15 min each], 100% [2X 15 min]).

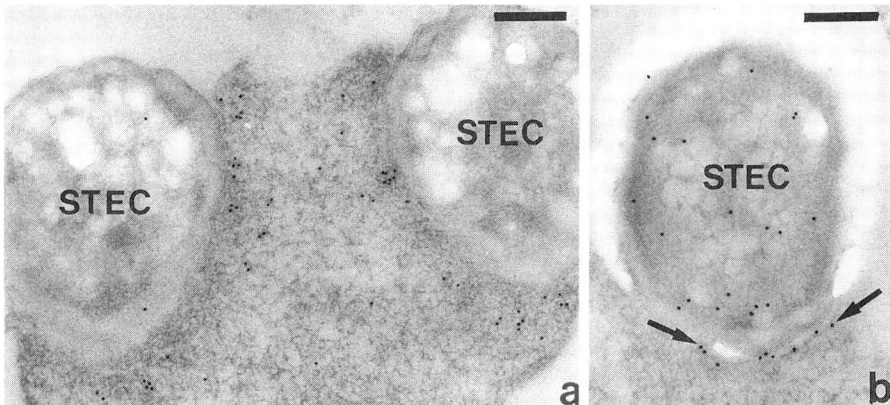

Fig. 4. Post-embed immunogold-labeling electron micrographs showing A/E lesion formation on Caco-2 intestinal cells. Sections were stained with anti-actin (**A**) and anti-Tir/EspE (**B**). Note that, in (**A**), the gold labels cytoskeletal elements that accumulate beneath the intimately attached bacteria, whereas, in (**B**), the gold labels apical cell membrane in the region of intimate bacterial attachment (arrows) and also the bacterial cytosol. Scale bar, 0.2 µm.

10. Dry samples from 100% alcohol using a Critical Point Drying apparatus.
11. Mount coverslips (cut to size if necessary) on appropriate SEM stubs using double-sided adhesive tape or epoxy glue.
12. Coat specimens with a thin layer of platinum using a sputter coating apparatus. Examine using SEM (*see* **Note 10**).
13. **Figure 5** shows the application of conventional and immuno-SEM to studies of A/E lesion formation. Conventional SEM (without immunolabeling) can be performed by eliminating **steps 1–6**.

4. Notes

1. Blocking buffers are used to minimise nonspecific antibody binding. A routine blocking buffer consists of PBS containing 0.2% BSA, but the concentration of BSA can be increased to 1% if necessary. Normal serum, again up to 1%, is also sometimes included. For example, when using secondary antibodies raised in goats, the addition of 0.2–1% normal goat serum would be included. Some blocking buffers also include low concentrations of nonionic detergent such as Tween-20.
2. High-titer primary antibodies, preferably raised against highly purified antigen, are essential for successful immunolabeling. Appropriate dilutions of the antibody have to be found empirically by finding the highest dilution consistent with good specific labeling and a low background: 1 : 100 is probably a good starting point.
3. A high-quality second antibody is essential for labeling the primary antibody specifically and with low background. Such antibodies conjugated to a variety of different fluorochromes are now widely available commercially (e.g., Sigma,

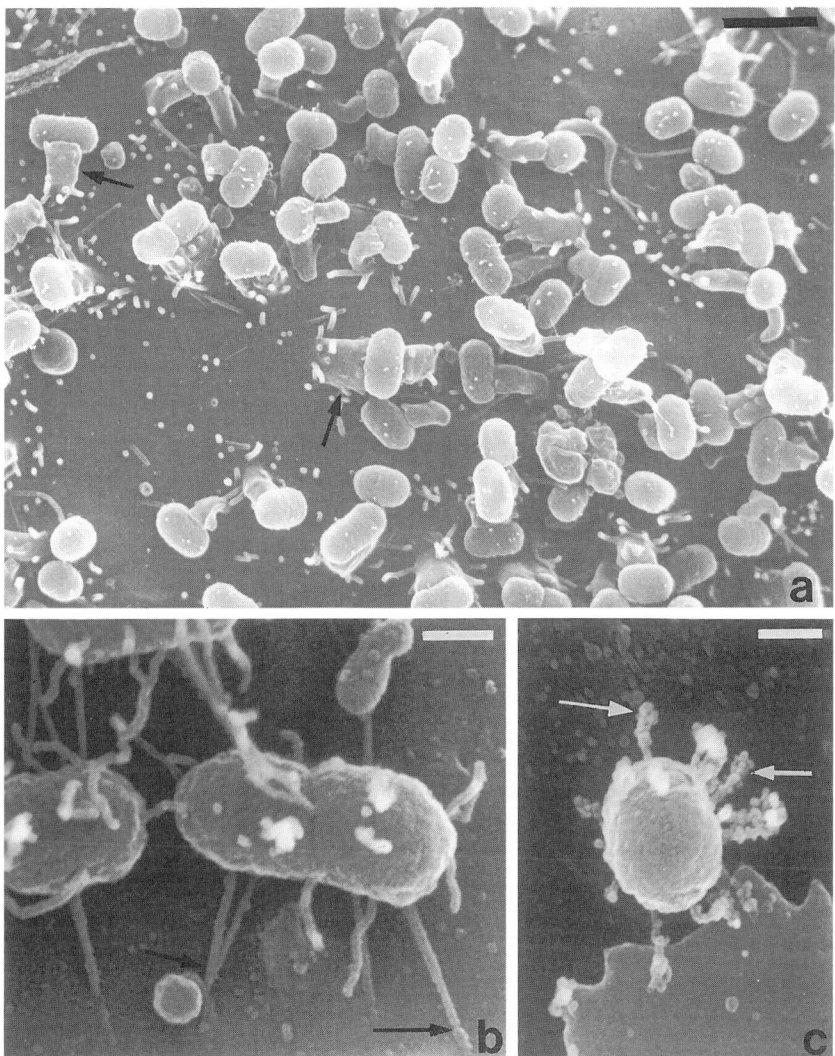

Fig. 5. Conventional (**A**, **B**) and immunogold-labeling (**C**) scanning electron micrographs showing stages of A/E lesion formation. In (**A**), bacteria have formed A/E lesions and many are seen intimately attached to raised pedestal-like structures (arrows). Prior to intimate attachment filamentous structures are seen to connect bacteria and the epithelial cell surface (**B**, arrows) and these structures are confirmed as EspA filaments by immunogold labeling (**C**); note that the filaments are coated with the large gold particles (**C**, arrow). Scale bars: (**A**) 1 μm; (**B, C**) 0.2 μm.

Molecular Probes), although green (FITC, Alexa$_{488}$) and red fluorochromes (TRITC, Alexa$_{594}$, Texas red) remain the most common. A general DNA stain such as DAPI, which fluoresces blue, is often used to visualize bacteria by fluorescence.

4. Phalloidin, when conjugated to a fluorochrome, provides a highly specific stain for cellular actin. It is, however, highly toxic and so should be handled with extreme care. When purchased lyophilized, it should be reconstituted, aliquoted into useful amounts (e.g., sufficient for 10 coverslips) and stored frozen at –20°C. Always wear gloves and a face guard when handling this reagent.

5. To support bacteria, electron microscopy (EM) grids need to be coated with a thin film of either plastic (formvar) or carbon. Coated grids or the expertise to produce them should be available in any specialized EM facility.

6. High-quality gold conjugated second antibodies are widely available commercially and can be purchased with various sized gold particles (5 nm, 10 nm, 20 nm, 30 nm, etc.). Smaller gold particles give higher labeling intensities; therefore, for high-magnification work, 5- or 10-nm particles are the preferred size. We routinely use 10-nm particles for most TEM studies. For SEM studies, because of the lower resolution of the microscope, we use 30-nm particles.

7. Electron-microscopy-grade glutaraldehyde is essential for good fixation for all EM procedures. It is usually purchased concentrated in sealed vials under nitrogen and should be made fresh just prior to use. We purchase vials of 8% aqueous glutaraldehyde from Polysciences, Inc.

8. It is important that bacteria be grown under appropriate conditions so that they express the antigens of interest. For example, the LEE genes that encode A/E lesion formation are not expressed when bacteria are grown in routine bacteriological media but they are expressed when grown in cell culture medium (MEM/FCS) for 3–4 h at 37°C.

9. Unicryl resin is a largely hydrophilic acrylic embedding resin ideal for both preservation of tissue structure and antigenicity. To preserve antigenicity of the sample Unicryl is best polymerized using UV irradiation according to the manufacturer's instructions. The resin remains liquid down to –50°C and so can be polymerized at low temperature if required. We routinely polymerize at –10°C in a cryostat, in which case, alcohol dehydration can be carried out by progressive lowering of temperature (e.g., 30%, 50% at 4°C; 70%, 95%, and 100% at –10°C).

10. Depending on the SEM available it may not be necessary to sputter-coat the sample (**step 12**); examination of uncoated samples is possible with modern field-emission-gun SEMs operating at low voltages. Gold labeling is best imaged using a backscattered rather than a secondary-electron detector if available because the signal is proportional to the atomic number of the element (i.e., a strong signal is obtained from the gold particles) *(6)*.

Acknowledgments

This work was supported by the Wellcome Trust.

References

1. Kaper, J. B. (1998) Enterohemorrhagic *Escherichia coli. Curr. Opin. Microbiol.* **1,** 103–108.

2. Knutton, S., Baldini, M. M., Kaper, J. B., and McNeish, A. S. (1987) Role of plasmid-encoded adherence factors in adhesion of enteropathogenic *Escherichia coli* to HEp-2 cells. *Infect. Immun.* **55,** 78–85.

3. Frankel, G., Phillips, A. D., Rosenshine, I., Dougan, G., Kaper, J. B., and Knutton, S. (1998) Enteropathogenic and enterohaemorrhagic *Escherichia coli*: more subversive elements. *Mol. Microbiol.* **30,** 911–921

4. Knutton, S., Baldwin, T., Williams, P. H., and McNeish, A. S. (1989) Actin accumulation at sites of bacterial adhesion to tissue culture cells: basis of a new diagnostic test for enteropathogenic and enterohemorrhagic *Escherichia coli*. *Infect. Immun.* **57,** 1290–1298.

5. Knutton, S., Adu-Bobie, J., Bain, C., Phillips, A. D., Dougan, G., and Frankel, G. (1997) Down regulation of intimin expression during attaching and effacing enteropathogenic *Escherichia coli* adhesion. *Infect. Immun.* **65,** 1644–1652.

6. Ebel, F., Podzadel, T., Rohde, M., Kresse, A. U., Kramer, S., Deibel, C., et al. (1998) Initial binding of Shiga toxin-producing *Escherichia coli* to host cells and subsequent induction of actin rearrangements depend on filamentous EspA-containing surface appendages. *Mol. Microbiol.* **30,** 147–161.

7. Knutton, S., Rosenshine, I., Pallen, M. J., Nisan, I., Neves, B. C., Bain, C., et al. (1998) A novel EspA-associated surface organelle of enteropathogenic *Escherichia coli* involved in protein translocation into epithelial cells. *EMBO J.* **17,** 2166–2176.

8. Rosenshine, I., Ruschkowski, S., Stein, M., Reinscheid, D. J., Mills, S. D., and Finlay, B. B. (1996) A pathogenic bacterium triggers epithelial signals to form a functional bacterial receptor that mediates actin pseudopod formation. *EMBO J.* **15,** 2613–2624.

9. Batchelor, M., Knutton, S., Caprioli, A., Huter, V., Zanial, M., Dougan, G., et al. (1999) Development of a universal intimin antiserum and PCR primers. *J. Clin. Microbiol.* **37,** 3822–3827.

13

Detection and Characterization of EHEC-Hemolysin

Herbert Schmidt and Roland Benz

1. Introduction

The ability of particular *Escherichia coli* strains to lyse erythrocytes of several mammalian species was first described by Kayser in 1903 *(1)*. This phenomenon was termed *hemolysis* and the bacterial determinants involved were termed *hemolysins*. The best characterized *E. coli* hemolysin is α-hemolysin, which is mainly produced by uropathogenic *E. coli* (UPEC). When blood agar plates are inoculated with α-hemolysin-producing bacteria, large clear zones of hemolysis are observed that occurred after 4–6 h of incubation at 37°C *(2)*. This chapter aims to describe three protocols useful for studying the hemolysin of enterohemorrhagic *E. coli* (EHEC): detection of the toxin, measurement of hemolytic activity, and, finally, determination of the pore-forming activity of the toxin. EHEC are a subgroup of Shiga toxin-producing *E. coli* (STEC, and are associated with severe human disease. It should be noted here that the enterohemolytic phenotype is not a characteristic property of all EHEC/STEC. Whereas almost all EHEC O157:H7 are enterohemolytic, the sorbitol-fermenting *E. coli* O157:H⁻ are nonhemolytic. EHEC O111 differs in their capacity to produce E-Hly and EHEC O103:H2 overproduces E-Hly so that an α-hemolysin-like phenotype occurs on blood agar plates *(3,4)*. In STEC, strains isolated from a variety of human and nonhuman sources also differ in their ability to produce EHEC-hemolysins.

1.1. Detection of EHEC-Hemolysin

Enteropathogenic *E. coli* (EPEC) were reported in 1988 to produce a distinct type of hemolysin, termed *enterohemolysin (5)*. The so-called enterohemolytic phenotype is characterized by small turbid zones of hemolysis occurring after incubation at 37° for 18–24 h and can only be detected on

From: *Methods in Molecular Medicine, vol. 73: E. coli: Shiga Toxin Methods and Protocols*
Edited by: D. Philpott and F. Ebel © Humana Press Inc., Totowa, NJ

blood agar plates containing defibrinated, washed red blood cells. A close association of EHEC and this certain kind of enterohemolytic phenotype was described in 1989 *(6)* and it was proposed that this could be used as a diagnostic marker for the identification of EHEC strains. (Such enterohemolysin agar plates can be easily prepared in the laboratory but can be also purchased from several companies.)

Some years later, we demonstrated that this hemolytic phenotype is encoded by a large plasmid of *E. coli* O157:H7, termed pO157 *(7)*. The EHEC-hemolysin (E-Hly) of *E. coli* O157:H7 is located from position kb 7.83 to position kb 15 in the physical map of plasmid pO157 *(8)*. Four genes, E-*hlyC*, E-*hlyA*, E-*hlyB*, and E-*hlyD* constitute an operon responsible for the synthesis, activation, and transport of the protein out of the bacterial cell *(8,9)*. The E-Hly genes are related to the genes encoding α-hemolysin and other so-called RTX operons *(9)*. E-Hly is encoded by the E-*hlyA* gene and has a molecular weight of approx 110 kDa. Whereas α-hemolysin is exported via the HlyB/HlyD proteins from the cell in high amounts, E-Hly remains mostly cell-associated and is exported only in low amounts, which can only be detected with highly sensitive methods *(9,10)*. E-Hly is activated by acylation mediated by the E-HlyC protein. RTX toxins may be divided in two categories: the hemolysins, which affect a variety of cell types, and the leukotoxins, which are cell-type-specific and species-specific. In addition to their toxic action on target cells, these molecules offer an interesting model for the targeting, insertion, and translocation of aqueous proteins into lipid membranes *(11)*. Hemolysins and other RTX toxins have been implicated in pathogenesis caused by the respective bacteria. Cell death mechanisms such as necrosis or apoptosis have been reported and were shown to be concentration dependent. E-Hly of *E. coli* O157:H7 is a typical RTX protein, which is able to lyse erythrocytes and other eukaryotic cells of various origins *(12)*.

1.2. Measurement of Hemolytic Activity

Investigation of some properties of E-Hly needs larger amounts of E-Hly. This is simply achieved by precipitation of culture supernatants of strains that produce large amounts of E-Hly. This is usually not possible with E-Hly from wild-type strains, as the amount of E-Hly secreted from such strains is too low. However, laboratory strains, carrying a cloned E-*hly* operon (i.e., pEO40) *(7)* and, in addition, a recombinant plasmid carrying the HlyB/HlyD proteins from the *E. coli* α-hemolysin operon (such as pRSC6), secrete large amounts of protein in the supernatant under laboratory conditions *(9)*. Also, EHEC of serotype O103:H2 secrete large amounts of E-Hly *(3)*.

Hemolysin from culture supernatants can be incubated with erythrocytes and the hemolytic activity can be determined by measuring the optical density of the reaction sample in a spectral photometer.

1.3. Determination of the Pore-Forming Activity of E-Hly

We could show that E-Hly has the capacity to form pores in lipid bilayer membranes, similar to that of α-hemolysin. The determination of biophysical characteristics of such pores demonstrated that the size of the pore mediated by E-Hly is approx 2.6 nm and that the channel is cation-selective *(10)*. For such measurements in lipid bilayers, purified hemolysin is needed. This can easily be achieved by excising hemolysin bands from sodium dodecyl sulfate–polyacrylamide gel electrophoresis (SDS-PAGE) as described in **Subheading 3.3.1.**

2. Materials

2.1. Detection of E-Hly on Blood Agar Plates

1. Defibrinated sheep erythrocytes (Oxoid) (*see* **Note 1**).
2. Tryptose blood agar base (Difco Laboratories).
3. Sterile $CaCl_2$ solution, usually 100 mM.
4. Petri dishes (90 mm).
5. Falcon tubes (50 mL).
6. Phosphate-buffered saline (PBS).
7. Membrane filter with a pore size of 0.2 µm (Schleicher & Schüll).

2.2. Measurement of Hemolytic Activity

1. Todd–Hewitt broth (THB) (Difco).
2. Sorvall centrifuge, GSA rotor, and corresponding 250-mL beaker.
3. Membrane filter with a pore size of 0.2 µm (Schleicher & Schüll).
4. Gentamycin (stock: 10 mg/mL), chloramphenicol (stock: 30 mg/mL), and ampicillin (stock 50 mg/mL).
5. Polyethylenglycol 4000.
6. Glycerol.
7. Toothpicks.
8. Washed sheep erythrocytes (*see* **Subheading 3.1.1.**).
9. A 96-well microtiter plate (Nunc).
10. Microplate reader (e.g., Dynex).

2.3. Determination of the Pore-Forming Properties of E-Hly

2.3.1. Purification of E-Hly

1. Sodium dodecyl sulfate (SDS) acrylamide/bisacrylamide, ammonium persulfate, and *N,N,N',N'*- tetramethylethylendiamine (TEMED) (Serva).
2. Mini protein gel apparatus.
3. 10 mM Tris-HCl, pH 7.0.
4. Urea.
5. Tabletop centrifuge for Eppendorf tubes.

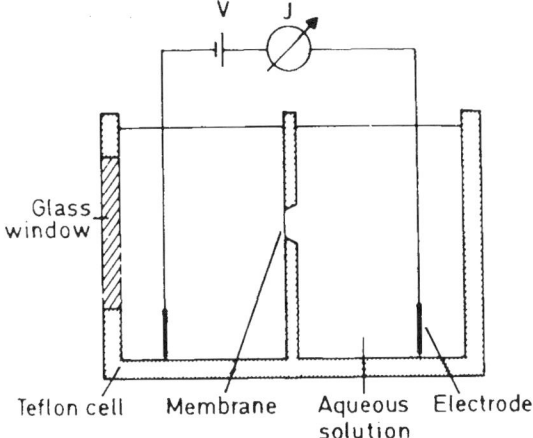

Fig. 1. Scheme of lipid bilayer instrumentation.

2.3.2. Lipid Bilayer Membrane Technology

1. Teflon chamber (self-made, *see* **Fig. 1**).
2. Asolectin (soybean lecithin type IV-S from Sigma, contains, in addition to lecithin, many different lipids).
3. *n*-Decane (high purity).
4. *n*-Butanol (high purity).
5. High-impedance electrometer (Keithley 617 Electrometer, Keithley Instruments Inc., Cleveland OH).
6. Current-to-voltage converter (Keithley 427 current amplifier, Keithley Instruments Inc.).
7. Two Ag/AgCl electrodes with salt bridges (Metrohm, Herisau, Switzerland).
8. Voltage source made on the basis of a voltage-divider circuit (output voltage 5–550 mV).
9. Radio cables for the connection between membrane cell and electrometer or current amplifier.
10. Salt solutions: 0.01 M KCl; 0.03 M KCl; 0.1 M KCl; 0.3 M KCl; 1 M KCl; 0.1 M LiCl; 0.1 M KAc, pH 6.
11. Microscope lamp for membrane illumination and telescope for membrane inspection in the reflected light.

3. Methods

3.1. Detection of EHEC-Hemolysin

3.1.1. Preparation of Erythrocytes

1. Mix 25 mL of the defibrinated sheep erythrocytes with 15 mL phosphate-buffered saline (PBS) and stir carefully with a baton made of glass (*see* **Note 2**).
2. Centrifuge for 15 min at 950*g* in a Sorvall SS34 rotor. Discard the supernatant, again add 15 mL of PBS, mix carefully, and repeat the procedure several times until the supernatant is clear (usually three times).

3. Discard the supernatant and resuspend the pellet with 15 mL of PBS. Store washed erythrocytes at 4°C not longer than 48 h.

3.1.2. Preparation and Inoculation of Blood Agar Plates (see **Note 1**)

1. Dissolve 26.4 g tryptose blood agar base in 760 mL bidistilled water and autoclave for 15 min at 121°C. Transfer the vessel to a water bath and cool it to 48°C.
2. Add filter-sterilized $CaCl_2$ to a final concentration of 10 mM.
3. Add the washed erythrocytes (approx 40 mL) into the blood agar base medium and mix gently with a glass pipet. (The temperature of the medium must be between 48°C and 50°C !!)
4. Pour a thin layer of the medium into Petri dishes (20 mL/Petri dish) (*see* **Note 3**).
5. Inoculate one of the plates (as a quality control) with *E. coli* O157:H7 strain EDL933, an α-hemolysin producer (i.e., WAF100/pSF4000), and *E. coli* K-12 derivative C600.
6. Inoculate the plates for testing of the enterohemolytic phenotype by single-colony streaking. The enterohemolytic phenotype occurs after 18–24 h of incubation at 37°C as small, turbid zones of hemolysis. In contrast, the α-hemolytic phenotype occurs after 4–6 h of incubation at 37°C as large, clear zones of hemolysis. *E. coli* C600 is nonhemolytic. Do not forget to perform quality controls in each experiment.

3.2. Measurement of Hemolytic Activity

3.2.1. Precipitation of Extracellular E-Hly

1. Transform *E. coli* HB101/pRSC6 with 500 ng of plasmid pEO40 and select transformants on blood agar plates containing chloramphenicol (30 µg/mL) and ampicillin (50 µg/mL).
2. Select a single colony that shows a large, clear zone of hemolysis and transfer it to a 300-mL Erlenmeyer flask containing 100 mL prewarmed Todd–Hewitt broth, 5 mM $CaCl_2$, 30 µg/mL chloramphenicol, and 50 µg/mL ampicillin.
3. Grow the bacterial suspension for 11 h at 180 rpm at 37°C in a rotary shaker, until it reaches an optical density (OD_{600}) of 2.5.
4. Transfer the culture to a Sorvall GSA beaker and centrifuge for 15 min at 10,400g at 4°C in a Sorvall GSA rotor.
5. Transfer the culture supernatant to a fresh GSA beaker, mix with gentamycin to a final concentration of 20 µg/mL, and incubate at room temperature for 10 min.
6. Adjust supernatant to 20% (w/v) PEG 4000 and 3% (v/v) glycerol; mix well by stirring slowly at 4°C for 1 h.
7. Centrifuge for 20 min at 13,200g at 4°C in a Sorvall GSA rotor, discard the supernatant, and dissolve the pellet in 0.5 mL of 0.85% (w/v) NaCl solution. Carefully rinse the walls of the beaker to completely dissolve the sediments.
8. Prepare aliquots and store them immediately at –20°C not longer than 7 d or use the E-Hly preparation freshly (within 3–4 h) stored in an ice bath. Storage for longer times will result in decreased activity.

3.2.2. Hemolysin-Assay

1. Dilute erythrocytes 1:10 and 1:50 in 0.85% NaCl.
2. Use the erythrocytes for the following mixtures undiluted and 1:10, and 1:50 diluted:
   Mix A: Measurement of total lysis
   　　180 μL of distilled water
   　　20 μL washed sheep erythrocyte suspension
   　　1 μL $CaCl_2$
   Mix B: Background lysis
   　　180 μL of a 0.85 % NaCl solution
   　　20 μL washed sheep erythrocyte suspension
   　　1 μL $CaCl_2$
   Mix C: Hemolysis
   　　160 μL of a 0.85 % NaCl solution
   　　20 μL washed sheep erythrocyte suspension
   　　20 μL of the E-Hly precipitate
   　　1 μL $CaCl_2$
3. Incubate the mixtures for 1 h at 37°C, centrifuge, and determine the amount of released hemoglobin spectrophotometrically at A_{570} nm in a microplate reader. Run samples in duplicate.
4. Compare values of different dilutions, use values with absorption values under 2, and calculate percentage of lysis as follows:

(A_{570} of total lysis- A_{570} of background A_{570} of sample – A_{570} of background) × 100

3.3. Determination of the Pore-Forming Activity of E-Hly

3.3.1. Purification of E-Hly

1. Load eight lanes of a 7% SDS/polyacrylamide minigel (acrylamide:bisacrylamide = 44:0.8) with 30 μL of the E-Hly precipitate.
2. Electrophorese with 150 V for 45 min.
3. Excise the 110-kDa E-Hly bands unstained (*see* **Note 4**).
4. Pool the gel slices and incubate them overnight at 7°C in 0.5 mL of a solution composed of 6 *M* urea and 10 m*M* Tris-HCl, pH 7.0.
5. Centrifuge at highest speed in a minicentrifuge for 5 min.
6. Use supernatants freshly or store at –20°C.

3.3.2. Lipid Bilayer Technique

The formation of ion-permeable channels by E-Hly and other membrane-damaging toxins can be studied using artificial lipid bilayer membranes formed according to the Mueller–Rudin method (solvent containing membranes) *(13)* (*see* **Note 5**).

1. Use a Teflon chamber (45 mm long, 25 mm wide, and 25 mm high with two compartments, each with a volume of 5 mL) for membrane formation across the circular hole (area about 0.5 mm^2) between the two compartments.

2. Treat the hole with 5 μL of 1% solution of asolection in chloroform for higher membrane stability and let the chloroform evaporate.
3. Fill the two compartments with 5 mL salt solution each.
4. Paint with a teflon loop 5 μL 1% asolectin solution in *n*-decane/n-butanol (10/1, v/v) across the hole between the compartments to form a lipid lamella.
5. Inspect the lamella in the reflected light with the telescope until Newton's colors disappear and the lipid bilayer membrane appears to be optically black in the reflected light of the microscope lamp.

3.3.3. Determination of Channel-Forming Activity of E-Hly

1. Add small amounts of purified E-Hly (about 500 ng/mL for multichannel experiments and about 10–50 ng/mL for single-channel recordings) while stirring to one or both sides of the membrane *(10,15)* (*see* **Note 6**).
2. Apply a membrane potential of about 20 mV to the asolectin membrane.
3. Channels appear within 2–3 min after the addition of the toxin.
4. Analysis of data is as follows:

   The addition of smaller amounts of E-Hly to lipid bilayer membranes allows the resolution of step increases in conductance as shown in **Fig. 2**. This means that the membrane activity described above is caused by the formation of ion-permeable channels in the membranes. One step reflects the opening of one conductive unit (i.e., of one channel in the membrane). These conductance steps are specific to the presence of E-Hly. They are not observed when only concentrated supernatants of cells lacking the toxin are added to the aqueous phase. This means that the channels are not formed by porins or other proteins of the corresponding *E. coli* cells. **Figure 2** shows that the conductance steps observed in the presence of E-Hly have a limited lifetime (mean lifetime about 20 s) and usually decayed back to the small conductance state. The occurrence of two types of pore caused by hemolysin can also be seen in **Fig. 3**, which shows a histogram of the conductance fluctuations observed under the conditions of **Fig. 2** (20 mV membrane potential). The most frequent value for the single-channel conductance of E-Hly in 0.15 M KCl (the conditions of **Fig. 2**) was about 550 pS. A comparison of the macroscopic conductance data derived from multichannel experiments shows that a channel density of more than 10^6 channels/cm^2 can be obtained in reconstitution experiments. This suggests that the formation of E-Hly channels is not a rare event and is definitely not an artifact.

   The lipid bilayer technique allows excellent access to both sides of the membrane. As a consequence, it is possible to perform single channel experiments in different salts and concentrations. A large variety of different ions were permeable through the E-Hly channels. Furthermore, the single-channel conductance of E-Hly was not a linear function of the specific conductance of the bulk aqueous phase. Obviously, the channel contains negatively charged groups, which influence the ion concentration at the channel opening and, thus, the single-channel conductance. The dependence of the single-channel conductance on the bulk aqueous salt concentration allows one to estimate the size of the E-Hly channel *(10)*. According to this estimate, it has a diameter of about 2.6 nm.

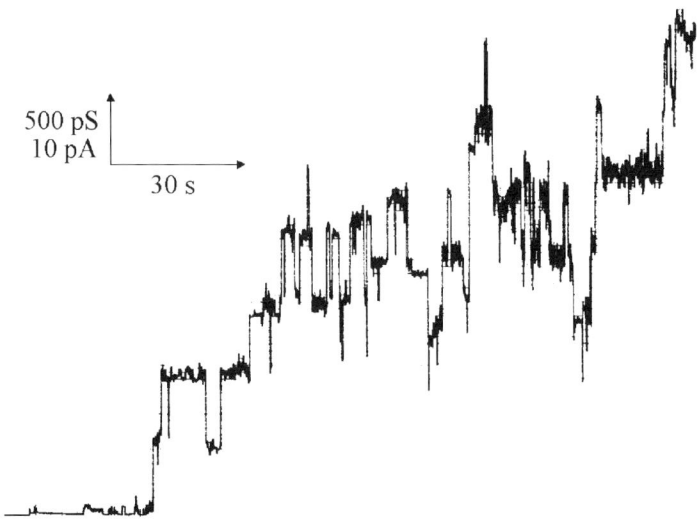

Fig. 2. Single-channel recording of an asolectin membrane in the presence of 50 ng/mL E-Hly from *E. coli* O157. The aqueous phase contained 150 m*M* KCl (pH 6.0). The applied membrane potential was 20 mV; temperature = 20°C.

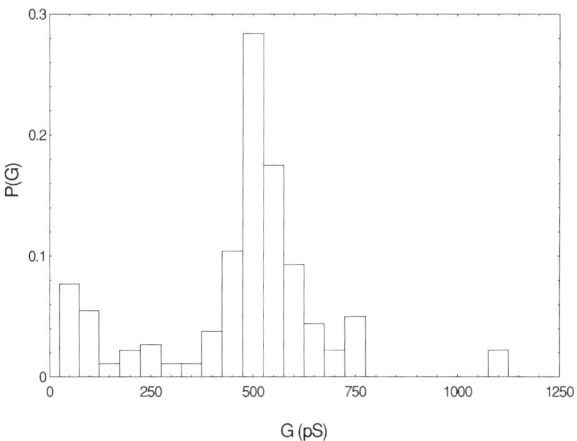

Fig. 3. Histogram of the probability $P(G)$ for the occurrence of a given conductivity unit observed with membranes formed of asolectin/n-decane in the presence of 50 ng/mL E-Hly from *E. coli* O157. $P(G)$ is the probability that a given conductance increment G is observed in the single-channel experiments. It was calculated by dividing the number of fluctuations with a given conductance increment by the total number of conductance fluctuations. The aqueous phase contained 150 m*M* KCl. The applied membrane potential was 20 mV; temperature = 20°C. The average single-channel conductance was 550 pS for 483 single-channel events.

The lipid bilayer experiments allow also the estimation of the diameter of E-Hly channel from the dependence of the single-channel conductance on the radii of the hydrated cations. The estimation is based on the same assumptions, which have been used previously for the derivation of the Renkin correction factor *(16)*. It can be applied to the channel conductance when the entering of a hydrated ion into effective area, A, of the channel mouth (total cross-section A_0) is the rate-limiting step (and not the diffusion of the hydrated ion through the channel itself).

The Renkin correction factor for the movement of a substrate into an aqueous channel is given by

$$A/A_0 = (1 - a/r)^2 [1-2.104 (a/r) + 2.09(a/r)^3-0.95(a/r)^5] \qquad (1)$$

A is the effective area of the channel mouth, A_0 is the total cross-sectional area of the channel, r is the radius of the channel, and a is the radius of the hydrated solute or ion. For the application of the Renkin correction factor to the single-channel conductance, we have to know the radii of the hydrated ions and their diffusion coefficients in the aqueous phase. The radii of the hydrated ions can be calculated from the limiting molar conductivities, λi, by using the Stokes equation:

$$a = Fezi^2/6\pi\nu\lambda i \qquad (2)$$

F (96500 As/mol) is the Faraday constant, e (1.602×10^{-19} A · s) is the elementary charge, zi is the valency of the ions, and ν (1.002×10^{-3} kg/(m · s)) is the viscosity of the aqueous phase. The free-diffusion coefficient of the hydrated ions is also given implicitly by the limiting molar conductivity, λi, of the different ions. It is not necessary to calculate it (although this is possible) because both the Renkin correction factor times the diffusion coefficient and the single-channel conductance may be given relative to the values for one given ion, the hydrated Rb$^+$ ion. It is noteworthy that the Renkin correction factor can only be applied if the conductance of the channel is limited by the movement of one ion species. However, this does not represent a serious restriction in the case considered here because the E-Hly channel conducts preferentially cations.

Figure 4 shows the best fit of the single-channel conductance of the E-Hly channel as a function of the hydrated ion radii (calculated according to **Eq. 2**) with the Renkin correction factor times the aqueous diffusion coefficient of the corresponding cation. The data are given relative to the data for Rb$^+$ (relative permeability equal to unity) and the best fit of the single-channel conductances was obtained with $r = 1.3$ nm, which means that the diameter of the E-Hly channel is around 2.6 nm. The data lie within the range from $r - 1.0$ nm to $r = 1.6$ nm, as shown in **Fig. 4**. A diameter of 2.6 nm agrees very well with our own protection experiments, showing that dextran 4 (molecular mass 4000 Dalton) was not permeable through the E-Hly channel *(10)*. Such a diameter is also very similar to that of the *E. coli* HlyA channel determined previously by osmotic protection experiments *(17)*.

3.4. Concluding Remarks

The methods described here can be used to study functional and molecular properties of pore-forming toxins. In particular, the black lipid bilayer tech-

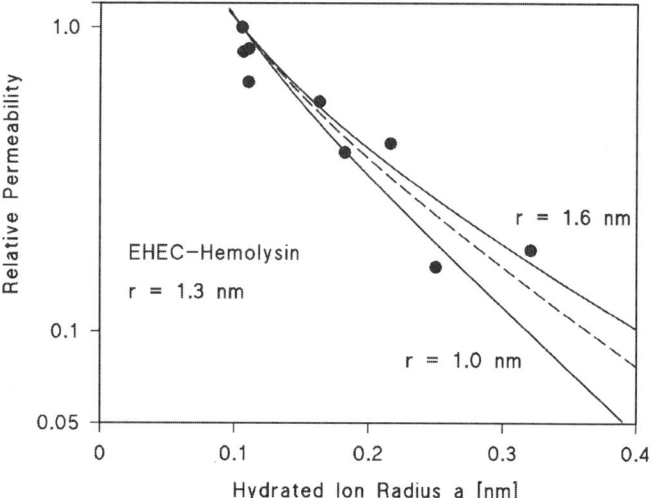

Fig. 4. Fit of the single-channel conductance data of the E-Hly channels by using the Renkin correction factor times the aqueous diffusion coefficients of the different cations. The values were normalized to 1 for $a = 0.105$ nm (Rb⁺). Single-channel conductances were normalized to the ones of Rb⁺ and plotted vs the hydrated ion radii taken from **ref. 8**. The single-channel conductances correspond to Li⁺, Na⁺, K⁺, NH₄⁺, Cs⁺, N(CH₃)₄⁺, N(C₂H₅)₄⁺ and Tris⁺, which were all used for the pore-diameter estimation. The fit (solid lines) is shown for E-Hly channels with $r = 1.6$ nm (upper line) and $r = 1.0$ nm (lower line). The best fit of all data was achieved with $r = 1.3$ nm (diameter = 2.6 nm), which corresponds to the broken line.

nique is a powerful method for studying biophysical properties of pore-forming proteins. It may help to understand the mechanism of pore formation and the attributes of the toxin in membranes composed of different lipids. The strength of the pore-forming activity in vitro may suggest a stronger in vivo activity but comparative analyses have so far not yet been performed. Future studies on E-Hlys should include systematic experiments with cell lines of different origins and in vivo studies.

4. Notes

1. The easiest way to prepare blood agar plates for detection of EHEC-hemolysin is to use defibrinated sheep erythrocytes or to purchase commercial enterohemolysin agar plates. However, in our experience, human erythrocyte concentrates works best. Such concentrates may be obtained from an institute of transfusion medicine from a university hospital. These concentrates can be defibrinated by a simple protocols and give good results. The disadvantage is that these washed human erythrocytes can be stored only up to 24 h in a refrigerator.

2. The erythrocyte washing procedure has to be performed very careful, because washed erythrocytes are fragile and may undergo lysis before incubation with hemolysins. The best way is to use a 20-mL glass pipet and stir carefully by hand.

3. Pouring only small amounts of blood agar in Petri dishes is a critical feature. Enterohemolysis can hardly be detected on agar plates containing usual amounts of agar (e.g., 30–50 mL). Best results are obtained by pouring 15–20 mL into the Petri dishes.

4. Excision of unstained E-Hly bands from an SDS gel is critical for using in the lipid bilayer technique. After electrophoresis, the two left most lanes containing the molecular-weight standard and one E-Hly lane are excised, separated from the rest of the gel and stained regularly. After staining, both parts of the gel are brought together and the unstained bands are labeled with a scalpel and excised unstained.

5. The lipid bilayer technique allows the sensitive detection of current through the membrane. It is, however, not very well suited for the study of fluxes of uncharged solutes. For sensitive electrical measurements, the membrane cell has to be surrounded by a Faraday cage to avoid the 50- or 60-cycle noise of the line and other perturbations caused by electric fields. It is also necessary to insulate the membrane cell against mechanical oscillations. Two Ag/AgCl electrodes are inserted into the aqueous phases on both sides of the membrane. Electrodes with salt bridges have to be used when the aqueous salt solutions do not contain chloride or when salt gradients are established across the across the membrane. The electrodes are switched in series with a voltage source (output voltage 5–250 mV) and an electrometer (Keithley 617). In the case of the single-channel recordings, the electrometer is replaced by a current amplifier (Keithley 427 or a current amplifier based on an operational amplifier). The amplified signal is monitored with a storage oscilloscope and recorded with a tape or a strip-chart recorder. The sensitivity of the method is on the order of 0.1 pA (which corresponds to a flux of about 5×10^5 ions/s) or a few pS (10^{-12} A/V). The lipid bilayer technique allows good access from both sides of the membrane. This means that, with this technique, the ionic composition on both sides of the membrane can be controlled. It is possible to establish salt gradients across the membrane by the addition of concentrated salt solution to one side of the membrane. Zero-current membrane potential measurements allow the measurement of the ionic selectivity of channels if the membrane contains a sufficient number of channels (*18*).

6. After a short lag time of about 2 min, the membrane current starts to increase by many orders of magnitude within 15–20 min (*16*). This process is similar for E-Hly and for HlyA of *E. coli* and of other enteric bacteria. It indicates the insertion of membrane-active, ion-permeable material into the membranes. After about 30 min, the membrane conductance (i.e., the current per unit voltage) increases at a much slower rate. This slow conductance increase continues usually until the lipid bilayer membrane are mechanically disrupted. When the rate of conductance increase is relatively slow (as compared to the initial one), it is possible to study the membrane conductance as a function of the protein concentration. Interest-

ingly, the slope of the specific membrane conductance vs the E-Hly concentration is much steeper than unity *(10)*. This result indicated that more than one toxin molecule is involved in channel formation. In fact, it has been demonstrated for *E. coli* HlyA that several hemolysin molecules are needed to form a channel *(19)*.

References

1. Kayser, H. (1903) Über Bakterienhämolysine, im Besonderen das Colilysin. *Z. Hyg. Infektionskr.* **42,** 118–138.
2. Beutin, L. (1991) The different hemolysins of *Escherichia coli*. *Med. Microbiol Immunol. Berl.* **180,** 167–182.
3. Schmidt, H., Geitz, C., Tarr, P. I., Frosch, M., and Karch, H. (1999) Non-O157:H7 pathogenic Shiga toxin-producing *Escherichia coli*: phenotypic and genetic profiling of virulence traits and evidence for clonality. *J. Infect. Dis.* **179,** 115–123.
4. Schmidt, H. and Karch, H. (1996) Enterohemolytic phenotypes and genotypes of Shiga toxin-producing *Escherichia coli* O111 strains from patients with diarrhea and hemolytic-uremic syndrome. *J. Clin. Microbiol.* **34,** 2364–2367.
5. Beutin, L., Montenegro, M., Zimmermann, S., and Stephan, R. (1986) Characterization of hemolytic strains of *Escherichia coli* belonging to classical enteropathogenic O-serogroups. *Zentralbl. Bakteriol. Mikrobiol. Hyg.* A **261,** 266–279.
6. Beutin, L., Montenegro, M.A., Orskov, I., Orskov, F., Prada, J., Zimmermann, S., et al. (1989) Close association of verotoxin (Shiga-like toxin) production with enterohemolysin production in strains of *Escherichia coli*. *J. Clin. Microbiol.* **27,** 2559–2564.
7. Schmidt, H., Karch, H., and Beutin, L. (1994) The large-sized plasmids of enterohemorrhagic *Escherichia coli* O157 strains encode hemolysins which are presumably members of the *E. coli* alpha-hemolysin family. *FEMS Microbiol. Lett.* **117,** 189–196.
8. Schmidt, H., Kernbach, C., and Karch, H. (1996) Analysis of the EHEC hly operon and its location in the physical map of the large plasmid of enterohaemorrhagic *Escherichia coli* O157:H7. *Microbiology* **142,** 907–914.
9. Schmidt, H., Beutin, L., and Karch, H. (1995) Molecular analysis of the plasmid-encoded hemolysin of *Escherichia coli* O157:H7 strain EDL 933. *Infect. Immun.* **63,** 1055–1061.
10. Schmidt, H., Maier, E., Karch, H., and Benz, R. (1996) Pore-forming properties of the plasmid-encoded hemolysin of enterohemorrhagic *Escherichia coli* O157:H7. *Eur. J. Biochem.* **241,** 594–601.
11. Lally, E. T., Hill, R. B., Kieba, I. R., and Korostoff, J. (1999) The interaction between RTX toxins and target cells. *Trends Microbiol.* **7,** 356–361.
12. Bauer, M. E. and Welch, R. A. (1996) Characterization of an RTX toxin from enterohemorrhagic *Escherichia coli* O157:H7. *Infect. Immun.* **64,** 167–175.
13. Mueller, R., Rudin, D. O., Tien, H. T., and Wescott, W. C. (1962) Reconstitution of excitable membrane structure in vivo and its transformation into an excitable system. *Nature* **194,** 979–981.
14. Dilger, J. P. and Benz, R. (1985) Optical and electrical properties of thin monoolein membranes. *J. Membr. Biol.* **85,** 181–189.

15. Benz, R., Schmid, A., Wagner, W., and Goebel, W. (1989) Pore formation by the haemolysin of *Escherichia coli*: evidence for an association–dissociation equilibrium of the pore-forming oligomers. *Infect. Immun.* **57,** 887–895.

16. Renkin, E. M. (1954) Filtration, diffusion, and molecular sieving through porous cellulose membranes. *J. Gen. Physiol.* **38,** 225–243.

17. Bhakdi, S., Mackmann, N., Nicaud, J. M., and Holland, I. B. (1986) *Escherichia coli* hemolysin may damage target cell membranes by generating transmembrane pores. *Infect. Immun.* **52,** 63–69.

18. Benz, R., Janko, K., and Läuger, P. (1979) Ionic selectivity of pores formed by the matrix protein (porin) of *Escherichia coli. Biochim. Biophys. Acta* **551,** 238–247.

19. Ludwig, A., Benz, R., and Goebel, W. (1993) Oligomerization of *Escherichia coli* haemolysin (HlyA) is involved in pore formation. *Mol. Gen.* **241,** 89–96.

14

Shiga Toxin Receptor Glycolipid Binding

Pathology and Utility

Clifford A. Lingwood

1. Introduction

The tissue specificity and age-related etiology of Shiga toxin (Stx)-induced pathology strongly implicate receptor binding as a major determinant of Stx-induced hemolytic uremic syndarome (HUS) *(1)*. In this review, Shiga toxin receptor binding is considered in relation to the following: (1) the multivalency and multiple valencies of the Shiga toxin B- subunit pentamer and the molecular basis of its specificity, including the important role played by the lipid moiety of globotriaosyl ceramide (Gb_3) and its plasma membrane microenvironment; (2) the internalization of the toxin–receptor complex and subsequent intracellular trafficking; (3) the role of Gb_3 in cell signaling pathways; (4) the upregulation of Gb_3 synthesis and Stx sensitivity in cancers, particularly drug-resistant cancers and the involvement of the P-glycoprotein in Gb_3 biosynthesis; (5) the ability of Gb_3 synthesis to be upregulated by cytokines or short-chain fatty acids to increase Stx susceptibility; 6) Gb_3 as a risk factor for the development of HUS, because *in summa*, these properties define the mechanism by which Shiga toxemia results in clinical sequellae. However, these properties also imbue Stx with characteristics that can be utilized as tools in cell biology to measure aspects of homeostasis and to provide new methods for immunization and DNA transfection.

1.1. Basis of Shiga Toxin Receptor Binding Specificity

The glycolipid Gb_3 is the functional receptor for all Shiga toxins *(2)*, even Stx2e *(3)*, which preferentially binds to Gb_4 *(4)*. Although the binding is medi-

From: *Methods in Molecular Medicine, vol. 73: E. coli: Shiga Toxin Methods and Protocols*
Edited by: D. Philpott and F. Ebel © Humana Press Inc., Totowa, NJ

ated via the terminal Galα1-4Gal carbohydrate of Gb$_3$ *(5,6)*, the hydroxyl groups involved in binding for each of the different Shiga toxins is different *(7)*, indicating that the different members of the Shiga toxin family recognize the galabiose moiety differently. In addition, although the carbohydrate contains the crucial determinants of binding, the lipid-free oligosaccharide sequence is not recognized *(8)*, indicating the important role played by the lipid moiety *(9)*. Multivalency plays an important role in receptor binding, not only because of the pentameric array of receptor-binding B-subunits but because each B subunit potentially has more than one binding site. Glycolipids in membranes can associate into multimolecular and multivalent domains for toxin binding. Similarly, although the oligosaccharide itself does not bind, when coupled to diatomaceous earth *(10)* or conjugated in a pentameric "star-fish" array *(11)*, effective toxin binding is observed. The manner by which Shiga toxins bind their glycolipid receptors is extremely complex *(12)* and an understanding of the optimal monomeric receptor presentation would be of advantage prior to construction of multivalent dendrimers.

Molecular modeling predicted two Gb$_3$ binding sites *(7,13)*: one in the cleft between adjacent B-subunit monomers, such that the β-galactose of the Gb$_3$ oligosaccharide stacked against Phe 30, and the α-galactose formed a hydrogen-bonding network with the aspartate loop *(7)*. Binding in this site also accommodated aminoGb$_4$ in which the terminal sugar is GalNH$_2$, which is the preferred receptor for all Shiga toxins *(7)*. The amine of aminoGb$_4$ formed a salt bridge with the Asp16. Binding in this site also provided an explanation for the lack of Gb$_4$ binding by Stx1 because of the unfavorable apposition of the acetamido carboxyl group with the carboxyl group of Asp16. The acetamido group cannot reorient because it is involved in a intramolecular hydrogen bond to the hydroxyl at the 3' position of the terminal *N*-acetyl galactosamine sugar residue. In Stx2e, which binds Gb$_4$ in addition to Gb$_3$ *(4)*, there is a cleft between Glu15 and Asp16 sufficient to accommodate the acetyl group. [The lack of Stx1/Gb$_4$ binding is not explained in the modeling of Stx2e/Gb$_4$ binding based on the Stx1/Gb$_3$ sugar cocrystal structure *(14)* (*see* below), however.] Our model also explains the site specific mutagenesis studies in which the receptor-binding specificity of Stx2e was altered to bind Gb$_3$ only, by mutagenesis of both Lys66 and Gln64 to Gln and Glu, respectively, because, together, these mutations allow the reorientation of Gln64 to block the Gb$_4$ acetamido group *(7)*.

A second higher-energy (lower-affinity) docking minimum was proposed in a shallow depression on the other side of Phe30 *(7)*, different thermodynamic conformers about the anomeric linkage of Gb$_3$ *(15)* being preferentially docked into this site II and the previous site I. Homology modeling was used to dock Gb$_3$ in site I of each the Stx species. These model docking experiments agreed very well with results of binding to deoxy Gb_3 analogs for each Stx, predicting four differences in binding for each of the four Stxs assayed (Stx1, Stx2, Stx2c, and Stx2e) *(16)*.

The crystal structure of the Stx1B-subunit pentamer complex with the lipid-free globotriaosyl oligosaccharide has been solved by X-ray analysis *(17)* and the solution structure of the same complex solved by nuclear magnetic resonance (NMR) *(18)*. The cocrystal structure found three sites of density as a result of the sugar. Site I corresponded to the site I we predicted, with some variation in the terminal αGal positioning. Site II also corresponded to the site II we predicted, however, the oligosaccharide was oriented in a different position from that predicted and site III was identified as primarily Trp34. This was not identified by modeling and, indeed, was not found in the solution structure complex. In the cocrystal structure, site II showed the highest oligosaccharide density. Similarly, in the solution structure, site II was identified as the oligosaccharide-binding domain. We speculate, however, that the structures of both these complexes, because they involve the lipid-free oligosaccharide, do not represent the preferred binding site for Gb_3 glycolipid. Our recent studies using fluorescent resonance energy transfer support this conclusion *(19)*. We synthesized a coumarin-Gb_3 analog and measured the distance from the coumarin to the single Trp34 when bound in the StxB pentamer. Although the coumarin-Gb_3 is a modification of the native Gb_3 species, it binds Stx as well as Gb_3 by thin-layer chromatography overlay *(19)*. The efficiency of energy transfer between the bound coumarin-Gb_3 and the Trp34 is a direct measure of the distance between these two chromaphores. This distance was calculated from FRET to be 13.2 Å. The equivalent distance for a coumarin-Gb_3 docked according to the cocrystal structure in either site I, II, or III was far greater than this measurement. However, the distance measured according to the theoretical docking in site II, as we had proposed, was approximately equal to this FRET-calculated distance. We had proposed that site II was preferentially utilized by Stx2c *(7)* and found, indeed, that the coumarin-Gb_3 was preferentially bound by Stx2c, as opposed to other Stxs, and therefore suggested that by generation of the coumarin derivative we have generated a Gb_3 homolog which preferentially docked in site II, as modeled.

The Stx2e mutant, GT3, in which Gb_4 binding was removed *(20)* has recently been cocrystalized with the lipid-free Gb_3 sugar *(21)*. In this first crystal structure of the Stx2 series, only site I and site II receptor occupancy was seen with similar docking parameters to the Stx1B-Gb_3 sugar cocrystal. It was suggested that although the role of site III in Gb_3 binding in GT3 was much less than for Stx1, nevertheless, in addition to site I, site III may play a role in Stx2e binding Gb_4, because of the appropriate position of Asn18 to interact with the terminal GalNAc (Asp18-Asn induces Stx1/Gb_4 binding) *(20)*.

Although our studies indicate that site I is the high-affinity Gb_3-binding site, it is still possible to use blockage of the lower-affinity site II as a mechanism of toxin inactivation. At high concentrations, lipid-free Gb_3 sugar analogs preferentially dock in site II *(17,21)*. When multimerized, the Stx affinity for such analogs is greatly increased. The binding of a cleverly designed "pentamer of dimers" can crosslink between Stx pentamers via site II linkage to strongly inactivate both Stx1 and Stx2 in cell culture *(11)*. The multimer, however, was not found to simultaneously bind to the different sites in each monomer of a given pentamer.

However, our approach has been to optimize the monomer structure for site I docking prior to the generation of multivalent Gb_3 analogs. In this regard, we have generated a water-soluble Gb_3 analog *(22)* by substituting the fatty acid of Gb_3 with a rigid hydrocarbon ring structure, α-adamantan. This conjugate preferentially partitions into water and retains high-affinity Stx binding. The total synthesis of adamantyl-Gb_3 *(23)* confirmed the inhibitory activity of this Gb_3 derivative (Lingwood, unpublished). Understanding the mechanism by which the aglycone moiety of such species promotes the receptor binding of Shiga toxin is a goal of our future research.

Site-specific mutagenesis studies have implicated both site I and site II as important in Gb_3 binding and cell cytotoxicity *(24)*. Although the mutagenesis of Trp 34 (site III) itself only had a marginal effect on Gb_3 and cell binding, mutation of Trp 34 in combination with a mutation that blocks either site I or site II had a major inhibitory effect *(25)*. The site-specific mutant in which site I and site III had been specifically blocked and which showed a major decrease in Gb_3, cell surface binding and cytotoxicity still showed a fully occupied site II in the cocrystal structure with the lipid-free Gb_3 oligosaccharide. These mutational studies concluded that both site II and site I were important for Gb_3 binding and that site II alone was insufficient for high-affinity Gb_3 recognition. Such a conclusion is consistent with our original modeling studies. Analysis of kinetic and equilibrium Stx1/Gb_3 binding provides strong evidence for two binding sites *(12)*. However, only one dissociation constant was observed. In addition, evidence of cooperative binding depending on the Gb_3 concentration was obtained.

2. Intracellular Stx Trafficking

Stx1 undergoes a process of retrograde transport following endocytosis via clathrin-coated pits *(26,27)*. Electron microscopic (EM) studies show toxin within the Golgi cysterni and, in highly Stx1-sensitive cells, within the endoplasmic reticulum (ER) and nucleus *(28,29)*. Cells of moderate Stx sensitivity showed intracellular trafficking of Stx to the Golgi only *(29)*. Such cells that target the toxin/receptor complex to the Golgi were found to contain relatively lower levels of short fatty acid chain Gb_3 isoforms, whereas these isoforms were elevated for cells of high Stx sensitivity that targeted the receptor complex further along the retrograde transport pathway to the ER and nucleus *(29)*. Cells of moderate Stx sensitivity could be sensitized by treatment with sodium butyrate *(29,30)* and this resulted in the induction of retrograde transport of the Stx receptor complex to the ER/Golgi *(31)* and the induction of the increased synthesis of short (C:16, C:18) fatty acid isoforms of Gb_3 *(29)*. ER targeting was confirmed by colocalization with the ER chaperone, BIP and ERGIC 53, a marker of the intermediate compartment (between Golgi and ER) *(29)*.

The additional targeting of the ER/nucleus in cells of high Stx sensitivity suggests that Stx may have an additional effect to inhibition of cytosolic ribosomal protein synthesis *(33)*. Stx can be coprecipitated with the B23 nucleolar antigen *(32)*. Thus, the restricted location of Stx within the nucleus *(29)* is the nucleolus. The RNA glycanase activity of the StxA-subunit may therefore affect RNA processing within the nucleolus. Such a mechanism may relate to the observed increase in specific mRNA half-lives following treatment of some cells with low Stx concentrations *(34)*.

The mechanism by which short-chained Gb_3 fatty acid isoforms are preferentially utilized to target the more distal elements of the retrograde transport pathway has yet to be elucidated but may involve the tendency of vesicularization to reduce the bilayer width and, thus, the shorter-chain fatty acid isoforms will be selected on a thermodynamic basis *(35)*. The StxA- or B-subunit does not contain a KDEL or related sequence and, thus, the Golgi/ER trafficking is not based on the recognition of this motif.

The mechanism of Stx internalization varies between cell types. For Daudi B lymphoma cells *(27)* and HeLa cells *(26)*, the pathway is via clathrin-coated vesicles, but for Vero cells, a fraction (approx 50%) of the surface-bound FITC-StxB is internalized by a caveoli-dependent pathway *(36)*. Unlike many other bacterial toxins, internalization and subsequent cytotoxicity of Stx, was found to be independent of pH gradients. When such gradients were collapsed with monensin, retrograde transport of StxB from the cell surface was unaffected *(36)*. At high Stx concentrations and following prolonged incubation (24 h), caveoli-mediated Stx1 endocytosis becomes more significant and inhibitors of this pathway must be utilized to protect cells from Stx cytotoxicity under such conditions *(37)*. Caveoli-mediated endocytosis to early endosomes provides a mechanism for direct access to the endoplasmic reticulum without Golgi transit *(38)*.

Several cell types (e.g., monocytes, macrophages, PMNs, dendritic cells) express Gb_3, bind Stx, and yet are not killed but rather stimulated to produce cytokines *(32,39–41)*. Johannes' group have recently established that in such cells, although Stx is internalized, the receptor/Stx complex does not undergo retrograde transport, rather the Stx is internalized to lysosomes, partially degraded and released into the cytoplasm *(32)*. Glycolipids can be organized on the cell surface in detergent-insoluble domains or "rafts" containing high-cholesterol, glycosphingolipids and GPI-anchored cell surface proteins *(42,43)*, thought to be sites of transmembrane signaling *(44)*. In macrophages that target Stx to lysosomes, Gb_3 was not present in such domains, in contrast to cells that underwent retrograde Stx transport *(32)*. Thus, cell surface organization of Gb_3 may dictate subsequent intracellular trafficking. Intracellular Stx could be colocalized with the B23 nucleolar antigen in monocyte-derived cells *(32)*. Thus, it is likely that Stx can access the nucleus via both the caveoli-mediated transport pathway and a clathrin-coated vesicle-mediated retrograde transport pathway through the secretory endomembrane system. Thus, nuclear access from the cytoplasm via nuclear pores must occur in monocyte cells. In cells in which the nucleus is targeted after retrograde transport, nuclear access could similarly be achieved via membrane translocation from the ER to the cytosol. Alternatively, because the nuclear membrane is continuous with the ER, the Stx may be able to flip directly into the nucleus. Because the B-subunit targets the nucleolus, this raises the question as to whether this nuclear organelle contains Gb_3. Certainly, the nucleolus does not have a classical membrane. In cells of high Stx sensitivity, it has been shown that internalized Stx is colocalized with the BIP ER chaperone *(29)*. In the HeLa cells, BIP can also be colocalized with Stx but, as would be expected, not in monocyte cells *(32)*. **Figure 1** is a schematic summary of these intracellular Stx 1/Gb_3 trafficking routes.

Monocytes have long been implicated in the mechanism of Stx-induced HUS *(45)* and it is entirely feasible that Stx-stimulated cytokine release from monocytes serves to upregulate (local?) Gb_3 synthesis in target cells within the renal glomerulus. Several reports have shown cytokine-mediated increases in endothelial cell sensitivity to Stx resulting from increased α-galactosyl transferase activity responsible for Gb_3 synthesis *(46–48)*. This enzyme has recently been cloned by three groups *(49–51)* and it will be of interest to determine the expression of Gb_3 synthase within target glomerular cells during renal ontogeny. The availability of RNA probes for this enzyme will allow more precise definition of the roles of cytokines in Stx-induced disease.

3. Gb_3 and Signal Transduction

Gb_3 is involved in the signal transduction of two eukaryotic cell surface proteins, the $α_2$-interferon receptor IFNAR1 *(52)* and the B-cell differentiation antigen CD19 *(53)*. Both of these proteins show sequence similarity to the B-subunit of Stx in their N-terminal extracellular domains *(54,55)*. This sequence similarity aligns with site I, not site II, from our modeling studies *(55)*. Signal transduction of both interferon *(52,56,57)* and CD19 ligation

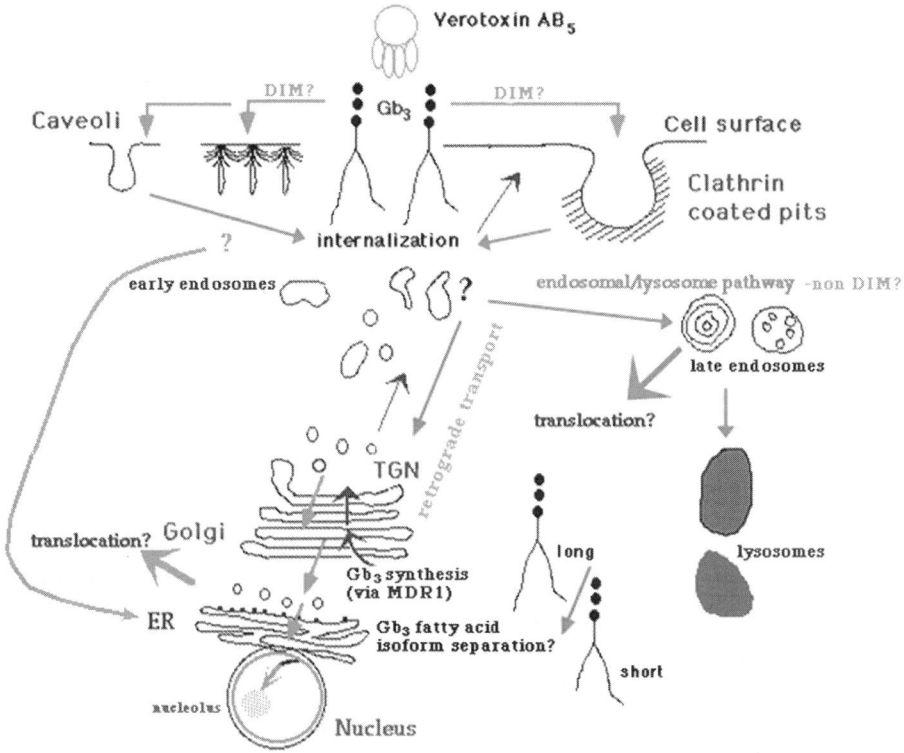

Fig. 1. Gb₃ synthesis and Stx intracellular traffic.

(53,58) has been shown to be compromised in cells deficient in Gb_3. Interferon binding to the α_2-interferon receptor can initiate two processes: growth inhibition and an antiviral activity. Both of these activities are Gb_3 dependent. Ligation of CD19 on B-cells can have a variety of effects, including the induction of homotypic adhesion and B-cell apoptosis. These two processes are also Gb_3 dependent. CD19 surface expression may also be Gb_3 dependent *(57)*. In Gb_3-containing B-cells, ligation of CD19 results in the internalization of this antigen and intracellular trafficking to the nuclear envelope *(58)*. In Gb_3-deficient cells, the CD19 is internalized (more slowly) but not to the nuclear envelope, and apoptosis *(58)* and homotypic adhesion *(53)* are not induced. Therefore, the topology of CD19-induced signaling cascades is likely important to the successful outcome of CD19 signaling. Internalization of Gb_3 by StxB treatment protects anti-CD19-treated cells from apoptosis *(58)* and prevents CD19 internalization. Thus, CD19 endocytosis is Gb_3 dependent. In turn, CD19 ligation prevents some StxB internalization *(58)*, arguing for the common usage of Gb_3. B-Cell CD19-ligation-dependent homotypic adhesion *(53)* and interferon-

induced growth inhibition *(57)* were both recovered following reconstitution of Gb_3-negative cells with Gb_3.

Following α_2-interferon binding to the interferon receptor complex, the cytosolic kinases Tyk2 and Jak1 are activated to phosphorylate two signal transducers, Stat1 and Stat2. These recruit a third protein to form ISGF3, the interferon-stimulated gene factor 3 complex, a DNA-binding element that then translocates to the nucleus to effect the upregulation of interferon-sensitive genes *(59)*. This complex is not effectively assembled to bind to the ISGF3 recognition sequence in Gb_3 negative cells *(52)*. Nuclear translocation of Stat1 was defective in Gb_3-deficient cells, and increased expression of the α_2-interferon-inducible double-stranded RNA-dependent protein kinase (PKR) was also not seen in such cells. Rather, PKR became unstable and was broken down after α_2-interferon stimulation. Addition of StxB together with α_2-interferon also prevented α_2-interferon-induced Stat1 nuclear translocation and induced PKR breakdown *(56)*.

Interferon-mediated growth inhibition, like CD19 signal transduction, requires internalization of the ligand/receptor complex. The antiviral activity induced by interferon binding to its receptor, however, does not require internalization and the role of Gb_3 in $\alpha2$-interferon signaling in antiviral activity is mediated at the cell surface *(56)*. Whereas short-chain Gb_3 isoforms are preferentially utilized to mediate intracellular Stx trafficking in highly Stx-sensitive cells *(29)* (presumably for growth inhibition by $\alpha2$-interferon and signal transduction by CD19 ligation also), the longer-chain Gb_3 isoforms are sufficient to carry out this cell surface accessory function of Gb_3 in $\alpha2$-interferon antiviral activity *(56)*. These studies for the first time demonstrate a distinct functional discrimination between different fatty acid isoforms of Gb_3, and they suggest that in receptor-mediated endocytosis and subsequent retrograde trafficking of Gb_3 binding ligands, there is a gradation of Gb_3 isoforms, in that the short-chain species are preferentially utilized or preferentially reside in the more distal targets (i.e., the endoplasmic reticulum and nuclear envelope), whereas the longer-chain species are preferentially retained at the cell surface or internalized only as far as the Golgi. The demonstrated ability of Gb_3 ligation to induce transmembrane signaling to activate cytosolic kinase activity *(60)* and the differential organization of Gb_3 within or excluded from detergent-insoluble membrane microdomains throughout intracellular trafficking *(32)* indicates a large potential for regulated, topologically distinct Gb_3-associated signal transduction pathways.

Shiga toxin is a potent inducer of apoptosis *(61–67)*. Gb_3 ligation can also induce apoptosis in B lymphoid cells *(68)*. However, most Stx-sensitive cells are relatively unaffected by the StxB subunit, although it is internalized in a similar manner. It has recently been shown in gastrointestinal cell lines that

Stx-induced apoptosis increases proapoptotic Bax and is prevented by overexpression of antiapoptotic Bcl-2 *(69)*. This induction of apoptosis is intimately related to the binding of Gb_3, because internalization of Stx into Gb_3-negative T84 cells (by an unknown mechanism) *(70)* had no effect *(69)*.

Transmembrane signaling to activate cytosolic *src* kinase activity has been shown to occur following Stx-binding cell surface Gb_3 *(60)* leading to apoptosis. Bound Gb_3 was present in rafts, and following activation, the *scr* kinase was released from association with the raft, presumably to allow the phosphorylation of distal substrates. Kinase activation was inhibited by filipin, which binds cholesterol to disrupt rafts. The fact that filipin poorly protects other cells against Stx *(37)* indicates that Stx signaling pathways, like intracellular trafficking, can vary according to cell type.

4. Antineoplastic Activity of Stx

The antitumor activity of Stx induced by the increased expression of Gb_3 in certain tumor cells lines, particularily drug-resistant forms *(71)*, also provides data consistent with this trafficking pathway. Gb_3 is upregulated in primary ovarian tumors, but particularly in their metastases and drug-resistant cases *(72)*. Gb_3 expression was overall less in the more differentiated tumor samples and this correlated with reduced FITC-Stx staining of frozen sections. Drug-resistant differentiated tumor sections were the exception and showed high Stx staining. Gb_3 from such tumors migrates more slowly on thin-layer chromatography, consistent with a higher content of short fatty acid isoforms *(72)*. Gb_3 expression and Stx sensitivity were dramatically increased *(73)* following transfection of MDCK cells with the MDR1 gene encoding the Pgp pump responsible for many instances of multidrug-resistant cancer phenotype. This Gb_3 increase and Stx sensitivity was prevented in the presence of cyclosporin A, a selective inhibitor of the Pgp pump. In the transfected cells, Stx was internalized via retrograde transport to the ER/nuclear envelope and even within the nucleus, as we had previously shown for highly Stx-sensitive Gb_3 positive cells *(29)*. This correlated with the fact that the elevated Gb_3 observed after MDR1 transfection was exclusively short-chain fatty acid isoforms. Both LC, and to some extent GC, were increased following transfection and the conversion of GC to LC was prevented by cyclosporin. We therefore proposed *(73)* that Pgp acts as a glycolipid "flippase" in the Golgi to flip glucosyl ceramide from the cytosolic surface of the Golgi, where it is made *(74)*, to the lumen for access to the lactosyl ceramide synthase and subsequent elongation glycolipid glycosyl transferases. Other studies have previously shown that Pgp was a nonspecific lipid flippase capable of flipping glucosyl ceramide to the cell surface in polarized epithelial cells *(75)*. Our subsequent work indicates (DeRosa and Lingwood, unpublished) that Pgp is similarly involved in the Gb_3 synthesis of many cell types. In the MDR1 transfected cells, Pgp was mainly observed in the Golgi *(73)*. Some cells do exhibit a multidrug-resistant phenotype by the action of a Golgi located Pgp, pumping drugs from the cytosol into the Golgi lumen to avoid cytotoxic effect *(76)*. Pgp was expressed on the cell surface also, and in a percentage of cells, this colocalized with the surface expression of Gb_3. Inhibition of Gb_3 synthesis by cyclosporin resulted in the lack of surface Gb_3 expression, but, interestingly, there was a corresponding lack of surface Pgp expression *(73)*. Current studies are investigating the possible role of Gb_3 in MDR function. These studies provide a molecular explanation for the hypersensitivity of multiple-drug-resistant cells we have previously observed *(71)* and further supports our contention that, despite its involvement in HUS, Shiga toxin provides an attractive option for antineoplastic therapy.

Intratumoral Stx injection has been shown to be a effective treatment for astrocytoma xenografts in mice *(66)*. Both tumor cells and their neovasculature were induced to undergo apoptosis. Similar studies underway for other Gb_3-positive tumor xenograft models are promising. Malignant meningioma cranial xenografts and their vasculature are responsive to Stx *(76a)*. Both in primary ovarian *(72)* and glioblastoma sections *(66)*, blood vessels within the vicinity of the tumor were Stx reactive, further suggesting that Stx has an antiangiogenic as well as antineoplastic effect *(77)*. Stx also selectively eliminates tumor cells from stem cell preparations for autologous stem cell transplantation *(78,79)*. Prior to aggressive chemo/radiotherapy, hematopoetic stem cell preparations are isolated from cancer patients to repopulate the lymphoid system after tumor treatment. Elimination of contaminating tumor cells in such preparations should increase the efficacy of this treatment approach.

We have now shown that Stx also targets the invasive malignant lymphoma form of posttransplant lymphoproliferative disease *(80)*. This is a systemic monoclonal Ebstein–Barr virus (EBV)[+] B-cell lymphoma that can develop following prolonged immunosupression after organ transplant. Interestingly, the Gb_3 of EBV-infected B-lymphoma (highly Stx sensitive) cell lines used as a model of this malignancy was primarily of the short fatty acid isoform.

5. Gb₃ as a Risk Factor for HUS

It is possible, even likely, that Gb_3 fatty acid isoforms expression plays a role as a risk factor for HUS. Studies analyzing the Gb_3 fatty acid content of red cells of patients who develop HUS following gastrointestinal Shiga toxin-producing *E. coli* (STEC) infection and show a significant difference *(81)*. Although it is unlikely that red blood cell binding plays a significant role in Stx-mediated pathology, it is possible that this provides a genetic marker of a predisposition for synthesis of restricted Gb_3 isoforms in the more relevant cell types. It is of interest to note that cell surface Stx binding did not correlate with the increased susceptibility of cells containing the shorter acid Gb_3 isoforms *(29)*. In in vitro binding assays for Stx1, Gb_3 recognition increases as a function of carbon chain length up to C:20 *(82)*. Although this is an interesting thermodynamic observation and provides much of the impetus for examining the role of the lipid moiety on glycolipid receptor function, it is obvious that toxin binding to Gb_3 within the plasma membrane environment is a more complex event because cells that showed elevated short-chain fatty acid Gb_3 isoforms did not bind any more (or less) toxin than cells preferentially expressing long-chain Gb_3 fatty acid isoforms *(29)*. However, it is also possible that plasma membrane Gb_3 isoforms in both such cell types are equivalent and only the intracellular Gb_3 species are distinct, in terms of their fatty acid isoform content. The lower-affinity binding to these shorter fatty acid Gb_3 isoforms might play a role in cell susceptibility (e.g., if the toxin were more able to dissociate from these Gb_3 isoforms intracellularly).

From our data on MDR regulation of Gb_3 synthesis *(73)*, it is likely relevant that the nephron is a major site of Pgp expression in normal cells *(83)* and that clinical use of its inhibitor, cyclosporin can result in the induction of HUS *(84)* and increases Pgp *(85)*.

Thrombocytopenia is one of the cardinal clinical symptoms of hemolytic uremic syndrome. Gb_3 has been found on platelets *(86)*; however, the addition of Stx has been shown to be without effect *(87)*, indicating that the toxin is not directly responsible for this effect. Although endothelial cells are the primary target, as defined by histological studies of HUS renal samples *(86)*, mesangial cells also express Gb_3 and are sensitive to Stx *(88)*. In this regard it is of interest to note that cultured pediatric mesangial cells are far more sensitive than those cultured from adult kidneys *(65,89)*, suggesting that this may, indeed, be a relevant target cell in vivo. The mesangial cells are not killed outright by Stx, but, rather, their ability to respond to mitogenic stimulation is eliminated *(88,90)*, suggesting that there may be even yet a third type of cellular response to Stx.

There is still considerable uncertainty as to the mechanism by which Stx passes from the gastrointestinal tract. Mucosal cells within the colon do not express Gb_3 *(91,92)*. The status of Gb_3 expression of gastrointestinal vascular endothelial cells has not been studied to any extent. Our studies from a single sample showed no Stx reactivity *(72)*; certainly, Stx binding to gastrointestinal endothelial cells in animal models of STEC-induced disease has been demonstrated *(93)*. Therefore, it may be that regulation of gastrointestinal endothelial cell Gb_3 is also necessary for the development of hemorrhagic colitis. Sodium butyrate has been shown to upregulate Gb_3 synthesis in cultured cells *(29,94,95)* and is a metabolic product of many bacteria. Thus, it is possible that fluctuations in gastrointestinal butyrate could serve to effect an increase in local endothelial Gb_3 content. Stxs have been shown to undergo trancytosis in several in vitro model systems of the mucosal epithelial cells *(96)*. One such model is T84 cells. These cells are entirely Gb$_3$ negative *(70)*. CaCo2 cells are another model. They express Gb$_3$, and in this model, trancytosis was shown to be saturable, but, strangely, distinct pathways were inferred for the trancytosis of Stx1 and Stx2 *(97)*. In this case, it is possible that the trancytosis was the result of an unusual intracellular trafficking pathway for Gb$_3$.

It is possible that bacterial attachment may facilitate Stx epithelial transit. The binding of the enterohemorrhagic *E. coli* might facilitate Shiga toxemia by increasing the mucosal permeability *(98)* or via a type 3 secretion-dependent process. Alternatively, the attachment of such bacteria to epithelial cells has been shown to induce apoptosis *(67)* via the recognition of a phosphatidyl ethanolamine *(99)* and such an increased cellular turnover may compromise the mucosal barrier. Gastrointestinal mucosal trancytosis of Stx represents the most amenable target for prophylaxis of STEC-induced HUS. However, clinical studies using insoluble Gb$_3$ dietary analogs to mop up gastrointestinal toxin in patients at risk have not been overly successful *(100)*. This suggests that free toxin within the gastrointestinal tract may not play an important role in the subsequent development of disease. Although systemic administration of effective soluble Gb$_3$ analogs represents a feasible prophylactic approach *(11,22)*, this represents a more complex and invasive strategy. Studies with the pig edema disease toxin using gastrointestinal loops in vivo were unable to demonstrate the disease following toxin administration into the lumen of such loops. Only when high concentrations of bile acids were included with the toxin were disease symptoms apparent *(101)*. Perhaps bile plays a similar role in the etiology of human STEC-induced disease.

6. Shiga Toxin as a Tool

In addition to the utility of the antineoplastic activity of Stx, the B-subunit, as a result of its intracellular routing, is useful in several biological systems:

6.1. Measurement of Homeostasis–Intraorganelle pH

Targeting of the FITC-labeled B-subunit to the Golgi of living cells allowed the accurate pH determination of the Golgi lumenal pH *(102)*. Because the fluorescein fluor is pH sensitive, quantitating the fluorescent image within the Golgi by digital image analysis and then requantitating the fluorescence when the pH gradient maintained by the Golgi membrane was collapsed to an external pH standard curve allowed the measurement of the intra-Golgi pH in Vero cells to be measured as 6.35. The acidic pH of the Golgi was maintained by a baflomycin-sensitive vacuolar proton pump. Using a recombinant B-subunit to which a KDEL sequence had been attached to the C-terminus *(103)* and which Johannes' group had shown, now became, in part, a resident ER protein *(102)*, the intralumenal pH of the endoplasmic reticulum was similarly measured and found to be neutral *(104)*. Thus, if the A-subunit does translocate to the cytoplasm from the ER, this is not mediated by a pH-induced conformational change *(105,106)*.

If other indicators can be coupled to StxB, it is possible that, other properties of these organelles can be measured. In addition, for highly Stx-sensitive cells, these parameters could be measured in the nucleus.

6.2. Antigen Presentation

As mentioned earlier, dendritic cells are Gb_3 positive but Stx resistant. Dendritic cells are the primary site for antigen processing and presentation to T-cells for immune response *(107)*. In these cells, Stx targets the endosome/lysosomal pathway *(32)*. StxB C-terminus peptide chimeras can be used to target peptide antigens to dendritic cells, internalize to access the antigen processing pathway and deliver the peptide to the class 1 MHC system within the endoplasmic reticulum for appropriate surface presentation for T-cell recognition *(108)*. The process requires internalization of the StxB chimera and can be prevented by inhibitors of the proteosome *(109)*. Thus this process bypasses the regulatory and degradative requirements of antigen presentation and provides a new targeted method for T-cell immunization.

6.3. DNA Transfection

We considered that the retrograde vesicular trafficking of Stx from the cell surface to the nucleus might provide a mechanism for transfection of cells with exogenous DNA. The major limits on the efficacy of transformation with exogenous DNA are the degradation of the DNA in the lysosome after uptake and further subsequent breakdown in the cytosol *(110)*. Vesicular transport to the nucleus should avoid such problems. We have therefore constructed a StxB chimera with λ*cro*, the smallest transcription factor. This chimera bound to Gb_3 and the λ*cro* 18-bp DNA recognition motif simultaneously and was able to efficiently transfer this fluorescent DNA motif to the nucleus of cells that so target Stx *(111)*.

Acknowledgments

Work from my laboratory has been supported by MRC grant MT13073, NCIC grant no. 11090, and NIH grant R01 DK52098.

References

1. Lingwood, C. A. (1994) Verotoxin-binding in human renal sections. *Nephron* **66,** 21–28.
2. Waddell, T., Cohen, A., and Lingwood, C. A. (1990) Induction of Verotoxin sensitivity in receptor deficient cell lines using the receptor glycolipid globotriosyl ceramide. *Proc. Natl. Acad. Sci. USA* **87,** 7898–7901.

3. Keusch, G., Jacewicz, M., Acheson, D., Donohue-Rolfe, A., Kane, A., and McCluer, R. (1995) Globotriasylceramide,Gb₃, is an alternative functional receptor for Shiga-like toxin 2e. *Infect. Immun.* **63,** 1138–1141.

4. DeGrandis, S., Law, H., Brunton, J., Gyles, C., and Lingwood, C. A. (1989) Globotetraosyl ceramide is recognized by the pig edema disease toxin. *J. Biol. Chem.* **264,** 12,520–12,525.

5. Lingwood, C. A., Law, H., Richardson, S., Petric, M., Brunton, J. L., DeGrandis, S., et al. (1987) Glycolipid binding of purified and recombinant *Escherichia coli*-produced Verotoxin in vitro. *J. Biol. Chem.* **262,** 8834–8839.

6. Waddell, T., Head, S., Petric, M., Cohen, A., and Lingwood, C. A. (1988) Globotriosyl ceramide is specifically recognized by the *E. coli* Verocytotoxin 2. *Biochem. Biophys. Res. Commun.* **152,** 674–679.

7. Nyholm, P. G., Magnusson, G., Zheng, Z., Norel, R., Binnington-Boyd, B., and Lingwood, C. A. (1996) Two distinct binding sites for globotriaosyl ceramide on verotoxins: molecular modelling and confirmation by analogue studies and a new glycolipid receptor for all verotoxins. *Chem. Biol.* **3,** 263–275.

8. Boyd, B., Zhiuyan, Z., Magnusson, G., and Lingwood, C. A. (1994) Lipid modulation of glycolipid receptor function: presentation of galactose α1–4 galactose disaccharide for Verotoxin binding in natural and synthetic glycolipids. *Eur. J. Biochem.* **223,** 873#-3878.

9. Lingwood, C. A. (1996) Aglycone modulation of glycolipid receptor function. *Glycoconj. J.* **13,** 495–503.

10. Armstrong, G. D., Fodor, E., and Vanmaele, R. (1991) Investigation of Shiga-like toxin binding to chemically synthesized oligosaccharide sequences. *J. Infect. Dis.* **164,** 1160–1167.

11. Kitov, P. I., Sadowska, J. M., Mulvey, G., Armstrong, G. D., Lingaw, H., Pannu, N. S., et al. (2000) Shiga-like toxins are neutralized by tailored multivalent carbohydrate ligands. *Nature* **403,** 669–672.

12. Peter, M. and Lingwood, C. (2000) Apparent cooperativity in multivalent verotoxin glogotriaosyl ceramide binding: Kinetic and saturation binding experiments with radiolabelled Verotoxin [125I]-VT1. *Biochim. Biophys. Acta* **1501,** 116–124.

13. Nyholm, P.-G., Brunton, J. L., and Lingwood, C. A. (1995) Modelling of the interaction of Verotoxin-1 (VT1) with its glycolipid receptor, globotriaosyl-ceramide (Gb₃). *Int. J. Biol. Macromol.* **17,** 199–205.

14. Cummings, M., Ling, H., Armstrong, G., Brunton, J., and Read, R. (1998) Modeling the carbohydrate-binding specificity of pig edema toxin. *Biochemistry* **37,** 1789–1799.

15. Nyholm, P.-G. and Pascher, I. (1993) Steric presentation and recognition of the saccharide chains of glycolipids at the cell surface: favoured conformations of the saccharide-lipid linkage calculated using molecular mechanics (MM3). *Int. J. Biol. Macromol.* **5,** 43–51.

16. Lingwood, C. A., Mylvaganam, M., Arab, S., Khine, A. A., Magnusson, G., Grinstein, S., et al. (1998) Shiga toxin (Verotoxin) binding to its receptor glycolipid, in *Escherichia coli* O157:H7 and Other Shiga Toxin-Producing *E. coli*

Strains (Kaper, J. B. and O'Brien, A. D., eds.), American Society for Microbiology, Washington, DC, pp. 129–139.

17. Ling, H., Boodhoo, A., Hazes, B., Cummings, M., Armstronmg, G., Brunton, J., et al. (1998) Structure of the Shiga toxin B-pentamer complexed with an analogue of its receptor Gb$_3$. *Biochemistry* **37**, 1777–17788.

18. Richardson, J. M., Evans, P. D., Homans, S. W., and Donohue-Rolfe, A. (1996) Solution structure of the carbohydrate-binding B-subunit homopentamer of Verotoxin VT-1 from *E. coli*. *Nat. Struct. Biol.* **4**, 190–193.

19. Picking, W. W., McCann, J. A., Nutikka, A., and Lingwood, C. A. (1999) Localization of the binding site for modified Gb$_3$ on Verotoxin 1 using fluorescence analysis. *Biochemistry* **38**, 7177–7184.

20. Tyrrell, G. J., Ramotar, K., Toye, B., Boyd, B., Lingwood, C. A., and Brunton, J. L. (1992) Alteration of the carbohydrate binding specificity of verotoxins from Galα1-4Gal to GalNAcβ1-3Galα1-4Gαl and vice versa by site-directed mutagenesis of the binding subunit. *Proc. Natl. Acad. Sci. USA* **89**, 524–528.

21. Ling, H., Pannu, N. S., Boodhoo, A., Armstrong, G. D., Clark, C. G., Brunton, J. L., et al. (2000) A mutant Shiga-like toxin IIe bound to its receptor Gb$_3$: structure of a group II Shiga-like toxin with altered binding specificity. *Structure* **8**, 253–264.

22. Mylvaganam, M. and Lingwood, C. (1999) Adamantyl globotriaosyl ceramide–a monovalent soluble glycolipid mimic which inhibits Verotoxin binding to its glycolipid receptor. *Biochem. Biophys. Res. Commun.* **257**, 391–394.

23. Hansen, H. C. and Magnusson, G. (1999) Synthesis of some aglycon analogs of globotriosylceramide. *Carbohydr. Res.* **322**, 181–189.

24. Bast, D. J., Banerjee, L., Clark, C., Read, R. J., and Brunton, J. L. (1999) The identification of three biologically relevant globotriaosyl ceramide receptor binding sites on the Verotoxin 1 B subunit. *Mol. Microbiol.* **32**, 953–960.

25. Soltyk, A., MacKenzie, C. R., Wolski, W., Hirama, T. and Brunton, J. A mutational analysis of the globotriaosyl ceramide binding sites of Verotoxin VT1. Submitted.

26. Sandvig, K., Olnes, S., Brown, J., Peterson, O., and van Deurs, B. (1989) Endocytosis from coated pits of Shiga toxin: a glycolipid-binding protein from *Shigella dysenteriae* 1. *J. Cell Biol.* **108**, 1331–1343.

27. Khine, A. A. and Lingwood, C. A. (1994) Capping and receptor mediated endocytosis of cell bound verotoxin (Shiga-like toxin) 1; chemical identification of an amino acid in the B subunit necessary for efficient receptor glycolipid binding and cellular internalization. *J. Cell. Physiol.* **161**, 319–332.

28. Sandvig, K., Garred, Ø., Prydz, K., Kozlov, J., Hansen, S., and van Deurs, B. (1992) Retrograde transport of endocytosed Shiga toxin to the endoplasmic reticulum. *Nature* **358**, 510–512.

29. Arab, S. and Lingwood, C. (1998) Intracellular targeting of the endoplasmic reticulum/nuclear envelope by retrograde transport may determine cell hypersensitivity to Verotoxin: sodium butyrate or selection of drug resistance may induce nuclear toxin targeting via globotriosyl ceramide fatty acid isoform traffic. *J. Cell. Physiol.* **177**, 646–660.

30. Sandvig, K., Ryd, M., Garred, O., Schweda, E., and Holm, P. K. (1994) Retrograde transport from the Golgi complex to the ER of both Shiga toxin and the nontoxic Shiga B-fragment is regulated by butyric acid and cAMP. *J. Cell Biol.* **126,** 53–64.

31. Lingwood, C. A., Khine, A. A., and Arab, S. (1998) Globotriaosyl ceramide (Gb$_3$) expression in human tumor cells: intracellular trafficking defines a new retrograde transport pathway from the cell surface to the nucleus, which correlates with sensitivity to verotoxin. *Acta Biochem. Polon.* **45,** 351–359.

32. Falguieres, T., Baron, C., Mallard, F., Lingwood, C., Goud, B., Salamero, J., et al. (2001) Cell type differences in sensitivity to Shiga toxin-mediated inhibition of protein biosynthesis can be correlated with the existence of two distinct toxin transport pathways. *Mol. Biol. Cell* **12,** 2453–2468.

33. Saxena, S. K., O'Brien, A. D., and Ackerman, E. J. (1989) Shiga toxin, Shiga-like toxin II variant, and ricin are all single-site RNA N-glycosidases of 28 S RNA when microinjected into *Xenopus* oocytes. *J. Biol. Chem.* **264,** 596–601.

34. Bitzan, M. M., Wang, Y., Lin, J., and Marsden, P. A. (1998) Verotoxin and ricin have novel effects on preproendothelin-1 expression but fail to modify nitric oxide synthase (ecNOS) expression and NO production in vascular endothelium. *J. Clin. Invest.* **101,** 372–382.

35. Lingwood, C. A. Glycolipids and bacterial pathogenesis, in *Oligosaccharides in Chemistry and Biology/\/A Comprehensive Handbook* (Ernst, B., Sinay, P. and Hart, G., eds.), Wiley–VCH, Weinheim, in press.

36. Schapiro, F., Lingwood, C. A., Furuya, W., and Grinstein, S. (1998) pH-independent targeting of glycolipids to the Golgi complex. *Am. J. Physiol.* **274,** 319–332.

37. Khine, A. A. PhD thesis, University of Toronto, Canada, 2000.

38. Smart, E. J., Ying, Y.-S., Conard, P. A., and Anderson, R. G. W. (1994) Caveolin moves from caveolae to the Golgi apparatus in response to cholesterol oxidation. *J. Cell Biol.* **127,** 1185–1197.

39. Ramegowda, B. and Tesh, V. L. (1996) Differentiation-associated toxin receptor modulation, cytokine production, and sensitivity to Shiga-like toxins in human monocytes and monocytic cell lines. *Infect. Immun.* **64,** 1173–1180.

40. van Setten, P., Monnens, L., Verstraten, R., van der Heuvel, L., and van Hinsberg, V. (1996) Effects of Verotoxin-1 on non adherent human monocytes: binding characteristics, protein synthesis, and induction of cytokine release. *Blood* **88,** 174–183.

41. Yamasaki, C., Natori, Y., Zeng, X.-T., Ohmura, M., Yamasaki, S., Takeda, S., et al. (1999) Induction of cytokines in a human colon epithelial cell line by Shiga toxin 1 (Stx1) and Stx2 but not by non-toxic mutant Stx1 which lacks N-glycosidase activity. *FEBS Lett.* **442,** 231–234.

42. Simons, K. and Ikonen, E. (1997) Functional rafts in cell membranes. *Nature* **387,** 569–572.

43. Hooper, N. (1999) Detergent-insoluble glycosphingolipid/cholesterol-rich membrane domains, lipid rafts and caveolae. *Mol. Membr. Biol.* **16,** 145–156.

44. Brown, D. and London, E. (1998) Functions of lipid rafts in biological membranes. *Annu. Rev. Cell. Dev. Biol.* **14,** 111–136.

45. van Setten, P., van Hinsbergh, V., van den Heuvel, L., Preyers, F., Dijkman, H., Assmann, K., et al. (1998) Monocyte chemoattractant protein-1 and interleukin-8 levels in urine and serum of patents with hemolytic uremic syndrome. *Pediatr. Res.* **43,** 759–767.

46. Louise, C. B. and Obrig, T. G. (1991) Shiga toxin-associated hemolytic–uremic syndrome: combined cytotoxic effects of Shiga toxin, interleukin-1{b}, and tumor necrosis factor alpha on human vascular endothelial cells *in vitro. Infect. Immun.* **59,** 4173–4179.

47. van der Kar, N. C. J., Kooistra, T., Vermeer, M., Lesslauer, W., Monnens, L. A. H., and van Hinsbergh, V. W. M. (1995) Tumor necrosis factor a induces endothelial galactosyl transferase activity and verocytotoxin receptors. Role of specific tumor necrosis factor receptors and protein kinase C. *Blood* **85,** 734–743.

48. Keusch, G. T., Acheson, D. W. K., Aaldering, L., Erban, J., and Jacewicz, M. S. (1996) Comparison of the effects of Shiga-like toxin 1 on cytokine-and butyrate pretreated human umbilical and saphenous vein endothelial cells. *J. Infect. Dis.* **173,** 1164–1170.

49. Steffensen, R., Carlier, K., Wiels, J., Levery, S. B., Stroud, M., Cederen, B., et al. (2000) Cloning and expression of the histo-blood group Pk UDP-galactose: Galbeta 1-4Glcbeta 1-Cer alpha 1, 4-galactosyltransferase. Molecular genetic basis of the p phenotype. *J. Biol. Chem.* **275,** 16,723–16,729.

50. Kojima, Y., Fukumoto, S., Furukawa, K. T. O., Wiels, J., Yokoyama, K., Suzuki, Y., et al. (2000) Molecular cloning of globotriaosylceramide/CD77 synthase, a glycosyltransfease that initiates the synthesis of globo series glycosphingolipids. *J. Biol. Chem.* **275,** 15,152–15,156.

51. Keusch, J. J., Manzella, S. M., Nyame, K. A., Cummings, R. D., and Baenziger, J. U. (2000) Cloning of Gb_3 synthase, the key enzyme in globo-series glycosphingolipid synthesis, predicts a family of of α1,4 glycosyltransferases conserved in plants, insects and mammals. *J. Biol. Chem.* **275,** 25,315–25,321.

52. Ghislain, J., Lingwood, C. A., and Fish, F. N. (1994) Evidence for glycosphingolipid modification of the type 1 IFN receptor. *J. Immunol.* **153,** 3655–3663.

53. Maloney, M. D. and Lingwood, C. A. (1994) CD19 has a potential CD77 (globotriaosyl ceramide)-binding site with sequence similarity to verotoxin B-subunits: implications of molecular mimicry for B cell adhesion and enterohemorrhagic *Escherichia coli* pathogenesis. *J. Exp. Med.* **180,** 191–201.

54. Lingwood, C. A. and Yiu, S. C. K. (1992) Glycolipid modification of α-interferon binding: sequence similarity between α-interferon receptor and the Verotoxin (Shiga-like toxin) B-subunit. *Biochem. J.* **283,** 25,26.

55. Lingwood, C. A. (1996) Role of Verotoxin receptors in pathogenesis. *Trends Microbiol.* **4,** 147–153.

56. Khine, A. A. and Lingwood, C. A. (2000) Functional significance of globotriaosylceramide in α_2 interferon/ type I interferon receptor mediated anti viral activity. *J. Cell. Physiol.* **182,** 97–108.

57. Maloney, M. D., Binnington-Boyd, B., and Lingwood, C. A. (2000) Globotriaosyl ceramide modulates interferon-α-induced growth inhibition and CD19 expression in Burkitt's lymphoma cells. *Glycoconj. J.* **16**, 821–828.

58. Khine, A. A., Firtel, M., and Lingwood, C. A. (1998) CD77-dependent retrograde transport of CD19 to the nuclear membrane: functional relationship between CD77 and CD19 during germinal center B-cell apoptosis. *J. Cell. Physiol.* **176**, 281–292.

59. Williams, B. R. G. and Haque, S. J. (1997) Interacting pathways of interferon signaling. *Semin. Oncol.* **24**, S9-70–S9-77.

60. Katagiri, Y., Mori, T., Nakajima, H., Katagiri, C., Taguchi, T., Takeda, T., et al. (1999) Activation of Src family kinase induced by Shiga toxin binding to globotriaosyl ceramide (Gb$_3$/CD77) in low density, detergent-insoluble microdomains. *J. Biol. Chem.* **274**, 35,278–35,282.

61. Inward, C. D., Williams, J., Chant, I., Crocker, J., Milford, D. V., Rose, P. E., et al. (1995) Verocytotoxin-1 induces apoptosis in vero cells. *J. Infect.* **30**, 213–218.

62. Arab, S., Murakami, M., Dirks, P., Boyd, B., Hubbard, S., Lingwood, C., et al. (1998) Verotoxins inhibit the growth of and induce apoptosis in human astrocytoma cells. *J. Neurol. Oncol.* **40**, 137–150.

63. Taguchi, T., Uchida, H., Kiyokawa, N., Mori, T., Sato, N., Horie, H., et al. (1998) Verotoxins induce apoptosis in human renal tubular epithelium derived. *Kidney Int.* **53**, 1681–1688.

64. Kiyokawa, N., Taguchi, T., Mori, T., Uchida, H., Sato, N., Takeda, T., et al. (1998) Induction of apoptosis in normal human renal tubular epithelial cells by *Escherichia coli* Shiga toxins 1 and 2. *J. Infect. Dis.* **178**, 178–184.

65. Williams, J., Boyd, B., Nutikka, A., Lingwood, C., Barnett-Foster, D., Milford, D., et al. (1999) A comparison of the effects of verocytotoxin-1 on primary human renal cell cultures. *Toxicol. Lett.* **105**, 47–57.

66. Arab, S., Rutka, J., and Lingwood, C. (1999) Verotoxin induces apoptosis and the complete, rapid, long-term elimination of human astrocytoma xenografts in nude mice. *Oncol. Res.* **11**, 33–39.

67. Foster, D. B., Abul-Milh, M., Huesca, M., and Lingwood, C. A. (2000) Enterohemorrhagic Escherichia coli induces apoptosis which augments bacterial binding and phosphatidylethanolamine exposure on the plasma membrane outer leaflet. *Infect. Immun.* **68**, 3108–3115.

68. Mangeney, M., Lingwood, C. A., Caillou, B., Taga, S., Tursz, T., and Wiels, J. (1993) Apoptosis induced in Burkitt's lymphoma cells via Gb$_3$/CD77, a glycolipid antigen. *Cancer Res.* **53**, 5314–5319.

69. Jones, N. L., Islur, A., Haq, R., Mascarenhas, M., Karmali, M. A., Perdue, M. H., et al. (2000) Escherichia coli Shiga toxins induce apoptosis in epithelial cells that is regulated by the Bcl-2 family. *Am. J. Physiol. Gastrointest. Liver Physiol.* **278**, G811–G819.

70. Philpott, D. J., Ackerley, C. A., Kiliaan, A. J., Karmali, M. A., Perdue, M. H., and Sherman, P. M. (1997) Translocation of Verotoxin-1 across T84 monolayers: mechanism of bacterial toxin penetration of epithelium. *Am. J. Physiol.* **273**, G1349–G1358.

71. Farkas-Himsley, H., Rosen, B., Hill, R., Arab, S., and Lingwood, C. A. (1995) Bacterial colicin active against tumor cells in vitro and in vivo is Verotoxin 1. *Proc. Natl. Acad. Sci. USA* **92** 6996–7000.

72. Arab, S., Russel, E., Chapman, W., Rosen, B., and Lingwood, C. (1997) Expression of the Verotoxin receptor glycolipid, globotriaosylceramide, in ovarian hyperplasias. *Oncol. Res.* **9,** 553–563.

73. Lala, P., Ito, S., and Lingwood, C. A. (2000) Retroviral transfection of MDCK cells with human MDR1 results in a major increase in globotriaosyl ceramide and 10^5–10^6-fold increased cell sensitivity to verocytotoxin. *J. Biol. Chem.* **275,** 6246–6251.

74. Lannert, H., Gorgas, K., Meissner, I., Wieland, F. T., and Jeckel, D. (1998) Functional organization of the Golgi apparatus in glycosphingolipid biosynthesis. *J. Biol. Chem.* **273,** 2939–2946.

75. van Helvoort, A., Smith, A., Sprong, H., Fritzsche, I., Schinkel, A., Borst, P., et al. (1996) MDR1 P-glycoprotein is a lipid translocase of broad specificity, while MDR3 P-glycoprotein specifically translocates phosphatidyl choline. *Cell* **87,** 507–517.

76. Molinari, A., Cianfriglia, M., Meschini, S., Calcabrini, A., and Arancia, G. (1994) P-Glycoprotein expression in the Golgi apparatus of multidrug-resistant cells. *Int. J. Cancer* **59,** 789–795.

76a. Salhia, B., Rutka, J. T., Lingwood, C., Nutikka, A., and Van Furth, W. The treatment of malignant meningioma with verotoxin. *Neoplasia* **4,** 304–311.

77. Lingwood, C. A. (1999) Verotoxin/globotriaosyl ceramide recognition: angiopathy, angiogenesis and antineoplasia. *Biosci. Rep.* **19,** 345–354.

78. LaCasse, E. C., Saleh, M. T., Patterson, B., Minden, M. D., and Gariépy, J. (1996) Shiga-like toxin purges human lymphoma from bone marrow of severe combined immunodeficient mice. *Blood* **88,** 1561–1567.

79. LaCasse, E. C., Bray, M. R., Patterson, B., Lim, W.-M., Perampalam, S., Radvanyi, L. G., et al. (1999) Shiga-like toxin I receptor on human breast cancer, lymphoma, and myeloma and absence from CD34+ hematopoietic stem cells: implications for ex vivo tumor purging and autologous stem cell transplantation. *Blood* **94,** 1–12.

80. Arbus, G. S., Grisaru, S., and Lingwood, C. A. (2000) Verotoxin targets lymphoma infiltrates of patients with post-transplant lymphoproliferative disease. *Leuk. Res.* **24,** 857–864.

81. Newburg, D., Chaturvedi, P., Lopez, E., Devoto, S., Feyad, A., and Cleary, T. (1993) Susceptibilty to hemolytic–uremic syndrome relates to erythrocyte glycosphingolipid patterns. *J. Infect. Dis.* **168,** 476–479.

82. Kiarash, A., Boyd, B , and Lingwood, C. A. (1994) Glycosphingolipid receptor function is modified by fatty acid content: Verotoxin 1 and Verotoxin 2c preferentially recognize different globotriaosyl ceramide fatty acid homologues. *J. Biol. Chem.* **269,** 11,138–11,146.

83. Ernest, S., Rajaraman, S., Megyesi, J., and Bello-Reuss, E. N. (1997) Expression of MDR1 (multidrug resistance) gene and its protein in normal human kidney. *Nephron* **77,** 284–289.

84. Grupp, C., Schmidt, F., Braun, F., Lorf, T., Ringe, B., and Muller, G. (1998) Haemolytic uraemic syndrome (HUS) during treatment with cyclosporin A after

renal transplantation{\}is tacrolimus the answer? *Nephrol. Dial. Transplant.* **13,** 1629–1631.

85. Jette, L., Beaulieu, E., Leclerc, J. M., and Beliveau, R. (1996) Cyclosporin A treatment induces overexpression of P-glycoprotein in the kidney and other tissues. *Am. J. Physiol.* **270,** F756–F765.

86. Cooling, L. L. W., Walker, K. E., Gille, T., and Koerner, T. A. W. (1998) Shiga toxin binds human platelets via globotriaosylceramide (P^k antigen) and a novel platelet glycosphingolipid. *Infect. Immun.* **66,** 4355–4366.

87. Yoshimura, K., Fujii, J., Yutsuda, T., Kikuchi, R. T. S., Shirahata, S., and Yoshida, S. (1998) No direct effects of Shiga toxin 1 and 2 on the aggregation of human platelets in vitro. *Thromb. Haemost.* **80,** 529–530.

88. Robinson, L. A., Lingwood, C., Hurley, R. M., and Matsell, D. G. (1994) The binding and biological effects of *E. coli* verotoxin (VT-1) on human paediatric glomerular mesangial cells (MC), in *Recent Advances in Verocytotoxin-Producing* Eschcerichia coli *Infections* (Karmali, M. A. and Goglio, A. G., eds.), Elsevier Science Amsterdam, pp. 361–364.

89. Simon, M., Cleary, T., Hernandez, J., and Abboud, H. (1998) Shiga toxin 1 elicits diverse biological responses in mesangial cells. *Kidney Intl.* **54,** 1117–1127.

90. van Setten, P. A., van Hinsbergh, V. W., Van der Heuvel, L. P., van der Velden, T. J., van de Kar, N. C., Krebbers, R. J., et al. (1997) Verocytotoxin inhibits mitogenesis and protein synthesis in purified human glomerular mesangial cells without affecting cell viability: evidence for two distinct mechanisms. *J. Am. Soc. Nephrol.* **8,** 1877–1888.

91. Holgersson, J., Ströberg, N., and Breimer, M. E. (1988) Glycolipids of human large intestine: difference in glycolipid expression related to anatomical localization, epithelial/non-epithelial tissue and the *ABO, Le* and *Se* phenotypes of the donors. *Biochimie* **70,** 1565–1574.

92. Holgersson, J., Jovall, P.-A., and Breimer, M. E. (1991) Glycosphingolipids from human large intestine: detailed structural characterization with special reference to blood group compounds and bacterial receptor structures. *Biochem. J.* **110,** 120–131.

93. Richardson, S. E., Rotman, T. A., Jay, V., Smith, C. R., Becker, L. E., Petric, M., et al. (1992) Experimental verocytotoxemia in rabbits. *Infect. Immun.* **60,** 4154–4167.

94. Louise, C. B., Kaye, S. A., Boyd, B., Lingwood, C. A., and Obrig, T. G. (1995) Shiga toxin-associated hemolytic uremic syndrome: effect of sodium butyrate on sensitivity of human umbilical vein endothelial cells to Shiga toxin. *Infect. Immun.* **63,** 2765–2769.

95. Sandvig, K., Garred, Ø., van Helvoort, A., van Meer, G., and van Deurs, B. (1996) Importance of glycolipid synthesis for butyric acid-induced sensitization to Shiga toxin and intracellular sorting of toxin in A431 cells. *Mol. Biol. Cell* **7,** 1391–1404.

96. Acheson, D. W. K., Moore, R., De Breucker, S., Lincicome, L., Jacewicz, M., Skutelsky, E., et al. (1996) Translocation of Shiga toxin across polarized intestinal cells in tissue culture. *Infect. Immun.* **64,** 3294–3300.

97. Hurley, B. P., Jacewicz, M., Thorpe, C. M., Lincicome, L. L., King, A. J., Keusch, G. T., et al. (1999) Shiga toxins 1 and 2 translocate differently across polarized intestinal epithelial cells. *Infect. Immun.* **67,** 6670–6677.

98. Philpott, D., McKay, D., Mak, W., Perdue, M., and Sherman, P. (1998) Signal transduction pathways involved in enterohemorrhagic *Escherichia coli*-induced alterations in T84 epithelial permeability. *Infect. Immun.* **66,** 1680–1687.

99. Foster, D. B., Philpott, D., Abul-Milh, M., Huesca, M., Sherman, P. M., and Lingwood, C. A. (1999) Phosphatidylethanolamine recognition promotes enteropathogenic *E. coli* and enterohemorrhagic *E. coli* host cell attachment. *Microb. Pathol.* **27,** 289–301.

100. Armstrong, G. D., McLaine, P. N., and Rowe, P. C. (1998) Clinical trials of synsorb-Pk in preventing hemolytic uremic syndrome, in Escherichia coli *O157:H7 and Other Shiga Toxin-Producing* E. coli *Strains* (Kaper, J. B. and O'Brien, A. D., eds.), American Society for Microbiology, Washington, DC, pp. 374–384.

101. Waddell, T. E. and Gyles, C. L. (1995) Sodium deoxylcholate facilitates systemic absorption of Verotoxin 2e from pig intestine. *Infect. Immun.* **63,** 4953–4956.

102. Kim, J. H., Lingwood, C. A., Williams, D. B., Furuya, W., Manolson, M. F., and Grinstein, S. (1996) Dynamic measurement of the pH of the Golgi complex in living cells using retrograde transport of the Verotoxin receptor. *J. Cell Biol.* **134,** 1387–1399.

103. Johannes, L., Tenza, D., Anthoy, C., and Goud, B. (1997) Retrograde transport of KDEL-bearing B-fragment of Shiga toxin. *J. Biol. Chem.* **272,** 19,554–19,561.

104. Kim, J. H., Johannes, L., Goud, B., Antony, C., Lingwood, C. A., Daneman, R., et al. (1998) Non-invasive measurement of the pH of the endoplasmic reticulum at rest and during calcium release. *Proc. Natl. Acad. Sci. USA* **95,** 2997–3002.

105. Saleh, M. T. and Gariépy, J. (1993) Local conformational change in the B-sub-unit of Shiga-like toxin 1 at endosomal pH. *Biochemistry* **32,** 918–922.

106. Saleh, M., Ferguson, J., Boggs, J., and Gariepy, J. (1996) Insertion and orienta-tion of a synthetic peptide representing the C-terminus of the A1 domain of Shiga toxin into phospholipid membranes. *Biochemistry* **35,** 9325–9334.

107. MacPherson, G., Kushnir, N., and Wykes, M. (1999) Dendritic cells, B cells and the regulation of antibody synthesis. *Immunol. Rev.* **172,** 325–334.

108. Lee, R. S., Tartour, E., van der Bruggen, P., Vantomme, V., Joyeux, I., Goud, B., et al. (1998) Major histocompatibility complex class I presentation of exog-enous soluble tumor antigen fused to the B-fragment of Shiga toxin. *Eur. J. Immunol.* **28,** 2726–2737.

109. Haicheur, N., Bismuth, E., Bosset, S., Adotevi, O., Warnier, G., Lacabanne, V., et al. (2000) The B-subunit of Shiga toxin fused to a tumor antigen elicits CTL and targets dendritic cells to allow MHC class I restricted presentation of pep-tides derived from exogenous antigens. *J. Immunol.* **165,** 3301–3308.

110. Lechardeur, D., Sohn, K.-J., Haardt, M., Joshi, P. B., Monck, M., Graham, R. W., et al. (1999) Metabolic instability of plasmid DNA in the cytosol: a poten-tial barrier to gene transfer. *Gene Ther.* **6,** 482–497.

111. Facchini, L. and Lingwood, C. A. (2001) Verotoxin 1 B subunit{\}lambda Cro chimeric protein specifically targets DNA to globotriaosyl ceramide. *Exp. Cell Res.* **269,** 117–129.

15

Methods for the Purification of Shiga Toxin 1

Anita Nutikka, Beth Binnington-Boyd, and Clifford A. Lingwood

1. Introduction

In addition to its involvement in clinical disease, Shiga toxin (Stx) is being appreciated as a useful tool in various aspects of cell physiology. These include defining signal transduction pathways *(1–5)*, intracellular monitors of homeostatic processes *(6–8)*, inducers of apoptosis *(9–15)*, antigen carriers for cellular immunization *(16–18)*, and as an antineoplastic agent *(19–24)*. These various fields of study require the availability of purified toxin and/or its subunits. In order to facilitate the expansion of such studies, in the absence of a commercial supplier, a detailed purification scheme for Stx1 is presented in this chapter.

Standard biochemical techniques have traditionally been used in the purification of significant quantities of biologically active toxin *(25–27)*, The first procedure presented is based on these standard procedures. The second method developed in our lab takes advantage of the Gb_3-binding properties of the toxin *(28)*. A Gb_3-celite affinity column provides a simple rapid method for the purification of Stx1.

2. Materials
2.1. Shiga Toxin 1 Purification

1. 3X 1 L LB broth: 10 g Bacto-tryptone, 5 g yeast extract, 10 g NaCl, in 1 L of water in a 2-L Erlenmyer flask. Autoclave (Difco Laboratories, Detroit, MI).
2. LB-amp plates: 10 g Bacto-tryptone, 5 g yeast extract, 10 g NaCl, 15 g agar, all in 1 L water. Autoclave. When solution cools to approx 50°C, add ampicillin to 100 µg/mL. Pour plates. Store inverted at 4°C.
3. Ampicillin: 1000X stock solution: 100 mg/mL in water, sterile filter, store at –20°C (Sigma, St. Louis, MO).
4. Amicon YM 30 membrane (Amicon, Danvers, MA).

From: *Methods in Molecular Medicine, vol. 73: E. coli: Shiga Toxin Methods and Protocols*
Edited by: D. Philpott and F. Ebel © Humana Press Inc., Totowa, NJ

5. Phosphate-buffered saline (PBS).
6. Phenylmethylsulfonylfluoride (PMSF): 100X stock solution: 100 mM in acetone, store in freezer, handle with caution. PMSF solutions are inactivated in aqueous solutions, so add this just before extraction (Sigma, St. Louis, MO).
7. Lysozyme: 1000X stock solution: 100 mg/mL in water (Pharmacia Biotech, Baie D'Urfe, Quebec).
8. Hydroxyapatite Bio-Gel HT Gel (Bio-Rad Laboratories, Hercules, CA).
9. Dialysis tubing, 12,000–14,000 MWCO (Fisher, Nepean, Ontario).
10. 0.5 M Potassium phosphate stock, pH 7.2: 17.42 g K_2HPO_4, 6.8 g KH_2PO_4, pH with KOH and make up to 300 mL. Dilute stock as necessary for 10 mM or 100 mM.
11. PBE 94 Polybuffer Exchanger (degas exchanger under vacuum prior to pouring column) (Pharmacia Biotech).
12. 0.025 M Imidazole-HCl, pH 7.4 (store at room temperature; degas before using). (Sigma).
13. Polybuffer 74 (Pharmacia Biotech) Dilute 1 : 7 and pH to 5.0 with HCl. Degas before using.
14. Affi-gel Blue, 50–100 mesh (Bio-Rad).
15. 10 mM Sodium phosphate, pH 7.2. Store at room temperature.
16. 10 mM Sodium phosphate, 0.5 M NaCl, pH 7.2. Store at room temperature.

2.2. Purification of Shiga Toxin 1 Using a Gb$_3$-Celite Affinity Column

1. Celite 545 (Fisher Scientific, Nepean, Ontario).
2. Gb$_3$.
3. Chloroform. Work with chloroform in a well-ventilated fume hood.
4. 50 mM TBS: 50 mM Tris-HC1, 154 mM NaCl, pH 7.4.
5. 1 M Tris-HCl, pH 9.5.
6. 1 M Tris, pH ~ 11–11.5, unadjusted.
7. 2 M HCl.
8. Sodium azide.

3. Methods

3.1. Purification of Shiga Toxin 1

The following method describes Stx1 holotoxin purification from strain JB28, which has the Shiga toxin 1 operon cloned in pUC19 *(29)*. Yields range from 10–30 mg of purified protein from a 3-L culture.

1. Grow JB28 on LB-amp plates and passage at least once. Grow overnight cultures from single colonies in 3 mL of liquid LB broth supplemented with 100 µg/mL of ampicillin (in triplicate).
2. Supplement each liter of LB broth with 100 µg/mL ampicillin. Add 1 mL of overnight seed culture to each of the 1-L LB flasks. Grow cultures for 24 h at 37°C with shaking at approx 250 rpm.
3. Centrifuge bacteria at 9000 rpm for 15 minutes at 4°C and retain pellets (*see* **Note 1**).

4. Scrape pellets into a beaker and extract with 250 mL of cold PBS containing 100 μg/mL lysozyme and 1 mM PMSF. Sonicate on ice using a probe sonicator until a uniform suspension is attained (approx 1 min). Transfer the solution to a 500-mL flask and shake at 250 rpm at 37°C for 1 h. Pellet the bacteria by centrifuging at 9000 rpm for 15 min at 4°C. Remove the supernatant and keep it on ice.

5. Extract the remaining pellet a second time with a similar volume of extraction buffer. Pellet the bacteria by centrifugation. Pool the extraction supernatants and continue (*see* **Note 2**).

6. The supernatants should be yellow and the bacterial pellet should become finer and more diffuse with each extraction step. Filter the combined supernatants through filter paper and then through a glass fiber filter to clarify. This step serves to remove debris and will greatly speed the concentration step.

7. Concentrate the combined supernatants to 30–50 mL or less at 50–70 psi (max) using an Amicon cell with a YM30 membrane filter. Stop the concentration if precipitate begins to appear.

8. Dialyze the concentrated extract using 12,000–14,000 MWCO dialysis membrane in at least 2 L of 10 mM potassium phosphate, pH 7.2, overnight at 4°C with at least one change of dialysis buffer. The final salt concentration of the extract should be very low before running the hydroxyapatite column (i.e., [NaCl] < 5 mM).

9. Prepare the hydroxyapatite column (approx 2.5 cm × 20 cm) and equilibrate the column with at least 3 column volumes of 10 mM potassium phosphate, pH 7.2 (*see* **Note 3**).

10. Load the dialyzed sample onto the column and wash with 10 mM potassium phosphate, pH 7.2. Monitor the absorbance at 280 nm using a quartz cuvet. Continue to wash the column until the absorbance reading is close to zero.

11. Elute the toxin containing fraction from the column with approx 2 column volumes of 100 mM potassium phosphate, pH 7.2 collecting approx 3 mL fractions. Measure the absorbance of the fractions at 280 nm using a quartz cuvet. Pool the peak fractions that elute with the 100 mM phosphate (usually the yellow fractions). Dialyze the pooled fractions using 12,000–14,000 MWCO membrane in at least 2 L of 0.025 M imidazole-HCl, pH 7.4 overnight, at 4°C with one change of dialysis buffer.

12. While the sample is dialyzing pour the degassed chromatofocusing column (1.5 cm × 50 cm) and equilibrate the column with 0.025 M imidazole buffer, until the column effluent is at pH 7.4 (*see* **Note 4**).

13. Load the sample onto the column. When sample has been loaded, follow immediately with degassed polybuffer-HCl, pH 5.0 (50 mL Polybuffer 74 + 350 mL distilled water, a 1 : 7 dilution, with the pH adjusted to 5.0 with HCl).

14. Collect 1- to 2-mL fractions and test them for absorbance at 280 nm and pH. Measuring the pH of the fractions is imperative to ensure that the desired pH gradient has formed. Plot the absorbance at 280 nm and pH versus fraction number. Pool the peak fraction at about pH 6.7 for Stx1. If necessary, monitor the fractions by sodium dodecyl sulfate–polyacrylamide gel electrophoresis (SDS-PAGE) to verify the presence of Stx1 in any peaks. The Stx1 is 80–90% pure at this stage (*see* **Note 5**) (**Fig. 1**).

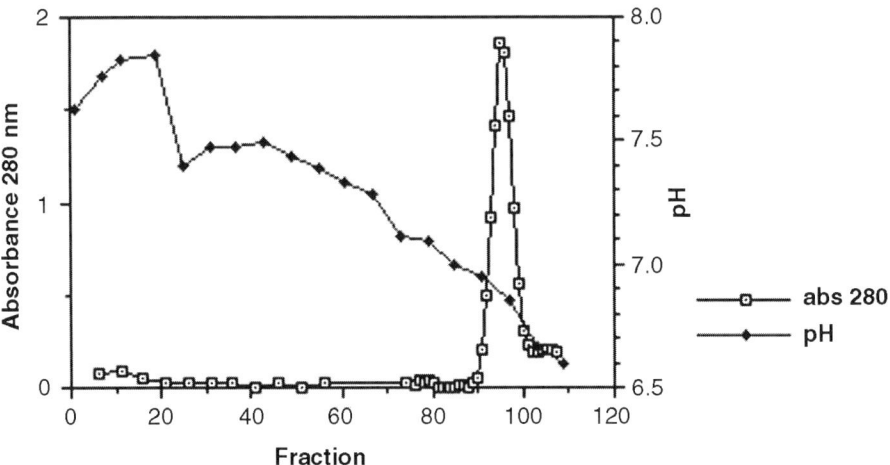

Fig. 1. Chromatofocusing elution profile. A typical elution profile of Stx1 from a chromatofocusing column. The Stx1 peak is typically symmetrical. Other peaks may be detected, depending on the extent of bacterial lysis.

15. Clean the column with 100 mL of 1 M NaCl. If a yellow band remains on the column, clean it with 100 mL of 1 M HCl, but quickly equilibrate it with 25 mM imidazole, pH 7.4. Store the column indefinitely in 20% ethanol in 25 mM imidazole.
16. Equilibrate a Affi-gel Blue column with 10 column volumes of 10 mM sodium phosphate buffer, pH 7.2 (*see* **Note 6**).
17. Load the fractions pooled from the chromatofocusing column onto the Affi-gel Blue column and wash the column with 10 mM sodium phosphate, pH 7.2, until the absorbance of the eluate at 280 nm is zero; at least 5 column volumes.
18. Elute the Shiga toxin 1 with 0.5 M NaCl in 10 mM sodium phosphate buffer and collect 2-mL fractions.
19. Measure the absorbance at 280 nm of each fraction and pool the main peak fractions that elute with 0.5 M NaCl.
20. Clean the column with 25 mL each of 8 M urea in wash buffer and 1 M NaCl in wash buffer. Re-equilibrate column with 10 mM phosphate containing 0.1% azide for storage.
21. Dialyze peak fractions against 2 L of PBS, or desired buffer, pH 7.4, at 4°C with one change. If necessary, concentrate the toxin to between 500 and 1000 µg/mL and aliquot into 50- to 200-µL portions and store at –20°C. The toxin is stable under these conditions for several years (*see* **Note 7**).

3.2. Purification of Stx1 Using a Gb$_3$-Celite Affinity Column

1. Dry 20 g of Celite 545 in a 100–130°C oven for 1 h. Cool and then store sealed in a dry place until needed.
2. Resuspend the celite in several volumes of chloroform in a glass (vacuum) flask and allow the celite to settle. Pour off any fine material that does not settle within

a few minutes. Repeat the procedure at least once or until the majority of the fines have been removed; this will ensure a rapid column flow rate. When the fines have been removed, add enough chloroform so that the celite can be thoroughly swirled into a uniform suspension. Degas the celite slurry under vacuum until air bubbles are no longer seen.

3. Resuspend 10 mg of Gb_3 in 5 mL of chloroform. Add the Gb_3 dropwise to the celite while swirling to ensure an even distribution of Gb_3 in the celite.

4. When all of the Gb_3 has been added, remove the chloroform by one of two methods: (1) Transfer the celite-Gb_3 slurry to a round-bottom flask and remove most of the chloroform using a rotary evaporator. Do not attempt to completely dry the celite by this method (*see* **Note 8**). When the celite is almost dry, transfer the celite to a large glass beaker and air-dry in the fume hood overnight. (2) If a rotovap is not available, transfer the celite slurry to a large glass beaker and allow to air-dry in a fume hood (*see* **Note 9**). When all of the chloroform has evaporated, put the Gb_3-celite in a 100°C oven for 5 min. After cooling, the affinity matrix may be stored indefinitely in a dessicator in the freezer.

5. To prepare the Gb_3-celite column resuspend the Gb_3-celite in several volumes of TBS (50 mM Tris-buffered saline, pH 7.4) in a vacuum flask. Thoroughly degas the celite under vacuum, with swirling (*see* **Note 10**). When the Gb_3-celite is thoroughly hydrated and degassed, pour it into a column (2.5 cm in diameter). Wash the column with 1 L of TBS containing 0.05% sodium azide at 4°C.

6. Prepare the crude toxin extract as usual (*see* **Subheading 3.1.**). Filter the extract to remove debris and add 0.05% sodium azide. In the cold room, pass the extract through the Gb_3-celite column very slowly (<1 mL/min) or twice with a faster flow rate (*see* **Note 11**).

7. Wash the unbound proteins from the affinity column with 1 L of TBS (maximum flow rate) followed by 0.5 L of 50 mM Tris containing 1 M NaCl.

8. Elute the toxin with 1 M Tris-HCl, pH 9.5 (no NaCl). Collect approx 2 mL fractions. Immediately identify toxin-containing fractions by reading the absorbance at 280 nm. Neutralize the pooled fractions using 2 M HCl and dialyze extensively against an appropriate buffer (i.e., PBS) (*see* **Note 12**).

9. Elute any tightly-bound toxin using 1 M Tris (pH unadjusted, approx 11 or 11.5). Again, neutralize immediately prior to dialysis. Usually, the majority of the toxin elutes at the lower pH (*see* **Note 13**).

10. Re-equilibrate the column with TBS with 0.05% azide and store at 4°C.

4. Notes

1. These growth conditions are specific for JB28. Growth of bacteria should be adjusted for the clones or sources that are used. Bacterial pellets can be stored indefinitely at –20°C.

2. The extraction step is one of the most important in determining the final yield of purified toxin. Most bacterial extraction methods are suitable for this purpose. When a sonicator is not available, we have also used a standard laboratory blender to extract bacteria with the addition of 0.2 mg/mL colimycin M (Parke-Davis) to

the extraction buffer. (Take steps to prevent the spread of aerosols during blending.) The blended solution sometimes gets very foamy, so that the addition of Antifoam A (Sigma) is helpful. The extraction method is sometimes quite harsh and results in the lysis of bacteria, thereby releasing DNA. Sonication of the extracted bacterial supernatants helps to shear the DNA so that columns flow at a faster rate. An alternative to the PMSF is a protease inhibitor cocktail tablet that is available from Boehringer-Mannheim, Germany (cat. no. 1836153).

3. All of the columns can be run at room temperature. If a protease inhibitor is used during extraction, nicking of the A-subunit is kept to a minimum (<10%). Hydroxyapatite columns come in various mesh sizes. We have found the Bio-Rad Bio-Gel HT hydroxyapatite columns have reasonable flow rates for this procedure. To clean the hydroxyapatite column for storage, wash with 0.5 L of 500 mM potassium phosphate, pH 7.2, and re-equilibrate with 10 mM potassium phosphate + 0.05% sodium azide.

4. All chromatofocusing column packing material and buffers should be thoroughly degassed before using. The column and buffers should be prepared at the temperature at which they will be used. If the columns are run at room temperature, make sure to warm the dialyzed sample to room temperature prior to loading.

5. After pooling the chromatofocusing peak fractions it is necessary to remove the ampholytes in the polybuffer. This is done using an Affi-gel Blue column. Gel filtration can also be used to remove ampholytes.

6. If there is evidence of blue dye leaching from the column (i.e., after a prolonged period of storage), wash the column with several column volumes of 0.5 M NaCl prior to equilibration.

7. Characterization of the purified Stx1: The protein concentration of the final purified Stx1 can be determined using the BCA protein assay (Pierce) using bovine serum albumin (BSA) for a standard curve. The purity of the toxin can be analyzed by SDS-PAGE (**Fig. 2**). Under nonreducing conditions, the purified toxin will run as two bands of about 8 kDa (B-subunit) and 32 kDa (A-subunit). Under reducing conditions, the A-subunit may migrate as a major 32-kDa band and a minor (<10%) 28-kDa band because of partial proteolytic cleavage that occurs during or prior to purification. Biological activity of the toxin can be determined using a vero cell cytotoxicity assay (expect about $10^9 CD_{50}$/mg/mL) Specificity of toxin binding to Gb_3 can be tested using a thin-layer chromatographic overlay (i.e., binding to Gb_3 but not Gb_4).

8. A rotovap can be used to remove the bulk of the chloroform, thereby speeding up the preparation of the celite affinity matrix. As the chloroform is removed from the celite in a rotovap, the solution is likely to bump and some of the celite may be lost. Remove the celite from the rotary evaporator prior to complete dryness. The remaining residual chloroform can be dried in a well-vented fume hood overnight as in Option 2.

9. Whichever method is used, it is crucial that the celite be completely dry. A properly made affinity matrix can be reused several times; failure to completely dry the celite will result in loss of Gb_3 during the next step.

10. Some of the celite may be initially difficult to wet and may float to the surface of the solution when under vacuum. Continue to swirl the celite gently until all of the celite is wet.

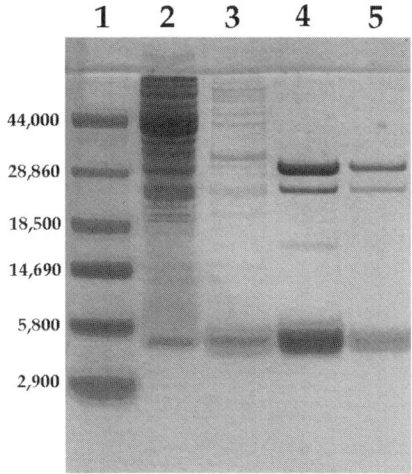

Fig. 2. 15% Tricine SDS-PAGE of Stx1 purification (Coomassie blue stained). (1) Prestained molecular-weight markers (approximate molecular weights as indicated), (2) JB28 extract, (3) hydroxyapatite peak, (4) Chromatofocusing peak, (5) Affi-Gel Blue peak (pure Stx1).

11. The Shiga toxin 1-containing extract should be loaded onto the Gb_3-celite column slowly to give the toxin a chance to bind to the Gb_3. The extract may also be run through the column more than once. If the binding capacity of the column is exceeded, there may also be Stx1 in the flowthrough fraction after loading. Check this solution before discarding it. The entire purification procedure can be repeated with the remaining extract.

12. The pH of the Stx1 should be adjusted as quickly as possible after eluting the toxin from the column.

13. The capacity of a column containing 10 mg Gb_3 is approx 5–10 mg of toxin. The capacity may decrease with each use of the column. If less Gb_3 is to be used to make the affinity matrix, scale down the amount of celite accordingly. Note that this method will also isolate any free B-subunit within the crude extract. We pass the affinity purified toxin (in 10 mM sodium phosphate) through a Cibachron Blue column to isolate the free B-subunit; the column binds the A-subunit of the holotoxin, which is eluted with 0.5 M sodium chloride. *See* **Subheading 3.1.**

References

1. Ghislain, J., Lingwood, C. A., and Fish, E. N. (1994) Evidence for glycosphingolipid modification of the type 1 IFN receptor. *J. Immunol.* **153,** 3655–3663.
2. Maloney, M. D. and Lingwood, C. A. (1994) CD19 has a potential CD77 (globotriaosyl ceramide)-binding site with sequence similarity to verotoxin B-subunits: implications of molecular mimicry for B cell adhesion and enterohemorrhagic *Escherichia coli* pathogenesis. *J. Exp. Med.* **180,** 191–201.

 3. Khine, A. A. and Lingwood, C. A. (2000) Functional significance of globotriaosylceramide in α2 interferon/type I interferon receptor mediated anti viral activity. *J. Cell Physiol.* **182,** 97–108.
 4. Khine, A. A., Firtel, M., and Lingwood, C. A. (1998) CD77-Dependent retrograde transport of CD19 to the nuclear membrane: functional relationship between CD77 and CD19 during germinal center B-cell apoptosis. *J. Cell Physiol.* **176,** 281–292.
 5. Katagiri, Y., Mori, T., Nakajima, H., Katagiri, C., Taguchi, T., Takeda, T., et al. (1999) Activation of Src family kinase induced by Shiga toxin binding to globotriaosyl ceramide (Gb$_3$/CD77) in low density, detergent-insoluble microdomains. *J. Biol. Chem.* **274,** 35278–35282.
 6. Kim, J. H., Lingwood, C. A., Williams, D. B., Furuya, W., Manolson, M. F., and Grinstein, S. (1996) Dynamic measurement of the pH of the Golgi complex in living cells using retrograde transport of the verotoxin receptor. *J. Cell Biol.* **134,** 1387–1399.
 7. Johannes, L., Tenza, D., Anthoy, C., and Goud, B. (1997) Retrograde transport of KDEL-bearing B-fragment of Shiga toxin. *J. Biol. Chem.* **272,** 19554–19561.
 8. Kim, J. H., Johannes, L., Goud, B., Antony, C., Lingwood, C. A., Daneman, R., et al. (1998) Non-invasive measurement of the pH of the endoplasmic reticulum at rest and during calcium release. *Proc. Natl. Acad. Sci. USA* **95,** 2997–3002.
 9. Mangeney, M., Lingwood, C. A., Caillou, B., Taga, S., Tursz, T., and Wiels, J. (1993) Apoptosis induced in Burkitt's lymphoma cells via Gb$_3$/CD77, a glycolipid antigen. *Cancer Res.* **53,** 5314–5319.
10. Inward, C. D., Williams, J., Chant, I., Crocker, J., Milford, D. V., Rose, P. E., et al. (1995) Verocytotoxin-1 induces apoptosis in vero cells. *J. Infect. Dis.* **30,** 213–218.
11. Taga, S., Carlier, K., Mishal, Z., Capoulade, C., Mangeney, M., Lécluse, Y., et al. (1997) Intracellular signaling events in CD77-mediated apoptosis of Burkitt's lymphoma cells. *Blood* **90,** 2757–2767.
12. Kiyokawa, N., Taguchi, T., Mori, T., Uchida, H., Sato, N., Takeda, T. et al. (1998) Induction of apoptosis in normal human renal tubular epithelial cells by *Escherichia coli* Shiga toxins 1 and 2. *J. Infect. Dis.* **178,** 178–184.
13. Taguchi, T., Uchida, H., Kiyokawa, N., Mori, T., Sato, N., Horie, H., et al. (1998) Verotoxins induce apoptosis in human renal tubular epithelium derived cells. *Kidney Int.* **53,** 1681–1688.
14. Williams, J., Boyd, B., Nutikka, A., Lingwood, C., Barnett-Foster, D., Milford, D., et al. (1999) A comparison of the effects of verocytotoxin-1 on primary human renal cell cultures. *Toxicol. Lett.* **105,** 47–57.
15. Gordon, J., Challa, A., Levens, J. M., Gregory, C. D., Williams, J. M., Armitage, R. J., et al. (2000) CD40 ligand, Bcl-2, and Bcl-xL spare group I Burkitt lymphoma cells from CD77-directed killing via verotoxin-1 B chain but fail to protect against the holotoxin. *Cell Death Different.* **7,** 785–794.
16. Noakes, K. L., Teisserenc, H. T., Lord, J. M., Dunbar, P. R., Cerundolo, V., and Roberts, L. M. (1999) Exploiting retrograde transport of Shiga-like toxin 1 for the delivery of exogenous antigens into the MHC class I presentation pathway. *FEBS Lett.* **453,** 95–99.

17. Lee, R. S., Tartour, E., van der Bruggen, P., Vantomme, V., Joyeux, I., Goud, B., et al. (1998) Major histocompatibility complex class I presentation of exogenous soluble tumor antigen fused to the B-fragment of Shiga toxin. *Eur. J. Immunol.* **28,** 2726–2737.

18. Haicheur, N., Bismuth, E., Bosset, S., Adotevi, O., Warnier, G., Lacabanne, V., et al. (2000) The B subunit of Shiga toxin fused to a tumor antigen elicits CTL and targets dendritic cells to allow MHC class I-restricted presentation of peptides derived from exogenous antigens. *J. Immunol.* **165,** 3301–3308.

19. Farkas-Himsley, H., Rosen, B., Hill, R., Arab, S., and Lingwood, C. A. (1995) Bacterial colicin active against tumour cells in vitro and in vivo is verotoxin 1. *Proc. Natl. Acad. Sci. USA* **92,** 6996–7000.

20. LaCasse, E. C., Saleh, M. T., Patterson, B., Minden, M. D., and Gariépy, J. (1996) Shiga-like toxin purges human lymphoma from bone marrow of severe combined immunodeficient mice. *Blood* **88,** 1561–1567.

21. Arab, S., Russel, E., Chapman, W., Rosen, B., and Lingwood, C. (1997) Expression of the Verotoxin receptor glycolipid, globotriaosylceramide, in Ovarian Hyperplasias. *Oncol. Res.* **9,** 553–563.

22. LaCasse, E. C., Bray, M. R., Patterson, B., Lim, W.-M., Perampalam, S., Radvanyi, L. G., et al. (1999) Shiga-like toxin I receptor on human breast cancer, lymphoma, and myeloma and absence from CD34+ hematopoietic stem cells: Implications for ex vivo tumor purging and autologous stem cell transplantation. *Blood* **94,** 1–12.

23. Arab, S., Rutka, J., and Lingwood, C. (1999) Verotoxin induces apoptosis and the complete, rapid, long-term elimination of human astrocytoma xenografts in nude mice. *Oncol. Res.* **11,** 33–39.

24. Arbus, G. S., Grisaru, S., Segal O., Dosch, M., Pop, M., Lala, P., et al. (2000) Verotoxin targets lymphoma infiltrates of patients with post-transplant lymphoproliferative disease. *Leuk. Res.* **24,** 857–864.

25. O'Brien, A. D. and LaVeck, G. D. (1983) Purification and characterization of a *Shigella-dysenteriae* 1-like toxin produced by *Escherichia coli. Infect. Immun.* **40,** 675–683.

26. Brown, J. E., Griffin, D. E., Rothman, S. W., and Doctor, B. P. (1982) Purification and biological characterization of shiga toxin from Shigella dysenteriae 1. *Infect. Immun.* **36,** 996–005.

27. Petric, M., Karmali, M. A., Richardson, S., and Cheung, R. (1987) Purification and biological properties of *Escherichia coli* verocytotoxin. *FEMS Microbiol. Lett.* **41,** 63–68.

28. Boulanger, J., Huesca, M., Arab, S., and Lingwood, C. A. (1994) Universal method for the facile production of glycolipid/lipid matrices for the affinity purification of binding ligands. *Anal. Biochem.* **217,** 1–6.

29. DeGrandis, S. A., Ginsberg, J., Toone, M., Climie, S., Friesen, J., and Brunton, J. (1987) Nucleotide sequence and promoter mapping of the *Escherichia coli* Shiga-like toxin operon of bacteriophage H-19B. *J. Bacteriol.* **169,** 4313–4319.

16

Methods for the Identification of Host Receptors for Shiga Toxin

Anita Nutikka, Beth Binnington-Boyd, and Clifford A. Lingwood

1. Introduction

The Shiga toxin receptor glycolipid Gb_3 (globotriaosylceramide) is also known as the p^k blood group antigen *(1)* and as CD77, a germinal center B-cell differentiation antigen *(2)*. Tissue localization of Gb_3 is associated with sites of Stx-induced pathology in animal models *(3–5)* and in humans *(6)* and, as such, is a major determinant of Stx1-induced disease.

The physiological role(s) of Gb_3 appears to be complex and has been recently implicated in a variety of cellular functions. Ligation of Gb_3 has been shown to activate cytosolic Src kinase in renal tubule cells *(7)*. Gb_3 regulates sensitivity of some cells to α_2-interferon *(8)* and plays a role in α_2 interferon signaling to induce both growth inhibition *(9)* and antiviral activity *(10)*. Binding of Stx1 to monocytes expressing Gb_3 stimulates cytokine production *(11–13)* but does not result in cell death. These results indicate that cell surface Gb_3 functions as a transducer of transmembrane signals. Furthermore, cellular levels of Gb_3 are modulated in response to various stimuli, including treatment with cytokines, lipopolysaccharide (LPS), butyrate, and fumonisin B1. Gb_3 has recently been shown to be elevated in many tumors *(14,15)*. The recent cloning of the α-Gal transferase responsible for Gb_3 synthesis will facilitate the elucidation of signaling pathways that regulate expression of Gb_3 *(16)*. However, Gb_3 function has been shown to be modulated by the structure of its lipid component, most notably the fatty acid chain length *(17,18)*. Therefore, methods for the detection and isolation of Gb_3 are important for studying the role of this glycolipid in Shiga toxin-induced pathology and regular cell function.

From: *Methods in Molecular Medicine, vol. 73: E. coli: Shiga Toxin Methods and Protocols*
Edited by: D. Philpott and F. Ebel © Humana Press Inc., Totowa, NJ

The distribution of Gb_3 in normal and malignant human tissues has been extensively studied with three different p^k-recognizing monoclonal antibodies. *(19–21)*. Although Gb_3 was found to be widely expressed in all three studies significant differences in tissue antibody staining were reported. These discrepancies were probably the result of the different epitope recognition and specificity of the various antibodies used and the differences in methodology. This demonstrates the general difficulty in assessing glycolipid antigens; commonly used solvent fixation and washing steps may cause redistribution or extraction of glycolipids.

The current report describes a peroxidase-based method for the specific staining of Gb_3 within frozen, unfixed tissue sections using purified Shiga toxin 1. The technique as described is free from background or nonspecific staining and results in excellent maintenance of tissue morphology. Methods for the extraction of glycolipids from cells and tissues are also provided; Gb_3 is identified within extracts separated by thin-layer chromatography (TLC) by Shiga toxin binding in a TLC overlay assay. This technique allows the qualitative assessment of Gb_3 quantity and fatty acid isoform content. Shiga toxin binding to Gb_3 is well characterized and is a specific and high-affinity interaction. Although Stx1 binding to Gb_3 may be modulated by changes in fatty acid content *(22)*, all Gb_3 species are generally recognized. Therefore, Shiga toxin binding is a powerful diagnostic tool for the identification of Gb_3 within tissues and cells.

2. Materials

2.1. Immunohistochemical Detection of Toxin Binding in Tissue Sections

1. Tissue-Tek Cryomold (Miles Inc., Elkhart, IN).
2. Tissue-Tek, OCT Compound (Sakura Finetek, Torrance, CA).
3. Surgipath Snowcoat X-TRA slides (Surgipath, Winnipeg, Manitoba).
4. Endogenous peroxidase blocker—freshly prepared: 1 mM sodium azide, 10 mM glucose, 1 U/mL glucose oxidase (all from Sigma, St. Louis, MO).
5. Biotin blocker (kit from Vector Laboratories Inc., Burlingame, CA).
6. Normal goat serum: 1 % in phosphate-buffered saline (PBS), sterile filtered; used for blocking and as a diluent for all antibodies (Jackson Immunoresearch Laboratories, Westgrove, PA). Store at 4°C.
7. Purified Shiga toxin 1: Store aliquots at 0.2–2.0 mg/mL at –70°C indefinitely. Prepare 200-ng/mL working solution fresh (*see* Chapters 3–7).
8. Anti-Shiga toxin antibody; for example, 13C4 (anti-Stx1 B-subunit monoclonal antibody, ATCC no. CRL 1794).
9. Biotinylated goat anti-mouse IgG (Jackson Immunoresearch Laboratories).
10. ABC Elite Reagent (Vector).
11. Diaminobenzidine (tablets from Sigma).

12. Mayer's Hematoxylin, Lillies Modification (Dako, Carpintera, CA).
13. 70% Ethanol, 95% ethanol, 100% ethanol, xylene (Sigma) These solutions can be reused several times.
14. Permount (Sigma).
15. PBS, sterile filter (10X solution, 1 L): 80 g NaCl, 2 g KCl, 11.5 g Na_2HPO_4, 1.36 g KH_2PO_4, pH 7.2–7.4.

2.2. Extraction of Glycolipids from Cells and Tissues

1. Chloroform:methanol, 2:1 (v/v). (All manipulations involving chloroform should be performed in a well-ventilated fume hood using solvent-resistant labware) (preferably glass). Store in a glass stoppered bottle at room temperature.
2. PBS (*see* **Subheading 2.1.** above).
3. Silica Gel (BioSil A, 200–400 mesh; Bio-Rad, Hercules, CA, or equivalent).
4. Glass wool (Fisher Scientific, Nepean, Ontario).
5. Acetone:methanol, 9:1, store at room temperature in a solvent-resistant bottle.
6. Screw-capped glass tubes with solvent resistant teflon-lined caps (Fisher).
7. 1 *N* NaOH in methanol: Dissolve NaOH pellets in methanol, store at room temperature.
8. Blender or homogenizer.
9. Separatory funnel.
10. Rotary evaporator.
11. Heating block.

2.3. Toxin Receptor by TLC Overlay

1. TLC plates (Machery-Nagel, Sil G UV plastic-backed (cat. no. 805023) or aluminum-backed (cat. no. 81813) silica TLC plates. Store at room temperature (*see* **Note 1**).
2. Chloroform:methanol:water (65:35:8, v/v/v).
3. Gelatin (bovine) (Sigma): 1% (w/v) gelatin in water; prepare fresh by heating; cool to 37°C prior to use.
4. TBS: 50 m*M* Tris-HCl, 154 m*M* NaCl, pH 7.4.
5. Purified Shiga toxin: Store aliquots at 0.2–2.0 mg/mL at –70°C indefinitely. Prepare 0.1 µg/mL working solution fresh.
6. Anti-Shiga toxin antibody (e.g., 13C4). Store aliquots at –70°C indefinitely. Prepare dilute working solution fresh.
7. Secondary antibody, HRP conjugated goat anti-mouse IgG (H+L) (Bio-Rad Laboratories, Hercules, CA).
8. Developer: Prepare a 3-mg/mL stock of 4-chloro-1 naphthol (Sigma) in methanol and store at –20°C. Prepare fresh developer by combining 1 vol of 4-chloro-1-naphthol solution with 5 vol of TBS and 1 µL of 30% hydrogen peroxide (Sigma) per 2 mL of solution.
9. 0.5% Orcinol (Sigma) in 2 *M* H_2SO_4. Store at room temperature protected from light. All spraying should be performed in a well-ventilated fume hood with protective clothing.

10. Neutral glycolipid standards. Gb_3 and Gb_4 are isolated in our lab from human kidney; neutral glycolipids are also available from Calbiochem, Matreya, or Sigma. Store the glycolipid solutions (0.1–0.5 mg/mL in chloroform : methanol) in screw-capped glass tubes at –20°C.

3. Methods

3.1. Immunohistochemical Detection of Toxin Binding

1. Embed tissue in permount and snap-freeze in liquid nitrogen. Then, 6-μm frozen sections should be cut from the tissue block and placed onto Surgipath X-TRA microscope slides.
2. Dry frozen sections overnight. All subsequent steps should be performed in a humid chamber (e.g., container with a wet paper towel) (*see* **Note 2**). Reserve one slide for a negative control (no Shiga toxin 1).
3. Prepare fresh endogenous peroxidase blocker (1 mM sodium azide, 1 U/mL glucose oxidase, 10 mM glucose). Block sections with endogenous peroxidase blocker at 35°C for 1 h. Rinse gently with PBS.
4. Block with Vector avidin solution for 15 min at room temperature. Rinse with PBS.
5. Block with Vector biotin solution for 15 min at room temperature. Rinse with PBS.
6. Block with 1% NGS/PBS for 20 min.
7. Incubate with 50–100 ng/mL of Shiga toxin 1 diluted in 1% NGS/PBS for 30 min. Rinse with PBS.
8. Incubate with 10 μg/mL of anti-Stx1 antibody diluted into 1% NGS/PBS for 30 min. Rinse with PBS.
9. Incubate with 1/500 dilution of biotinylated goat anti-mouse IgG diluted into 1% NGS/PBS for 30 min at room temperature. Rinse with PBS. Make up Vector ABC elite reagent in preparation for next step.
10. Incubate with ABC elite reagent in 1% NGS/PBS for 30 min. Rinse with PBS (*see* **Note 3**).
11. Develop with DAB for approx 4 min. Stop the reaction by immersing in distilled water.
12. If sections are to be counterstained, immerse in filtered hematoxylin solution for 30 s (or less depending on the amount of counterstaining desired), wash off excess by immersing in distilled water, and then "blue" by immersing in tap water for a few minutes (*see* **Note 4**).
13. Sections are then dehydrated through a series of ethanols from 70% to 100%. 2X 3 min in each of 70%, 95%, 100% and finishing with xylene. Sections can then be mounted with Permount.

3.2. Extraction of Shiga toxin Receptors from Cells and Tissues

3.2.1. Extraction of Gb$_3$ from Cells (see **Note 5**)

1. Isolate cells from at least two confluent T75 flask (approx 10^7/flask) of Vero cells by washing with PBS, trypsinizing, and then centrifuging.
2. Resuspend the cells in 0.75 mL PBS. (Cells can be frozen indefinitely for later use at this point.)

3. To a screw-capped solvent-resistant (i.e., glass) tube, add 12 mL of chloroform:methanol (2:1, v/v).

4. Add the cells to the chloroform:methanol mixture. Rinse the cell tube with an additional 0.25 mL of PBS and add it to the chloroform:methanol. (It is important to keep track of all volumes for the Folch partitioning in **step 6**.)

5. Cap the tubes and shake vigorously. Place on a shaker for at least 1 h to overnight.

6. Adjust the volumes to form a Folch partition, 2:1:0.6 (v/v/v), chloroform:methanol:water. In this example, add 1.4 mL of PBS to give (8:4:2.4). Let the phases separate (*see* **Note 6**).

7. Remove and discard the upper aqueous phase. Filter the lower organic phase through a Pasteur pipet that has been plugged with glass wool (a small plug at the bottom of the Pasteur pipet). This removes any insoluble debris in the sample. Rinse the tube and glass wool with a small amount of 2:1, chloroform:methanol.

8. Dry down the lower phase on a heating block under a gentle stream of nitrogen (or air) (*see* **Note 7**).

9. Resuspend the dried extract in chloroform:methanol (2:1), to approx 10^6 cells/20 μL for direct analysis by TLC, or further purify the sample by saponification (**Subheading 3.3.**) and/or silica gel chromatography (**Subheading 3.4.**).

3.2.2. Extraction of Gb₃ from Tissues (see **Note 8**)

1. Weigh the tissue and then chop into small pieces. Homogenize tissue with a blender or other homogenizer in 0.5 vol of PBS.

2. Slowly add the homogenate to 20 vol of chloroform:methanol, 2:1 (v/v), while stirring.

3. Allow the tissue to extract for 5–24 h with stirring at room temperature.

4. Filter the extracted homogenate through Whatman filter paper in a Buchner funnel.

5. Adjust the volumes of the filtrate to form a Folch partition (2:1:0.6, chloroform:methanol:water). Let the phases separate.

6. Remove the lower phase and dry down. Suspend the sample in chloroform:methanol to give about 0.5 mg of tissue/μL. The extract may be further purified by saponification (**Subheading 3.3.**) and silica gel chromatography (**Subheading 3.4.**) (*see* **Note 9**).

3.3. Saponification

1. Dry down the lower phase of the Folch partition and resuspend it in 1 *M* NaOH (made up in methanol, 2–5 mg/mL). Saponifying 10^7 cells in 0.5–1 mL is usually adequate. Solubilize the sample by vortexing and, if possible, by sonicating briefly in a bath sonicator. The solution may appear cloudy. As long as the sample is in suspension, the reaction will proceed.

2. Incubate at room temperature for 3–24 h.

3. Neutralize by adding an equivalent volume of 1 *N* HCl. Check the pH of the solution by spotting 1–2 μL onto a piece of pH paper. The solution should be slightly basic; adjust the pH as necessary.

4. Folch partition the sample, taking into account the amount of methanol and water already in the sample. The ratio should be 2:1:0.6 (*see* **Note 10**).

5. When the sample has separated into two phases, mark the level of both the upper and lower phases. Remove the upper phase and discard. Replace the upper phase with an equal volume of theoretical upper phase. 1:47:48 (chloroform: methanol:water) (*see* **Note 11**). Mix and again let separate. This helps to remove the salt from the sample. Repeat.

6. Again, dry down the lower phase and resuspend in an appropriate volume of chloroform:methanol (2:1) (e.g., 10^6 cells/20 µL).

3.4. Silica Gel Chromatography for Isolation of Neutral Glycolipids

1. Evaporate the solvent from a portion of the cell or tissue extract and dissolve the residue in chloroform.

2. Prepare a small silica column as follows: Make a glass wool plug in a small Pasteur pipet using a long Pasteur pipet to pack the glass wool tightly. Add 150 mg of activated BioSil A 200–400 mesh (Bio-Rad) to the plugged pipet per 50 mg of starting tissue or up to 10^8 cells (*see* **Note 12**). Add chloroform to the silica and suspend with the long pipet until no air bubbles remain. Wash the column briefly with chloroform; the column will not dry out when the chloroform drains to the top of the bed.

3. Place the column in a test tube. Add the sample to the column and let it drain in. Wash the sample tube with chloroform and add to the column.

4. Wash the column with 4 mL chloroform to elute nonpolar components.

5. Transfer the column to a fresh glass tube and elute the neutral glycolipids with 10 mL acetone:methanol 9:1. Evaporate the solvent and resuspend the residue in the original volume. A further elution with methanol will remove the remaining lipids from the column.

3.5 Detection of Toxin Receptor by TLC Overlay

1. Prepare TLC plates in duplicate: Plastic-backed TLC plates are cut to an appropriate size (e.g., 8 cm^2). Lipids are loaded in lanes that are marked with a pencil 1 cm from the bottom of the plate. Lanes should be separated by 0.5 cm and placed at least 1 cm from the edges of the plate.

2. Lipids are loaded using a Hamilton syringe. The use of a blow dryer speeds up the loading and helps to create a tighter band. Most cell lines require that at least 10^6 cells be run per lane. For tissue extracts, load between 5 and 20 mg equivalent of tissue. One lane should include standard lipids (approx 1 µg each).

3. Place a piece of filter paper in the solvent tank. Pour solvent (65:35:8 (chloroform:methanol:water, v:v:v) into the tank to a depth of less than 1 cm, making sure that the wick is also wet. Place the TLC plate in the tank and let the solvent front run up to the top of the plate. Remove from the tank and let air-dry.

4. One plate should be sprayed with orcinol (chemical detection of glycolipids); 0.5% orcinol in 3 M H_2SO_4, and heated at 110°C for 2–10 min (*see* **Note 13**). The other plate is placed into a blocking solution of 1% gelatin for 1–18 h at 37°C (*see* **Note 14**).

5. Rinse plate with TBS.

6. Incubate with 0.1 µg/mL of Stx1 (in TBS) for 1 h at room temperature. Wash with three rinses of TBS for 5 min each.
7. Incubate with primary antibody (13C4) at 1 µg/mL in TBS. Wash with 3X 5 min rinses with TBS.
8. Incubate with secondary antibody; 1/2000 in TBS. Wash with three rinses of TBS for 5 min each.
9. Incubate with 4-chloro-1-naphthol developer until bands reach sufficient intensity.
10. Rinse with water to stop color development before air-drying the plate (*see* **Fig. 1**).

4. Notes

1. The type of TLC plate used is important. Many brands of silica plates, although giving good separation of neutral glycolipids, do not stand up well to incubation and washing steps during the TLC overlay assay. The use of a plasticizing agent, polyisobutylmethacrylate (PIBM), is widely used to overcome this problem, however we have found that treatment of plates with PIBM results in nonphysiological binding of Stx1 to Gb_4 *(23)*. Our laboratory has found that Machery–Nagel plastic or aluminum-backed Silica Gel G plates give excellent resolution and remain intact throughout the overlay procedure without the need for PIBM treatment. We currently are using plates embedded with ultraviolet (UV) indicator; the binding agent in plates without indicator seems to be incompatible with methanol-containing solvent systems and the plates bubble while running. These plates can be run in an alternate solvent system; hexane:isopropanol:water, 10:10:1, which gives adequate resolution for TLC overlay assays.
2. Some tissues have high levels of endogenous peroxidase, making a negative control very important. The condition of the section is also critical for proper detection of Gb_3 in the sections. Tissues that have been poorly frozen show higher background binding and are more likely to come off during the extensive washing procedures. Tissues embedded in an embedding media, such as Tissue-Tek OCT compound, seem to stay on the slide better. If possible, include a positive control, such as a section of human kidney, which is abundant in Gb_3.

   Antibodies and other reagents are expensive. To minimize the amount of reagents used, a line can be drawn around the sections, with a wax pencil, to contain antibodies. A PAP pen, which provides a hydrophobic barrier, is available from DAKO.
3. This procedure gives enhanced detection. For a simpler though less sensitive method, remove the avidin/biotin blocking step and the biotinylated antibody step (i.e., omit **steps 4, 5**, and **9**). After **step 8**, simply use an HRP-linked secondary antibody (HRP goat anti-mouse) and develop. The amounts of Shiga toxin and subsequent antibody concentrations should be optimized for each batch. These dilutions should be used as a starting point.
4. Counterstaining with hematoxylin sometimes obscures detection of smaller amounts of Gb_3. Decreasing the counterstaining time to a minimum is one option. Omitting the counterstain allows best visualization of detected Gb_3, although correlation with tissue ultrastructure becomes much more difficult (**Fig. 2**).

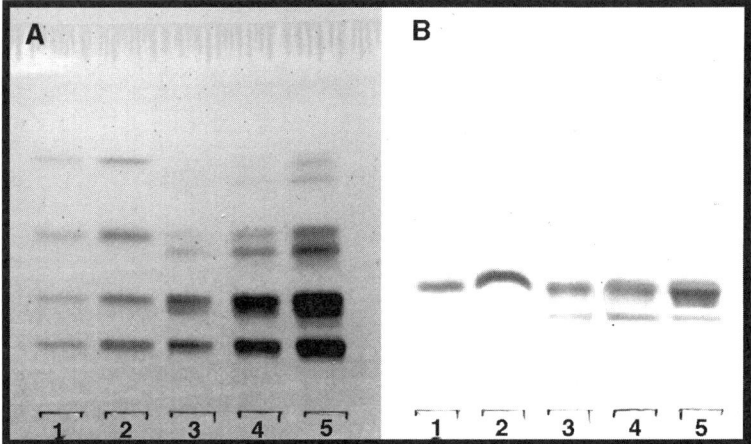

Fig. 1. Stx1 TLC overlay to detect Gb₃ in human kidney (**A**) Orcinol staining; (**B**) Stx1 overlay. Lanes 1 and 2: 0.5 and 1.0 µg of each glycolipid standard—from the top: galactosyl ceramide, lactosyl ceramide, Gb₃ and Gb₄. Lanes 3–5, doubling concentrations of acetone/methanol fraction from saponified human renal extract.

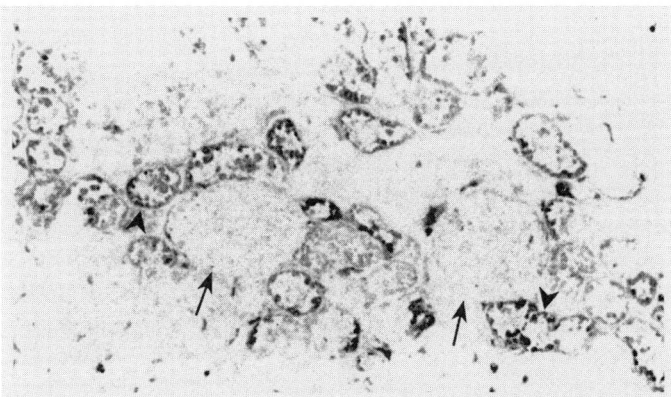

Fig. 2. Immunoperoxidase Stx1 staining of adult human kidney. Some proximal and distal tubules are stained by Stx (arrowheads), but glomeruli (arrowed) are unlabeled. Section has not been counterstained but imaged under DIC to permit black-and-white representation. (From **ref. 6**.)

5. The extraction protocols used in this method consists of resuspending cells/homogenizing tissues and subsequently extracting them in chloroform:methanol *(24)*. The extraction mixture is adjusted to a Folch partition and the lower neutral glycolipid containing phase is dried down and resuspended in a smaller more suitable volume. Gb₃ is found in the lower phase of the Folch partition with other

neutral glycolipids and phospholipids. As a general guide, a proper Folch partition will consist of approximately a two-thirds lower phase and a one-third upper phase.

6. Separation into lower and upper phases can take several hours. It is usually convenient to leave the sample to separate overnight. The separation of the phases can, however, be speeded up by centrifuging the sample at approx 3000 rpm for 15 min, or more if necessary. If the phases do not separate well in a reasonable time (i.e., there is a significant amount of white interface), it is best to increase the total volume of the Folch partition, mix well, and repeat the separation.

7. Drying down of the lower phase is best accomplished by placing the tube in a warm heating block under the a gentle stream of nitrogen (or air). Chloroform: methanol dries quickly; however, the lower phase of the Folch partition also contains a small amount of water, which will take a little longer to dry. Make sure that the sample is completely dry before continuing. The resuspension volume is based on Vero cells, which have a significant amount of Gb_3. The number of cells loaded can be adjusted as needed. It may be necessary to increase/decrease the resuspension volume. The only factor limiting the dilution is the solubility and the volume of sample that will need to be loaded on the TLC plate. Keeping the volumes small will make the loading much less tedious.

8. Extraction of tissues follows the same basic protocol as extraction of cells, but using larger volumes. When extracting larger amounts of tissue, 10 volumes of chloroform : methanol is usually sufficient to extract the Gb_3-containing neutral glycolipids from a well-homogenized tissue sample. However, if this is found to be insufficient, the tissue can be re-extracted with another similar volume of 2:1. Working with larger volumes of chloroform requires appropriate laboratory equipment. Large extraction funnels are recommended for these separations. Drying down of larger volumes of organic solvents is best accomplished using a rotary evaporator.

9. Some tissue (and cell) extracts contain significant amounts of cholesterol and phospholipid, which may affect Gb_3 migration on TLC or interfere with Gb_3 detection by TLC overlay. Adjusting the composition of the solvent in the running of the plate sometimes helps to separate the Gb_3 from the phospholipids. A more polar solvent, 60:40:9, or a less polar solvent, 65:25:4, sometimes separates the Gb_3 from the interfering phospholipids. It is, however, usually necessary to eliminate these interfering phospholipids by one of two methods; saponification (treatment with base) (*see* **Subheading 3.3.**) or silicic acid column chromatography (*see* **Subheading 3.4.**) Saponification removes base labile phospholipids from the sample. Some tissue extracts contain large amounts of nonpolar lipids, which makes saponification difficult. For these samples, it is best to use silicic acid chromatography. Silicic acid chromatography separates the lipids into three fractions: nonpolar lipids (i.e. cholesterol and fatty acids) eluted with chloroform, neutral glycolipids (including Gb_3) eluted with acetone:methanol (9:1), and long-chain glycolipids and phospholipids with methanol.

10. A Folch partition of 2:1:0.6, chloroform:methanol:water forms two phases. The theoretical upper phase consists of 3:48:47, chloroform:methanol:water and the

lower phase consists of 86:14:1, chloroform:methanol:water. If one phase is removed and replaced with an equal volume of the same theoretical phase, the proportions of the Folch partition are maintained. Therefore, the lower phase can be washed with the theoretical upper phase to remove the excess salts.

11. An alternative to Folch partitioning the neutralized sample is to dilute the sample with water and apply it to a C_{18} solid-phase extraction cartridge (i.e., SepPak, from Millipore-Waters) that has been preconditioned with methanol and equilibrated with water. The neutralized diluted sample is applied to the column under gravity, washed with approx 20 mL of water using a syringe as per the manufacturer's suggested protocol. The desalted neutral glycolipids are eluted with 20–30 mL of methanol. The sample is dried down and then resuspended in chloroform:methanol (2:1). A rotary evaporator is handy for drying down this larger volume of solvent.

12. Column size can be scaled up as necessary. The silica column can be poured in a larger glass chromatography column. However, most standard chromatography columns with polypropylene fittings can be used for brief periods of time with organic solvents. Do not store silica in these columns for extended periods of time. The organic solvents will eventually dissolve the adhesives that hold the column together.

13. Specific chemical detection of glycolipids with orcinol is preferred. Upon heating, glycolipids turn purple on a yellow background. Overdevelopment of the TLC plate will result in appearance of nonspecific bands. As an alternative, the TLC plate can also be put into a jar with iodine crystals. The iodine binds reversibly to unsaturated lipids, including fatty acids present in most glycolipids and phospholipids. This allows a general visualization of the lipid species in the extracted sample.

14. When the plates have been run in solvent and dried, it is best to prewet the plate. Immerse one edge of the plate in the overlay dish and let the water soak up slowly. This removes air bubbles from the silica and helps to decrease background binding of antibodies. This prewetting is especially important if using the aluminum-backed plates. Once the plate is wet, immerse it in the water, and then change to the gelatin-blocking solution.

References

1. Naiki, M. and Marcus, D. M. (1974) Human erythrocyte P and P^k blood group antigens: identification as glycosphingolipids. *Biochem. Biophys. Res. Commun.* **60,** 1105–1111.

2. Mangeney, M., Richard, Y., Coulaud, D., Tursz, T., and Wiels, J. (1991) CD77: an antigen of germinal center B cells entering apoptosis. *Eur. J. Immunol.* **21,** 1131–1140.

3. Boyd, B., Tyrrell, G., Maloney, M., Gyles, C., Brunton, J., and Lingwood, C. (1993) Alteration of the glycolipid binding specificity of the pig edema toxin from globotetraosyl to globotriaosyl ceramide alters in vivo tissue targetting and results in a Stx1-like disease in pigs. *J. Exp. Med.* **177,** 1745–1753.

4. Hertzke, D. M., Cowan, L. A., Schoning, P., and Fenwick, B. W. (1995) Glomerular ultrastructural lesions of idiopathic cutaneous and renal glomerular vasculopathy of greyhounds. *Vet. Pathol.* **32,** 451–459.
5. Zoja, J., Corna, D., Farina, C., Sacchi, G., Lingwood, C., Doyles, M., et al. (1992) Verotoxin glycolipid receptors determine the localization of microangiopathic process in rabbits challenged with verotoxin-1. *J. Lab. Clin. Med.* **120,** 229–238.
6. Lingwood, C. A. (1994) Verotoxin-binding in human renal sections. *Nephron* **66,** 21–28.
7. Katagiri, Y., Mori, T., Nakajima, H., Katagiri, C., Taguchi, T., Takeda, T., et al. (1999) Activation of Src family kinase induced by Shiga toxin binding to globotriaosyl ceramide (Gb$_3$/CD77) in low density, detergent-insoluble microdomains. *J. Biol. Chem.* **274,** 35,278–35,282.
8. Cohen, A., Hannigan, G. E., Williams, B. R. G., and Lingwood, C. A. (1987) Roles of globotriosyl- and galabiosylceramide in verotoxin binding and high affinity interferon receptor. *J. Biol. Chem.* **262,** 17,088–17,099.
9. Ghislain, J., Lingwood, C. A., and Fish, E. N. (1994) Evidence for glycosphingolipid modification of the type 1 IFN receptor. *J. Immunol.* **153,** 3655–3663.
10. Khine, A. A. and Lingwood, C. A. (2000) Functional significance of globotriaosylceramide in α_2 interferon/ type I interferon receptor mediated antiviral activity. *J. Cell. Physiol.* **182,** 97–108.
11. van Setten, P., Monnens, L., Verstraten, R., van der Heuvel, L. and van Hinsberg, V. (1996) Effects of verotoxin-1 on non adherent human monocytes: binding characteristics, protein synthesis, and induction of cytokine release. *Blood* **88,** 174–183.
12. Ramegowda, B. and Tesh, V. L. (1996) Differentiation-associated toxin receptor modulation, cytokine production, and sensitivity to Shiga-like toxins in human monocytes and monocytic cell lines. *Infect. Immun.* **64,** 1173–1180.
13. Sakiri, R., Ramegowa, B., and Tesh, V. L. (1998) Shiga toxin type 1 activates tumor necrosis factor-α gene transcription and nuclear translocation of the transcriptional activators nuclear factor-kB and activator protein-1. *Blood* **92,** 558–566.
14. Arab, S., Russel, E., Chapman, W., Rosen, B. and Lingwood, C. (1997) Expression of the Verotoxin receptor glycolipid, globotriaosylceramide, in ovarian hyperplasias. *Oncol. Res.* **9,** 553–563.
15. LaCasse, E. C., Bray, M. R., Patterson, B., Lim, W.-M., Perampalam, S., Radvanyi, L. G., et al. (1999) Shiga-like toxin I receptor on human breast cancer, lymphoma, and myeloma and absence from CD34+ hematopoietic stem cells: implications for ex vivo tumor purging and autologous stem cell transplantation. *Blood* **94,** 1–12.
16. Keusch, J., Manzella, S. M., Nyame, K. A., Cummings, R. D. and Baenziger, J. U. (2000) Cloning of Gb$_3$ synthase, the key enzyme in globo-series glycosphingolipid synthesis, predicts a family of α1,4 glycosyltransferases conserved in plants, insects and mammals. *J. Biol. Chem.* **275,** 25,315–25,321.
17. Sandvig, K., Garred, Ø., Prydz, K., Kozlov, J., Hansen, S., and van Deurs, B. (1992) Retrograde transport of endocytosed Shiga toxin to the endoplasmic reticulum. *Nature* **358,** 510–512.

18. Arab, S. and Lingwood, C. (1998) Intracellular targeting of the endoplasmic reticulum/nuclear envelope by retrograde transport may determine cell hypersensitivity to Verotoxin: sodium butyrate or selection of drug resistance may induce nuclear toxin targeting via globotriosyl ceramide fatty acid isoform traffic. *J. Cell. Physiol.* **177,** 646–660.

19. Kasai, K., Galton, J., Terasaki, P., Wakisaka, A., Kawahara, M., Root, T., et al. (1985) Tissue distribution of the Pk antigen as determined by a monoclonal antibody. *J. Immunogenet.* **12,** 213–220.

20. Pallesen, G. and Zeuthen, J. (1987) Distribution of the Burkitt's-lymphoma-associated antigen (BLA) in normal human tissue and malignant lymphoma as defined by immunohistological staining with monoclonal antibody 38.13. *J. Cancer Res. Clin. Oncol.* **113,** 78–86.

21. Oosterwijk, E., Kalisiak, A., Wakka, J., Scheinberg, D., and Old, L. J. (1991) Monoclonal antibodies against $Gal\alpha1$-4$Gal\beta1$-4Glc (P^k, Cd77) produced with a synthetic glycoconjugate as immunogen: reactivity with carbohydrates, with fresh frozen human tissues and hematopoietic tumors. *Int. J. Cancer* **48,** 848–854.

22. Kiarash, A., Boyd, B., and Lingwood, C. A. (1994) Glycosphingolipid receptor function is modified by fatty acid content: Verotoxin 1 and Verotoxin 2c preferentially recognize different globotriaosyl ceramide fatty acid homologues. *J. Biol. Chem.* **269,** 11,138–11,146.

23. Yiu, S. C. K. and Lingwood, C. A. (1992) Polyisobutylmethacrylate modifies glycolipid binding specificity of verotoxin 1 in thin layer chromatogram overlay procedures. *Anal. Biochem.* **202,** 188–192.

24. Folch, J., Lees, M., and Sloane-Stanley, G. H. (1957) A simple method for the isolation and purification of total lipids from animal tissues. *J. Biol. Chem.* **226,** 497–521.

17

Shiga Toxin B-Subunit as a Tool to Study Retrograde Transport

Frédéric Mallard and Ludger Johannes

1. Introduction

Shiga toxin, produced by *Shigella dysenteriae* and Shiga toxin-producing *Escherichia coli*, is a bacterial protein toxin composed of two subunits, A and B *(1)*. The A-subunit is the actual toxin whose ribosomal RNA *N*-glycosidase activity leads to the inhibition of protein biosynthesis in target cells. For binding to the cellular Shiga toxin receptor, the glycosphingolipid globotriaosylceramide (Gb$_3$ or CD77), the A-subunit depends on its noncovalent interaction with the B-subunit, termed here StxB, composed of a homopentamer of B-fragments. In a number of cell types, Shiga toxin is transported from the plasma membrane to the endoplasmic reticulum (ER), via early endosomes and the Golgi apparatus *(1–3)*, bypassing the late endocytic pathway *(4)*. It is generally admitted that this toxin, like other members of the Shiga family *(5)*, translocates from the lumen of the compartments to the cytosol at the level of the ER *(6)*. Apart from its importance for the entry of Shiga toxin into target cells, the study of the retrograde transport route also plays a fundamental role in our understanding of intracellular membrane homeostasis and the steady-state localization of cellular proteins. The fine dissection of the molecular mechanisms underlying the fundamental steps of the retrograde route requires experimental systems that allow the quantitative analysis of these steps. In this chapter, we summarize what we call the Shiga toolbox: mutant StxB and experimental setups that allow measuring endocytosis, arrival in the trans-Golgi network (TGN), and the ER. The basic element in the Shiga toolbox is StxB, which conserves the intracellular transport characteristics of the holotoxin, but which, in most cell types, does not have adverse effects.

From: *Methods in Molecular Medicine, vol. 73: E. coli: Shiga Toxin Methods and Protocols*
Edited by: D. Philpott and F. Ebel © Humana Press Inc., Totowa, NJ

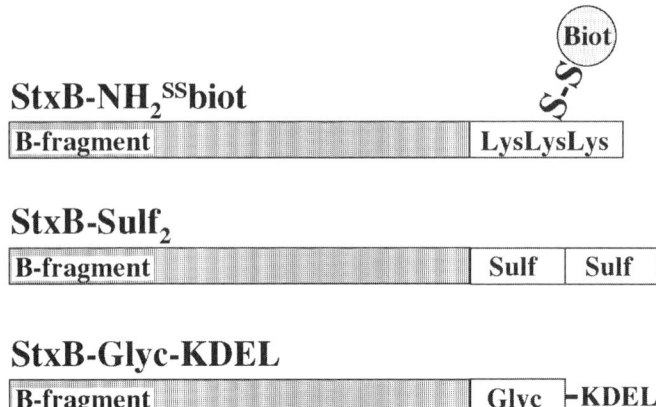

Fig. 1. StxB mutants, modified by C-terminally added Lys and biotinylated (StxB-NH$_2$^{SS}biot), by sulfation sites (StxB-Sulf$_2$), or by a glycosylation site and KDEL peptide (StxB-Glyc-KDEL).

The study of Shiga toxin endocytosis has been hampered by the fact that classical treatments to remove ligands from their plasma membrane-exposed receptors fail to do so in the case of StxB-Gb$_3$ interaction. In fact, StxB is very resistant to proteolytic cleavage, and washes at low pH do not dissociate it from its receptor. We have therefore constructed a StxB mutant, termed StxB-NH$_2$ (*see* **Fig. 1**), which is modified by three C-terminally added Lys. This mutant can be biotinylated, preserving its full activity, as opposed to wild-type StxB, which is rendered inactive by biotinylation (*7*). The experimental principle is the following: StxB-NH$_2$ is biotinylated by NHS-SS-biotin, introducing the biotin group, which is linked to StxB via a cleavable disulfide bond. StxB-NH$_2$^{SSbiot} is bound on ice to target cells, such as HeLa cells, a condition under which the protein interacts with its receptor without being internalized. After washing, the cells are shifted for variable times to 37°C, allowing for internalization to occur, replaced on ice, and biotin is then removed from cell surface exposed StxB-NH$_2$^{SSbiot} by treatment with mercaptoethanesulfonic acid (MESNA), a membrane, impermeable reducing agent. Control cells are mock treated to determine the total amount of cell associated StxB-NH$_2$^{SSbiot}. By Western blotting for biotin using alkaline phosphatase-coupled streptavidin (Stv-AP) on lysates of MESNA- and mock-treated cells, the fraction of internalized StxB can then be determined (*see* **Fig. 2**).

After internalization into early endosomes (EE), StxB is transported to the TGN/Golgi apparatus, bypassing late endosomes (*4*). Experimentally, StxB can be accumulated in EE by low-temperature incubation (19.5°C), a condition under which transport to the TGN is inhibited. Arrival in the TGN after the shift to 37°C can then be measured by following the sulfation of a mutant StxB,

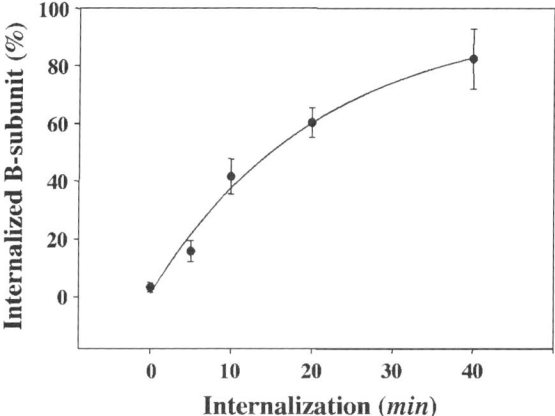

Fig. 2. Internalization kinetics of StxB-NH$_2$^{SSbiot} into HeLa cells. The half time of internalization is 15 min, and a plateau of 96% of internalized protein is reached, indicating that StxB enters cells rapidly and efficiently.

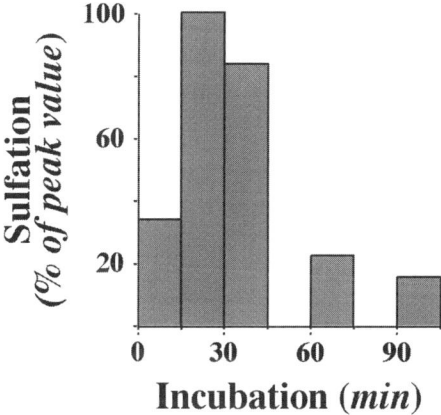

Fig. 3. Kinetics of StxB-Sulf$_2$ transport from EE to the TGN of HeLa cells. After StxB-Sulf$_2$ internalization into EE by low-temperature incubation, the cells were shifted to 37°C for the indicated times. Fifteen minutes before the end of each incubation period, radioactive sulfate was added for detection of passage through the TGN. Maximal passage was observed between 15 and 45 min. Thereafter, a new plateau was reached representing the steady-state presence of StxB-Sulf$_2$ in the TGN. (From **ref. 4**.)

termed StxB-Sulf$_2$ (*see* **Fig. 1**). In fact, sulfation is a TGN-specific posttranslational protein modification, catalyzed by sulfotransferase on a specific recognition sequence. StxB-Sulf$_2$ carries a tandem of such recognition sequences at its C-terminus. Bacterial purified protein is not sulfated because of the lack of sulfotransferase activity in bacteria. However, when purified StxB-Sulf$_2$ is

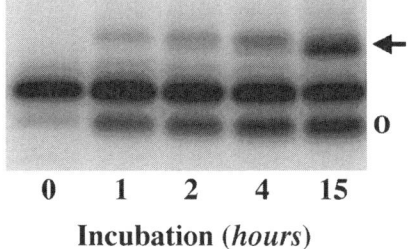

Incubation (*hours*)

Fig. 4. Kinetics of StxB-Glyc-KDEL glycosylation in the ER. Iodinated StxB-Glyc-KDEL was incubated with HeLa cells for the indicated times. The glycosylation product (arrow) and a proteolytical cleavage product (circle) are indicated. (From **ref. 8**.)

incubated with target cells, such as HeLa cells, in the presence of radioactive sulfate, the protein gets efficiently modified (*see* **Fig. 3**). At the end of a sulfation pulse, the target cells are lysed, StxB-Sulf$_2$ is immunoprecipitated, and sulfated protein is quantified after autoradiography using the PhosphorImager. The sulfation on endogenous proteins and proteoglycans is determined on the same lysates to control for sulfotransferase activity and cell number. Sulfation reaches a plateau value between 15 and 45 min shift to 37°C (*see* **Fig. 3**), indicating a maximum passage through the TGN during this time interval (*4*).

Finally, arrival of StxB in the ER can be followed quantitatively by biochemical means using the StxB mutant StxB-Glyc-KDEL (*see* **Fig. 1**) (*8*). This mutant carries a recognition sequence by oligosaccharyl-transferase (N-glycosylation site) and an ectopic retrieval signal at its C-terminus. Bacterially produced StxB-Glyc-KDEL is not glycosylated. The purified protein is iodinated in vitro, incubated with target cells at 37°C, and at the end of an incubation period, the target cells are lysed, and lysates are analyzed by sodium dodecyl sulfate–polyacrylamide gel electrophoresis (SDS-PAGE) on gradient gels that allow one to separate glycosylated from nonmodified StxB-Glyc-KDEL (*see* **Fig. 4**; the glycosylation product is indicated by an arrow). The percentage of glycosylated StxB-Glyc-KDEL is determined by the PhosphorImager and taken as a measure for the arrival of the protein in the ER. Note that the glycosylation of StxB-Glyc-KDEL, which increases linearly with time, is slower than the protein's arrival in the ER, as determined by morphological means (*8*), indicating that glycosylation on the folded StxB-Glyc-KDEL is a slow process. However, differences in glycosylation between incubation conditions (e.g., cells overexpressing different mutant proteins) remain equally constant over time, allowing one to deduce information on the efficiency of retrograde transport under the respective incubation conditions from the comparison of the glycosylation signals obtained after equivalent incubation times.

The presence of the ectopic retrieval signal, KDEL, is not necessary for retrograde transport as such, but it facilitates analysis of the glycosylation signal by retaining the protein in the ER, once having arrived in this compartment *(8)*.

2. Materials

2.1. Protein Purification

1. Six incubation bottles with 125 mL LB, six others with 375 mL.
2. Stock solution with 50 mg/mL ampicillin.
3. 10 m*M* Tris-HCl, pH 8.0; TSE buffer: 10 m*M* Tris-HCl, pH 8.0, 25% sucrose, 1 m*M* EDTA.
4. 1000X Protease inhibitor cocktail (1000X PIC): 1 mg/mL leupeptin, 1 mg/mL pepstatin, 1 mg/mL aprotinin, 1 mg/mL antipain, 1 M benzamidine, 40 mg/mL phenylmethylsulfonyl fluoride (PMSF).
5. Anion-exchange columns (QFF, MonoQ Amersham Pharmacia Biotech).
6. Bacterial incubator.
7. Rotary shaker.
8. Centrifuge.

2.2. Biotinylation

1. NHS-SS-biotin (Pierce), stock solution at 1 mg/mL in dimethyl sulfoxide (DMSO).
2. 1 *M* Carbonate buffer, pH 9.5.
3. Dialysis cassettes, molecular-weight cutoff (MWCO): 10 kDa (Pierce).
4. Nanosep 10-kDa microconcentration chambers for 0.1–0.5 mL (PallFiltron).

2.3. Internalization Assay

1. DMEM^{3+}: Dalbecoo's modified Eagle's medium (DMEM) (Gibco-BRL) supplemented with 10% SVF, 0,01% penicillin/streptomycin (Gibco-BRL), and 4 m*M* Gln (Gibco-BRL).
2. Buffers: 1 *M* Tris-HCl (pH 8.6); PBS^{++} (1X phosphate-buffered saline [PBS], 0.5 m*M* CaCl$_2$, 1 m*M* MgCl$_2$); 50 m*M* Tris-HCl (pH 8.0).
3. TNB buffer: 50 m*M* Tris-HCl, pH 8.6, 100 m*M* NaCl, 0.2% bovine serum albumin (BSA).
4. MESNA buffer: add 100 m*M* MESNA to TNB buffer.
5. RIPA: 1X PBS, 1% Nonidet P40 (NP-40), 0.5% deoxycholic acid, 0.5% SDS.
6. Other solutions: 5 *M* NaCl, 10% BSA, 100% methanol.
7. 2-Mercaptoethanesulfonic acid (MESNA, Sigma).
8. PIC (*see* **Subheading 2.1.4.**).
9. Antibodies: 13C4 mouse monoclonal anti-StxB from ATCC (cat. no. CRL 1794) at 1 mg/mL.
10. 3X SDS Sample Buffer without reducing agent (Bio-Rad).
11. Protein G–Sepharose beads 4 Fast Flow (Amersham Pharmacia Biotech) equilibrated in RIPA buffer.

12. Variation of Tris-tricine gels *(9)* (*see* **Note 1**): Anode buffer (0.2 M Tris-HCl, pH 8.9); cathode buffer (0.1 M Tris-HCl, pH 8.25, 0.1 M Tricine, 0.1% SDS); gel buffer (3 M Tris-HCl, pH 8.45, 0.3% SDS); polyacrylamide solution A (standard 30%/0.8% acrylamide/bisacrylamide solution plus 1.135% [w/v] of bisacrylamide); polyacrylamide solution B (standard 30%/0.8% acrylamide/bisacrylamide solution plus 0.138% (w/v) of bisacrylamide); stacking gel (for 12.5 mL: 1 mL polyacrylamide solution B, 3.1 mL gel buffer, 8.4 mL water, 10 µL TEMED (*N,N,N',N'*-tetramethylethylendiamine), 100 µL 10% ammonium persulfate [APS]); resolving gel (for 20 mL: 6.66 mL polyacrylamide solution A, 6.66 mL gel buffer, 2.66 g glycerol, 4.54 mL water, 6.6 µL TEMED, 66 µL of 10% APS).
13. Protein markers (Bio-Rad).
14. Whatman 3M paper, Immobilon P-PVDF paper (Millipore).
15. Transfer buffer: 3.03 g Tris-HCl, 1.44 g glycine, 200 mL ethanol, in 1 L of water.
16. Blotting buffer: 5% milk powder, 50 mM Tris-HCl, pH 8.0, 150 mM NaCl, 0.1% Tween-20.
17. Stv-AP (Jackson Laboratories).
18. Enhanced chemifluorescence (ECF) substrate (Amersham Pharmacia Biotech).
19. 24-well tissue culture test dishes, γ-sterilized, tissue culture-treated.
20. Materials: 37°C and 5% CO_2 incubator, 37°C water bath, Microfuge centrifuge (Beckman), microsyringe (Hamilton), Thermomixer (Eppendorf), vortex, electrophoresis power supply, semidry transfer apparatus (Bio-Rad), PhosphoImager (Molecular Dynamics), two transparency folders.

2.4. Sulfation Assay

1. DMEM[3+], RIPA buffer, Protein G–Sepharose beads 4 Fast Flow equilibrated in RIPA buffer, 13C4: monoclonal anti-StxB 13C4 (*see* **Subheading 2.3.**, **item 9**), 1000X Protease inhibitor cocktail, tissue culture test dishes (*see* **Subheading 2.3.**).
2. HBSS: 1X HBSS w/o calcium and magnesium (Gibco-BRL), 1 mM $MgCl_2$, 0.5 mM $CaCl_2$, 0.37% $NaHCO_3$ (Gibco-BRL), 1 g/L glucose (Gibco-BRL), pH 7.4.
3. MEM without sulfate: amino acids for RPMI 1640 (Gibco-BRL) (200 µg/mL L-Arg, 50 µg/mL L-Lys, 10 µg/mL L-Gly, 15 µg/mL L-His, 50 µg/mL L-Ile, 50 µg/mL L-Leu, 20 µg/mL L-Thr, 30 µg/mL L-Ser, 5 µg/mL L-Trp, 20 µg/mL L-Val, 15 µg/mL L-Phe; 5 µg/mL L-Cys, 1.5 µg/mL L-Met, 5 µg/mL L-Tyr, 2 mM Gln), 50 µg/L $Fe(NO_3)_2$, 6.7 mM NaH_2PO_4, 6.7 mM KCl, 1 mM $MgCl_2$, 0.5 mM $CaCl_2$, 11 mM NaCl, 0.37% $NaHCO_3$ (Gibco-BRL), 1 g/L glucose (Gibco-BRL), 1.1 g/L sodium pyruvate, 3.2 × 10^{-3}% Phenol Red, 4X vitamin mix for MEM (Gibco-BRL), pH 7.4.
4. Radioactive sulfate ($^{35}SO_4^{2-}$) (Amersham Pharmacia Biotech, cat. no. SJS-1). Radioactive concentration should not be lower than 30 mCi/mL.
5. 50 mM Tris-HCl (pH 8.0).
6. Microbeta Plus counter (Optiphase Supermix, cat. no. 1200-439).
7. 4 mL Plastic flasks (EG&G Wallac, cat. no. 3011-000).

2.5. Iodination

1. PD-10 columns (Amersham Pharmacia Biotech).
2. Iodination buffer: 10 mM NaCl, 100 mM Tris-HCl, pH 6.8.

3. Bovine serum albumin: 10 mg/mL stock solution stored at –20°C.
4. IodoBeads (Pierce).
5. Iodine ^{125}I (Amersham Pharmacia Biotech, catalog no. JMS-30, 5 mCi).

2.6. Glycosylation Assay

1. Tissue culture test dishes and DMEM^{3+} as in **Subheading 2.3.**
2. 10%/20% Acrylamide gradient gels.

3. Methods
3.1. Protein Purification

1. Inoculate six bottles with 125 mL LB/Amp each with 125 μL of an overnight culture grown at 30°C [StxB expression from pSU108 carrying the StxB coding region under control of a heat-inducible promotor *(10)*].
2. Grow overnight at 30°C.
3. Transfer into six bottles with 375 mL LB/Amp at 50°C; incubate 4 h at 42°C.
4. Centrifuge (6000*g*, 15 min, room temperature) to pellet cells.
5. Wash cells three times with 10 m*M* Tris-HCl, pH 8.0.
6. Resuspend cells in 200 mL TSE buffer; incubate at room temperature for 10 min; centrifuge to pellet cells.
7. Resuspend cells in 200 mL of ice-cold water containing 1X PIC; incubate on ice for 10 min.
8. Centrifuge at 4°C; collect supernatant; add 20 m*M* Tris-HCl, pH 8.0; snap-freeze in liquid nitrogen and store at –80°C.
9. The StxB has a rather acidic pK_i (for wild type StxB, the pK_i is 5.2). The protein can therefore easily be immobilized on anion exchangers, such as QFF and MonoQ. Wild-type StxB elutes from a MonoQ column at about 120–150 m*M* NaCl (*see* **Note 2**). To improve the purity of the preparation, protein may be passed through a gel filtration column (*see* **Note 3**).

3.2. Biotinylation

1. Dialyze StxB-NH$_2$ at 1 mg/mL against PBS (*see* **Notes 4–6**).
2. To 0.45 mL of dialyzed protein on ice, add 15 μL of 1 *M* carbonate buffer, pH 9.5, and 50 μL of NHS-SS-biotin at 1 mg/mL (*see* **Note 7**).
3. Incubate for 30 min on ice.
4. Add 50 m*M* NH$_4$Cl and dialyze against PBS to eliminate excess NHS-SS-biotin

3.3. Internalization Assay

1. About 12–15 h before the experiment, plate HeLa cells (*see* **Note 8**) in five 24-well dishes at 1×10^5 cells/well at about 80–90% confluence; for each condition (protein overexpression, drug treatment, etc.), 2 wells are needed (± MESNA treatment); each dish will be used for a single time-point corresponding to 0 min, 5 min, 10 min, 20 min, and 40 min (*see* **Fig. 1**).
2. Prepare the following solutions per well: StxB-NH$_2$^{SSbiot} at 1 μ*M* for binding in 0.25 mL DMEM^{3+} on ice, 2 mL DMEM^{3+} for washes on ice, 0.5 mL DMEM^{3+} at 37°C for internalization, 1.5 mL PBS^{++}; and 0.5 mL RIPA buffer.

3. Place cells on ice and wash once with ice cold DMEM^{3+}; label dishes corresponding to time-points.

4. Add StxB-NH$_2$^{SSbiot} in DMEM^{3+} to the cells; incubate on ice for 30 min.

5. Prepare TNB buffer (3 mL per couple of wells); MESNA buffer (1 mL per couple of wells) (*see* **Note 9**).

6. Wash cells three times with 0.5 mL DMEM^{3+} on ice.

7. Remove cells from ice (except for the 0-min dish), aspirate last wash, and add 0.5 mL of preheated DMEM^{3+} to begin internalization by transferring the 5-, 10-, 20-, and 40-min dishes to 37°C incubator (*see* **Note 10**).

8. At the end of each incubation period, place cells on ice, and wash each well once with 1 mL of TNB buffer from **Subheading 3.3.**, **step 5**; then add 1 mL of TNB buffer to one well and 1 mL of MESNA buffer to the other well; incubate on ice for 16 min.

9. In between incubation periods, prepare 1.5-mL Eppendorf tubes with 20 µL of 13C4 antibody at 1 mg/mL; chill tube rack and cover to 4°C.

10. Following MESNA treatment, wash cells three times with 0.5 mL of PBS^{++} on ice.

11. After final wash, lyse cells with 0.5 mL of RIPA buffer and add 1 µL of PIC; scrape the bottom of each well with a cut 1-mL pipet, and transfer lysed cells to Eppendorf tubes containing the antibody.

12. Add 100 µL of Protein G–Sepharose slurry to each tube (*see* **Note 11**).

13. Incubate tubes at 4°C for at least 90 min (end-over-end rotation).

14. Centrifuge tubes at 800*g*, 4°C, 1 min.

15. Aspirate supernatant and add 1 mL of 50 m*M* Tris-HCl, pH 8.0, to each tube; agitate beads gently, and recentrifuge as in **Subheading 3.3.**, **step 14**.

16. Aspirate supernatant, dry beads using a microsyringe; immediately add 30 µL of 1.5X SDS sample buffer without reducing agent.

17. Resuspend beads by vortexing; heat tubes for 5 min to 95°C in a Thermomixer.

18. Run samples on Tris-tricine gels for about 1.5 h at 25 mA per gel.

19. Cut the required surface of PVDF filter paper, three small pieces of Whatman paper of the size of the gel, and three large pieces of Whatman paper; immerse the three large pieces individually in transfer buffer, place in semidry transfer chamber, remove air bubbles with a roller; soak PVDF nitrocellulose membrane in concentrated methanol and lay on top of the three large pieces; place the gel on the membrane, remove air bubbles; immerse three smaller pieces in transfer buffer and layer on top of the gels, making sure that gels are completely covered; transfer at 10 V for 1 h.

20. Incubate membrane in blotting buffer under agitation for 1 h at room temperature or overnight at 4°C.

21. Dilute Stv-AP in 5 mL of blotting buffer and seal with PVDF membrane in plastic sack; incubate for at least 90 min at room temperature.

22. Perform two 10-min washes in blotting buffer, then two 10-min washes in blotting buffer without milk, followed by a 2- to 4-h wash in blotting buffer without milk to reduce background.

23. Set parameters on the PhosphorImager: blue phosphor screen, pixel size = 100 µm, PMT voltage = 800 V, press sample.

24. Develop membrane in 0.3 mL ECF by pipetting one line of ECF per gel in one transparency folder. Lay membranes with protein side face down on top of each line; close folder and smooth out air bubbles.

25. Transfer membranes to second transparency folder to reduce excess solution and smooth out air bubbles.

26. Place transparency folder in the PhosphorImager with protein side down at corner.

27. Begin acquisition of signal.

28. Data analysis: To determine the percentage of internalized $StxB-NH_2^{SSbiot}$ at each time-point and in each experimental condition, the signal obtained on the +MESNA sample (internalized) in divided by the signal obtained on the –MESNA sample (total cell associated).

3.4. Sulfation Assay

1. Seed cells in 24-well culture dishes 12–15 h before the experiment (10^5 cells in 1 mL $DMEM^{3+}$ per well, growth at 37°C in a 5% CO_2 incubator).

2. Depletion of endogenous sulfate: Wash cells three times with 0.5 mL HBSS at 37°C; replace the medium with 0.5 mL MEM without sulfate; incubate cells for 90 min at 37°C in a 5% CO_2 incubator.

3. Binding: Put cells on ice and replace medium with 0.25 mL ice-cold MEM without sulfate containing 1 μM $StxB-Sulf_2$; incubate on ice for 30 min (*see* **Note 12**).

4. Remove medium and wash three times with 0.5 mL cold MEM without sulfate.

5. Chase: Replace medium with 0.25 mL preheated MEM without sulfate containing 1 mCi/mL $^{35}SO_4^{2-}$ and incubate cells at 37°C in a 5% CO_2 incubator for the desired time (*see* **Notes 13–15**).

6. Put cells on ice and wash three times with 0.5 mL cold MEM.

7. Immunoprecipitation (*see* **Subheading 3.3.**, **steps 11–15**): Lyse cells two times in 0.5 mL RIPA containing 1X PIC at 4°C and pool the extracts; add 10 µg/mL 13C4 antibody and 100 µL Protein G–Sepharose slurry beads to the lysate; mix 90 min at 4°C by end-over-end rotation.

8. Collect 900 µL supernatant and wash beads three times with 1 mL of 50 mM Tris-HCl buffer, pH 8.0; dry beads with a Hamilton syringe and resuspend in sample buffer without dithiothreitol (DTT); boil samples for 5 min at 95°C.

9. Precipitate proteins in 900 µL supernatant from **Subheading 3.4.**, **step 8** with 10% trichloroacetic acid (TCA) for 1 h at 4°C prior to filtration on Glass Microfiber GF/C Filters (Whatman, catalog no. 1822-025); count filters by scintillation in a Microbeta counter (*see* **Note 16**).

10. Analysis (*see* **Fig. 2**): Run samples on a Tris-tricine mini-gel; dry the gel and expose it in a PhosphorImager cassette (12 h to 6 d); quantify bands corresponding to sulfated $StxB-Sulf_2$; amounts of sulfated $StxB-Sulf_2$ have to be normalized relative to total sulfation counts from **Subheading 3.4.**, **step 9** (*see* **Note 17**).

3.5. Iodination

1. Wash one IodoBead in 1 mL iodination buffer, dry the bead on filter paper, and transfer in a 1.5-mL Eppendorf tube containing 50 μL iodination buffer.
2. Add 1 mCi iodine (*see* **Note 18**) and incubate for 5 min at room temperature.
3. Add 100 μg recombinant B-Glyc-KDEL in 300 μL iodination buffer to the bead + iodine mix, incubate for 15 min at room temperature.
4. Separate iodinated protein from free iodine on a PBS-equilibrated PD-10 column; elute with PBS; collect two fractions of 1 mL, then four fractions of 500 μL.
5. For each fraction, add 2 μL iodinated protein to 450 μL BSA at 2 mg/mL; add 50 μL TCA 100% (10% final concentration); incubate for 1 h on ice; separate soluble and precipitated material and count both with a γ-counter; calculate free iodine (%) and the specific activity (cpm/ng StxB) for each fraction (*see* **Notes 19** and **20**).

3.6. Glycosylation Assay

1. Seed cells in 24-well culture dishes 12–15 h before the experiment (10^5 cells in 1 mL $DMEM^{3+}$ per well).
2. Binding: Place the cells on ice and replace the medium with 0.25 mL ice-cold $DMEM^{3+}$ containing 100 ng ^{125}I-StxB-Glyc-KDEL; incubate on ice for 30 min.
3. Remove unbound ^{125}I-B-Glyc-KDEL by washing cells three times with 0.5 mL cold $DMEM^{3+}$.
4. Chase: add preheated $DMEM^{3+}$ and incubate cells at 37°C in a 5% CO_2 incubator for the desired time.
5. Remove the medium and lyse cells two times in 100 μL sample buffer without reducing agent; pool extracts and boil samples at 95°C for 5 min (*see* **Note 21**).
6. Analysis: Run samples on a 10–20% gradient gel; fix and dry, then expose in a PhosphorImager cassette; quantify the glycosylation product in each condition (**Fig. 3** shows a typical result).

4. Notes

1. The classical Tris-tricine gel system is composed of three gel layers. For simplicity, we have eliminated the intermediate layer. As usual for SDS-PAGE, the stacking should be poured on the polymerized resolving gel.
2. The elution pattern of StxB mutants may vary. StxB-NH$_2$ elutes at 160 m*M* NaCl and StxB-Glyc-KDEL at 250 m*M*. StxB-Sulf$_2$ is very acidic and elutes in pH 6.0 solution (!) at 500 m*M*,
3. The purity of wild-type StxB, StxB-NH$_2$, and StxB-Sulf$_2$ is already satisfactory after ion exchange because of their elution at low or high ionic strength. For StxB-Glyc-KDEL, gel filtration improves purity.
4. Dialysis: The StxB monomer is a 7.7-kDa protein; however, the StxB pentamer is very stable. Therefore, dialysis membranes of MWCO of 10 kDa can be used without loss.
5. Freezing and storage: The purified StxB should be snap-frozen in liquid nitrogen and may then be kept at –20°C or for long term storage at –80°C.

6. Concentration: The StxB can be concentrated; we have found that micro-concentration chambers from PallFiltron allow one to minimize loss.

7. NHS-SS-biotin should be added under agitation to avoid high local concentrations.

8. In the laboratory, we work on a HeLa cell line that homogeneously and efficiently internalizes StxB. All protocols of this chapter have been established on these cells. However, they can be transposed to other cell lines.

9. MESNA should be dissolved as late as possible.

10. Internalization should be started with the 40-min dish; for best synchronization, medium may be added to the following dishes at intervals of 30 s.

11. Use 200-μL pipet with cut tip to avoid clogging leading to the addition of unequal amounts of Protein G–Sepharose to the tubes.

12. This step may be replaced by 40 min incubation at 19°C to accumulate StxB-$Sulf_2$ in early endosomes rather than at the cell surface.

13. It may be useful to briefly heat the dish in a 37°C water bath to have a better temperature shift.

14. The researcher is typically exposed to quantities of radioactive sulfate above 1 mCi; make sure to be adequately protected (work in radioprotection chamber).

15. Radioactive $^{35}SO_4^{2-}$ may be added at any given time-point after the shift to 37°C, allowing to sample continuous passage through the TGN or steady-state levels. **Figure 2** shows the result of a kinetics experiment.

16. The determination of sulfation on endogenous proteins and proteoglycans controls for sulfotransferase activity and cell number. For valid experiments, variations of the endogenous sulfation level obtained under different experimental conditions should be within 10%.

17. This step separates sulfated StxB-$Sulf_2$ from radioactive contaminants and metabolites. The gel can be fixed before complete migration to obtain sharp bands. It may be useful to cut the part of the gel corresponding to the migration front, as it contains nonfixable radioactive contaminants that may diffuse during gel drying and perturb subsequent quantification.

18. One millicurie of ^{125}I iodine, as typically used for StxB-Glyc-KDEL iodination, requires adapted protection (i.e., work in radioprotection chamber).

19. Typically, free iodine is below 5% and specific activity is around 5000 cpm/ng.

20. Iodinated StxB-Glyc-KDEL may be stored for up to 2 wk at 4°C without loss in binding capacity. However, after longer storage, at parasite band at 14 kDa and "smears" appear that make quantification difficult.

21. Cell lysates may be stored at –20°C. We noted, however, that analysis after storage results in "smears" on the gel and renders quantification difficult.

References

1. Sandvig, K. and van Deurs, B. (1996) Endocytosis, intracellular transport, and cytotoxic action of Shiga toxin and ricin. *Physiol. Rev.* **76,** 949–966.

2. Johannes, L. and Goud, B. (2000) Facing inward from compartment shores: how many pathways were we looking for? *Traffic* **1,** 119–123.

3. Johannes, L. and Goud, B. (1998) Surfing on a retrograde wave: how does Shiga toxin reach the endoplasmic reticulum? *Trends Cell Biol.* **8,** 158–162.

4. Mallard, F., Antony, C., Tenza, D., Salamero, J., Goud, B., and Johannes, L. (1998) Direct pathway from early/recycling endosomes to the Golgi apparatus relealed through the study of Shiga toxin B-fragment transport. *J. Cell Biol.* **143,** 973–990.

5. Wesche, J., Rapak, A., and Olsnes, S. (1999) Dependence of ricin toxicity on translocation of the toxin A-chain from the endoplasmic reticulum to the cytosol. *J. Biol. Chem.* **274,** 34,443–34,449.

6. Hazes, B. and Read, R. J. (1997) Accumulating evidence suggests that several AB-toxins subvert the endoplasmic reticulum-associated protein degradation pathway to enter target cells. *Biochemistry* **36,** 11,051–11,054.

7. Khine, A. A. and Lingwood, C. A. (1994) Capping and receptor-mediated endocytosis of cell-bound Verotoxin (Shiga-like toxin). 1: chemical identification of an amino acid in the B subunit necessary for efficient receptor glycolipid binding and cellular internalization. *J. Cell. Physiol.* **161,** 319–332.

8. Johannes, L., Tenza, D., Antony, C., and Goud, B. (1997) Retrograde transport of KDEL-bearing B-fragment of Shiga toxin. *J. Biol. Chem.* **272,** 19,554–19,561.

9. Schagger, H. and von Jagow, G. (1987) Tricine–sodium dodecyl sulfate–polyacrylamide gel electrophoresis for the separation of proteins in the range from 1 to 100 kDa. *Analy. Biochem.* **166,** 368–379.

10. Su, G.F., Brahmbhatt, H.N., Wehland, J., Rohde, M., and Timmis, K.N. (1992) Construction of stable LamB-Shiga toxin B subunit hybrids: analysis of expression in *Salmonella typhimurium aro*A strains and stimulation of B subunit-specific mucosal and serum antibody responses. *Infect. Immun.* **60,** 3345–3359.

18

Measuring pH Within the Golgi Complex and Endoplasmic Reticulum Using Shiga Toxin

Jae H. Kim

1. Introduction

Our understanding of pH regulation has been studied in-depth in the endocytotic pathway of the cell because of fluid phase and membrane-bound pH-sensitive markers that could be delivered and taken up exogenously by the cell *(1,2)*. Very little is understood, however, about the ionic composition of the two major compartments of the secretory pathway, the Golgi complex and the endoplasmic reticulum (ER). This has been largely the result of the relative inaccessibility of these compartments to exogenous probes, as the favored direction of flow of membrane and fluid-phase components in the secretory system is in the exocytotic direction.

The techniques described here harnesses the unique retrograde trafficking of Shiga toxin within the secretory pathway to permit noninvasive monitoring of organellar pH in mammalian cells. To begin with, the methods required for the noninvasive and continuous measurement of the luminal pH of the Golgi complex and ER hinge on the ease in which the purified recombinant B-subunit of Shiga toxin from Shiga toxin-producing *Escherichia coli* (STEC) can be labeled with a fluorescent dye that is not only bright enough for organellar detection but is also pH sensitive in the acidic range. First, the recombinant B-subunit is produced and labeled with either fluorescein isothiocyanate or DTAF (5-[4,6-dichlorotriazinyl]aminofluorescein) *(3,4)*. The resulting complex still retains its capacity to bind plasma membrane Gb_3 receptors on select cells and follow a retrograde pathway from plasma membrane to the Golgi complex. Here, the dye resides for several hours without significant degradation.

From: *Methods in Molecular Medicine, vol. 73: E. coli: Shiga Toxin Methods and Protocols*
Edited by: D. Philpott and F. Ebel © Humana Press Inc., Totowa, NJ

The cells are then able to be imaged in a digital fluorescence microscope system where fluorescence can be accurately detected and correlated with the pH of the lumen of the Golgi complex. The endoplasmic reticulum is accessed similarly except that the Shiga toxin B-subunit used is modified to carry a carboxyl-terminus KDEL segment, which confers ER localization *(5–7)*. Using these methods, the Golgi complex was found to be a moderately acidic organelle (pH ~6.5) that is acidified and tightly regulated by the vacuolar ATPase *(8,9)*. In contrast, the pH of the ER was neutral and had a high permeability to protons and, therefore, relied on the indirect regulation of its pH by plasmalemmal acid–base transporters *(10)*.

The basis of the fluorescence digital microscope setup is best explained by following the paths of light (*see* **Fig. 1** for diagrammatic scheme, *see* **refs.** *11* and *12* for more background reading on fluorescence microscopy). Excitation at 440 and 490 nm is provided by a xenon arc lamp via a computer-controlled shutter and filter wheel assembly (Sutter Instrument Co., CA), whereas continuous 620-nm illumination is achieved by filtering the transmitted incandescent source (halogen). The excitation light is attenuated by a neutral density filter and reflected to the cells by a dichroic mirror (510 nm), whereas the emitted fluorescence (>510 nm) and the transmitted red light (>620 nm) are separated by an emission dichroic mirror (580 nm). The red light is directed to a video camera, allowing continuous visualization of the cells, whereas the fluorescent light is directed onto a 520 ± 25 nm bandpass filter and imaged by a 512×512 frame-transfer cooled coupled charge device (CCD) camera (Princeton Instruments Inc., NJ). These images are captured digitally and transferred to a desktop computer that transfers the data into a fluorescence acquisition program.

The methods described here will primarily focus on the imaging techniques to make pH measurements with the fluorescently labeled Shiga toxin and Shiga toxin bearing the KDEL sequence. In summary, the use of Shiga toxin B-subunit as a pH probe has produced important insights into the regulation of organellar pH.

2. Materials

2.1. Hardware (see Note 1)

1. Microscope: inverted microscope, oil immersion lens 60X Plan Apo NA 1.4 (Leica DM IRB).
2. Xenon arc lamp 100 W and power source.
3. Optical filters (Omega Optical, VT):
   a. Heat filter
   b. 440-nm $\pm$ 10-nm and 490-nm $\pm$ 10-nm excitation bandpass filter
   c. 510-nm, 565-nm, and 580-nm dichroic mirror

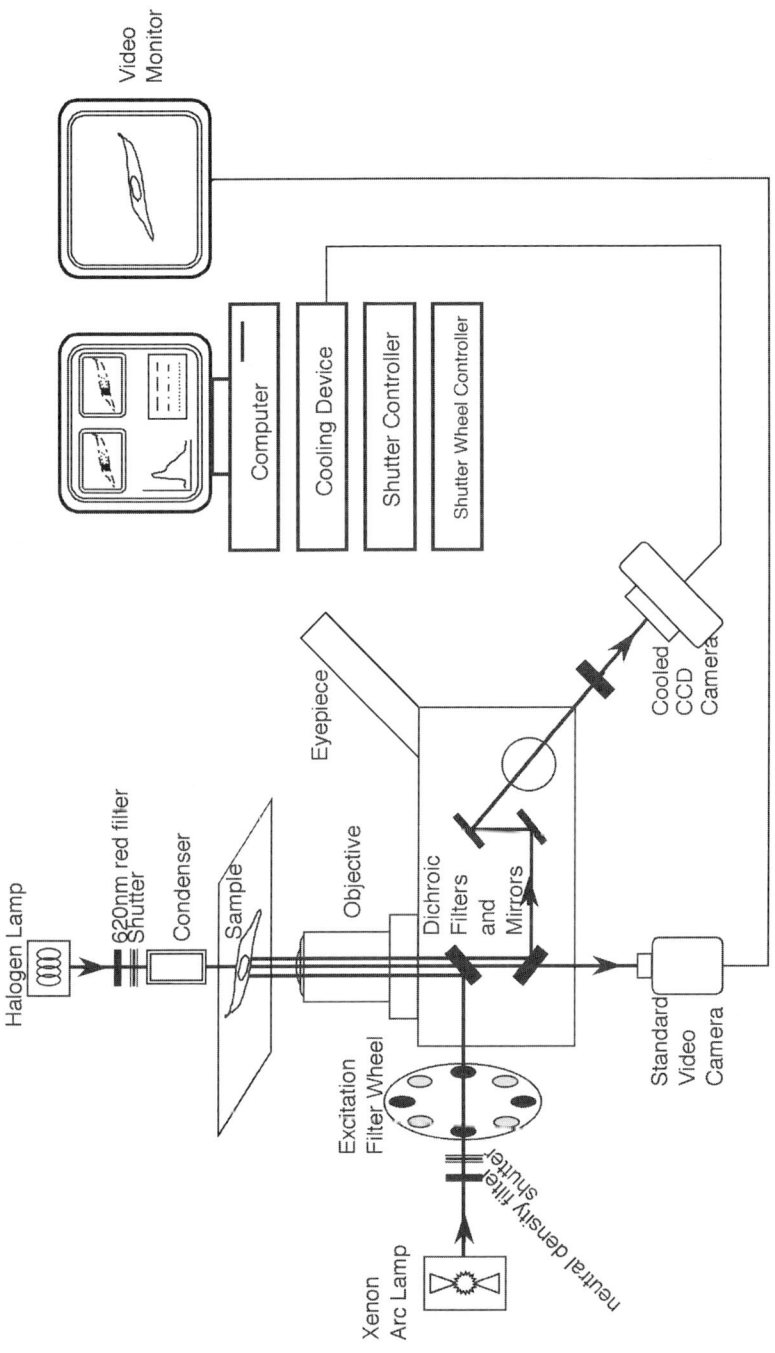

Fig. 1. Diagrammatic scheme of the digital fluorescence microscope setup.

     d. 520 ± 25-nm emission bandpass filter

     e. Neutral density filters (ND1, ND2, 30%)

 4. Rotating filter wheel (Sutter Instrument Co., CA).

 5. Shutter wheel controller (Lambda-10, Sutter Instrument Co., CA).

 6. Cooled CCD camera and detector controller (chip size, approx 1 M array, mono-chromatic, Princeton Instruments Inc., NJ).

 7. Standard video camera (CCD-72, MTI, IN).

 8. Shutter driver/timer (Uniblitz T132, MTI, IN).

 9. Computer: PC (minimum Pentium II 300 MHz), large hard drive (minimum 5 GB).

10. Media storage device:

     a. Iomega Zip drive

     b. CD Writer

11. Optical table (Technical Manufacturing Corp., MA).

12. Stainless-steel circular chambers (Atto, Molecular Probes, Inc., OR).

13. Open Perfusion Micro-Incubator (Medical Systems Corp., NY).

14. Circulating water bath.

2.2. Software

1. Universal Imaging Metafluor/Metamorph.

2. Microcal Origin.

3. Adobe Photoshop.

4. Microsoft Powerpoint.

2.3. Labeling StxB

1. Fluorescein isothiocyanate or DTAF [5-(4,6-dichlorotriazinyl) aminofluorescein; Molecular Probes, OR].

2. StxB (*see* Chapter 17 for purification procedure).

3. Carbonate–bicarbonate buffer: 0.5 M; prepare fresh by adding 5.8 mL of 5.3% Na_2CO_3 to 10 mL of 4.2% $NaHCO_3$), pH 9.5.

4. Centricon-10 filter units (Millipore).

2.4. Reagents and Cell Culture

1. Concanamycin (Kamiya Biochemicals, CA), 100 µM stock.

2. Nigericin (Sigma, MO), 5 mM stock.

3. Vero cells (ATCC) and culture medium (minimal essential medium [MEM] containing 5% fetal bovine serum (FBS), 20 mg/mL penicillin, and 20 mg/mL streptomycin) (*see* **Note 2**).

4. HeLa cells (ATCC) and culture medium (Dulbecco's modified Eagle's medium [DMEM] containing 10% FBS, 20 mg/mL penicillin, and 20 mg/mL streptomycin) (*see* **Note 2**).

5. Phosphate-buffered saline (PBS), pH 7.4.

6. PBS (pH 7.4) + 1 mM $CaCl_2$ and 1 mM $MgCl_2$.

7. Potassium-rich solutions of different pH's: 125 mM KCl, 20 mM NaCl, 10 mM HEPES, 10 mM MES, 0.5 mM $CaCl_2$, 0.5 mM $MgCl_2$ containing 5 mM nigericin and buffered to pH values ranging from 5.5 to 7.5 in steps of 0.5 pH units.

8. Glass coverslips, 25 mm in diameter.

3. Methods

3.1. Labeling StxB with FITC

1. Concentrate StxB (50–500 µg in carbonate–bicarbonate buffer) to <50 µL using a Centricon-10 filter.
2. Weigh FITC at Stx:FITC, 1:1, w/w in a small glass vial and dissolve in 1 mL of carbonate–bicarbonate buffer. Wrap in tin foil.
3. Add StxB to FITC solution. Shake the vial gently at room temperature for 1–2 h (or leave 18 h at 4°C).
4. Remove free FITC using a Centricon-10 filter repeating centrifugation until the filtrate is colorless and does not fluoresce under ultraviolet (UV) light.
5. Labeling usually results in 1–2 FITC molecules per B-subunit pentamer, determined by spectroscopy.

3.2. Measuring pH

1. Establish a separate dark room where a digital fluorescence imaging system (illustrated in schemata in **Fig. 1**) can be used to conduct these types of experiments.
2. Place the microscope setup on an optical (floating) table so that small movements will not impact the experiment.
3. Arrange two cameras (standard video camera for bright-field imaging and the cooled CCD camera for epifluorescence imaging [illustrated in **Fig. 1**]) so that concurrent imaging can take place and provide spatial referencing for the fluorescent images acquired.
4. Arrange the majority of the hardware equipment, including the computer and 17-in. video monitor on a tall tower adjacent to the table to optimize use of space.
5. Set up a 37°C circulating water bath nearby for warming solutions.
6. Grow the Vero or HeLa cells to approx 70% confluency on glass coverslips.
7. Wash the cells three times in cold PBS containing 1 mM CaCl$_2$ and 1 mM MgCl$_2$. To monitor the distribution of FITC-StxB, the cells were exposed to 10 µg/mL of the appropriately labeled Shiga toxin in the same PBS for 1 h at 4°C, to promote binding to the plasmalemmal receptors without endocytosis. Initiate internalization by incubating at temperature to 37°C for 2 h.
8. Carefully insert the coverslip into Atto Chamber within the thermostatted perfusion holder maintained at 37°C at the center of the field calibrated by a built-in thermocouple wire. The holder is a permanent adaptation fixed on the stage of the inverted microscope (*see* **Note 3**).
9. Do not tighten the top chamber excessively when securing the coverslip, as this may fracture the coverslip and render it unusable.
10. Bathe the cells in PBS with 1 mM CaCl$_2$ and 1 mM MgCl$_2$. Ensure that all solutions are mixed, pH balanced, and prewarmed to 37°C. Solutions in the chamber can either be static or dynamically exchanged. There will be greater volumes used in the latter, and warming of the solutions will have to be arranged.
11. Allow the buffer and cells to equilibrate for approx 5 min before starting the experiment.

12. Insert an appropriate neutral density filter to decrease the light exposure to the cells sufficient enough to provide a reasonable signal-to-noise ratio at less that 1-s acquisitions. Longer exposures will limit the rapidity of the experiments.

13. Obtain proper image acquisition software such as Metafluor software on a desktop computer with large primary hard drive (10 gigabytes) (*see* **Note 4**).

14. Set the microscope to view the fluorescence image and isolate an ideal region of interest based on tight organellar localization. Set it back to view the bright-light image and capture test acquisition images (*see* **Note 5**).

15. For an experiment acquire images for both wavelengths (490 nm and 440 nm) and save them. The computer will generate a third image representing ratiometric values for each pixel of fluorescence (490-nm image divided by the 440-nm image). These images are stored dynamically as the experiment is running (*see* **Note 6**).

16. To more precisely select the fluorescence originating from the Golgi complex or ER and not the whole cell, the area of localized fluorescence is outlined using a "region of interest" drawing tool. Juxtanuclear fluorescence is compatible with the morphology and location of the Golgi complex, whereas diffuse staining of the cytoplasm with fluorescence is consistent with ER localization.

17. Noncontributory fluorescence (background) emanating mostly from the dark current of the camera are digitally subtracted with each experiment. This is accomplished by acquiring background images (both wavelengths) without any sample on the microscope using the exact settings for acquisition necessary for the current experiment. The removal of such background fluorescence greatly enhances the signal-to-noise ratio (*see* **Note 7**).

18. Addition of 1 μL of concanamycin (final 100 n*M*) to the buffer with immediate mixing by pipetting up and down will result in rapid inhibition of the vacuolar ATPase (proton pump) and can be used to determine the dependence of the organelle on the pump.

19. At the end of each experiment, obtain a calibration curve of fluorescence vs pH *in situ* by sequentially perfusing the cells with potassium-rich media of different pH values. Allow approx 3–5 min for equilibration at each pH step.

20. Take the raw image data and rerun the experiment ratioing the 490-nm wavelength over the 440-nm wavelength data. This will generate values that reflect fluorescence data that are independent of concentration and leak of the dye. Background images from the experiment are subtracted by the program at this point. Select regions of interest again more precisely and correct for any shifts in the regions during the rerun as this may occur as a result of subtle movement during the experiment. The data stored are numerical average ratio values from the organellar regions selected (*see* **Note 8**).

21. Export the experimental data into a graphing program such as Microcal Origin and adjust the layout accordingly.

4. Notes

1. The exact configuration of the digital fluorescence microscope rig depends heavily on the needs of the user. The purchase and assembly of the fluorescence

equipment is best performed with the on-site assistance of experts from the various companies involved. Particular attention should be made toward compatibility of each component with each other, as this may lead to an enormous drain in time and resources. Ideally, one should seek companies that have both high quality equipment and superb servicing.

2. Only select cell types that have receptors for Shiga toxin with appropriate targeting to the secretory pathway. This poses one of the most difficult limitations of this method for organellar pH measurements. Despite this, however, we have found that Vero cells and HeLa cells (obtained from the American Tissue Culture Collection) have excellent labeling and are versatile cells with which to work.

3. The stainless-steel experimental chambers are rinsed in between experiments under hot water with additional spray rinsing using 70% ethanol to remove any residual hydrophobic agents, as any contamination may alter future experiments. The chambers are then air-dried before their next use. Pre-experimental measurements of the steady-state temperature at the center of the coverslip are necessary to ensure a stable temperature of 37°C. Use of an electrically heated objective lens will diminish the temperature gradient across the coverslip.

4. Depending on the size of the CCD chip in use and the size of the organelle of interest, you may choose to use a binning function to increase the signal-to-noise ratio. Increased binning allows for a larger area of photon collection but reduces the resolution as a direct consequence. Ideally, one attempts to achieve the maximum signal-to-noise ratio with a resolution that does not result in fragmentation of the fluorescent image of the organelle.

5. Regions are selected where at least five to six cells are in the field with optimal organellar staining.

6. The acquisition number should be kept to a minimum, as excessive exposure to the sample will result in photobleaching damage.

7. Experiments must be conducted in relative darkness, as external white light can influence the background of the experiment. The addition of a red filter (620 nm) will enable the use of red light during the experiment.

8. Invest in the fastest computer affordable, as this can often be a limiting factor, particularly when rerunning experiments take place and if image files are large. Nevertheless some planning is important as to the size of an individual experiment, as image files can accumulate quickly during a single experiment. Saving only the numerical ratio data is precarious, as there is no post hoc way to replay the experiment without the raw image data from both wavelengths. The other reason for keeping the raw image files is that the slight shifts that do occur during the experiment that result in false shifts of the average fluorescence of the region of interest. Replaying the experiment allows you to move the selected region and recalculate the ratio values.

References

1. Ohkuma, S. and Poole, B. (1978) Fluorescence probe measurement of the intralysosomal pH in living cells and the perturbation of pH by various agents. *Proc. Natl. Acad. Sci. USA* **75,** 3327–3331.

2. Maxfield, F. R. and Yamashiro, D. J. (1987) Endosome acidification and the pathways of receptor-mediated endocytosis. *Adv. Exp. Med. Biol.* **225,** 189–198.

3. Boulanger, J., Huesca, M., Arab, S., and Lingwood, C. A. (1994) Universal method for the facile production of glycolipid/lipid matrices for the affinity purification of binding ligands. *Anal. Biochem.* **217,** 1–6.

4. Khine, A. A. and Lingwood, C. A. (1994) Capping and receptor-mediated endocytosis of cell-bound verotoxin (Shiga-like toxin). 1: chemical identification of an amino acid in the B subunit necessary for efficient receptor glycolipid binding and cellular internalization. *J. Cell Physiol.* **161,** 319–332.

5. Johannes, L., Tenza, D., Antony, C., and Goud, B. (1997) Retrograde transport of KDEL-bearing B-fragment of Shiga toxin. *J. Biol. Chem.* **272,** 19,554–19,561.

6. Jackson, M. R., Nilsson, T., and Peterson, P. A. (1993) Retrieval of transmembrane proteins to the endoplasmic reticulum. *J. Cell Biol.* **121,** 317–333.

7. Miesenbock, G., and Rothman, J. E. (1995) The capacity to retrieve escaped ER proteins extends to the trans-most cisterna of the Golgi stack. *J. Cell Biol.* **129,** 309–319.

8. Kim, J. H., Lingwood, C. A., Williams, D. B., Furuya, W., Manolson, M. F., and Grinstein, S. (1996) Dynamic measurement of the pH of the Golgi complex in living cells using retrograde transport of the verotoxin receptor. *J. Cell Biol.* **134,** 1387–1399.

9. Stevens, T. H. and Forgac, M. (1997) Structure, function and regulation of the vacuolar (H+)-ATPase. *Annu. Rev. Cell Dev. Biol.* **13,** 779–808.

10. Kim, J. H., Johannes, L., Goud, B., Antony, C., Lingwood, C. A., Daneman, R., et al. (1998) Non-invasive measurement of the pH of the endoplasmic reticulum at rest and during calcium release. *Proc. Natl. Acad. Sci. USA* **95,** 2997–3002.

11. Bright, G. R., Fisher, G. W., Rogowska, J., and Taylor, D. L. (1989) Fluorescence ratio imaging microscopy. *Methods Cell Biol.* **30,** 157–192.

12. Rost, F. W. D. (1992) *Fluorescence Microscopy.* Cambridge University Press, Cambridge.

19

Detection of Shiga Toxin-Mediated Programmed Cell Death and Delineation of Death-Signaling Pathways

Nicola L. Jones

1. Introduction

Cells die by one of two morphologically distinct processes: necrosis or apoptosis. In general, necrosis is considered to be a pathologic process usually following a severe insult to the cell *(1)*. During necrosis, the plasma membrane loses selective permeability and the cell begins to swell. The organellar membranes also lose integrity and are unable to maintain normal function. Condensation of chromatin may occur, but it tends to be irregular. Leakage of cytoplasmic contents evokes an inflammatory response in the surrounding tissues. Overall, the configuration of the necrotic cell is maintained until its removal by professional phagocytes.

1.1. Morphologic Features of Apoptosis

Apoptosis is derived from an ancient Greek word for "falling off of leaves from trees" and is identified most readily by specific morphologic features (*see* **Fig. 1**). During apoptosis, the cell receives a stimulus that triggers a genetically programmed death-signaling cascade *(1)*. The cell begins to shrink in association with cytoplasmic condensation and vacuolation. The surface of the cell is altered with the loss of microvilli and the formation of blebs. Nuclear changes, including condensation and margination of chromatin around the nuclear envelope, are observed. The cell membrane surface constituents change such that phosphotidylserine is externalized. Ultimately, the cell separates into discrete, membrane-bound apoptotic bodies that are phagocytosed either by neighboring cells or professional scavengers. It is generally considered that, because of maintenance of cell membrane integrity, there is an absence of leakage of cytoplasmic constituents into the extracellular space. As a result of the

From: *Methods in Molecular Medicine, vol. 73: E. coli: Shiga Toxin Methods and Protocols*
Edited by: D. Philpott and F. Ebel © Humana Press Inc., Totowa, NJ

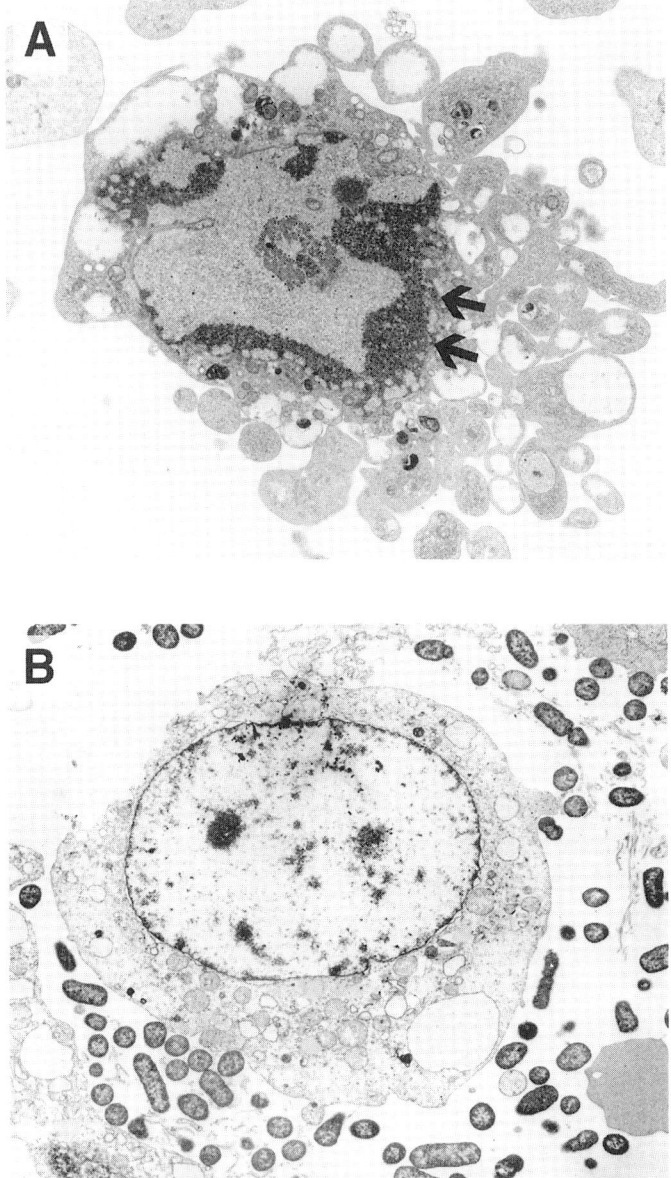

Fig. 1. Transmission electron micrographs of (**A**) an apoptotic HEp-2 cell and (**B**) a HEp-2 cell undergoing necrosis. (**A**) The apoptotic cell shows the characteristic features of apoptosis, including condensed and marginated nuclear chromatin (arrows) and cell membrane blebbing. In comparison, (**B**) the necrotic cell is swollen with loss of membrane integrity. Approximate magnification: ×46,000.

rapid removal of apoptotic bodies, programmed cell death is not associated with a marked degree of inflammation.

1.2. Biochemical Features of Apoptosis

A characteristic, although not universal, biochemical feature of apoptosis is fragmentation of nuclear DNA *(2)*. The DNA initially undergoes fragmentation into larger 50-kbp fragments and then 180- to 200-bp internucleosomal fragments. The presence of fragmented DNA can be exploited to detect apoptotic cells both *in situ* and in vitro *(3)*.

Identification of the morphologic features of apoptosis is generally considered the gold standard for detecting apoptotic cells *(1)*. However, early in the process of apoptosis, cells may not display these morphologic features. Therefore, additional techniques that identify biochemical features associated with apoptosis, such as DNA fragmentation or alteration of cell surface constituents, have been developed (*see* **Table 1**).

Commercially available immunoassays are available to detect DNA fragmentation. For example, histone-complexed oligonucleosomes, formed as a result of DNA fragmentation *(4)*, can be detected by immunoassays using antihistone antibodies, thereby allowing quantitation of the degree of apoptosis. Other commercial kits that are available include the cellular DNA fragmentation ELISA (enzyme-linked immunosorbent assay). DNA gel electrophoresis, in which a characteristic ladder formation is observed in cells undergoing apoptosis, can also be employed. Immunoassays that detect alterations in cell surface constituents such as exposure of phosphotidylserine by Annexin V labeling are also available. It is recommended that more than one technique that confirms the induction of programmed cell death be employed.

1.3. Signaling Pathways Mediating Apoptosis

A number of different stimuli and signaling pathways are able to initiate the final common process of programmed cell death *(5)*. Many of these pathways interact with each other to modulate the cell death signal response. The multiple signaling events also demonstrate both the complexity and redundancy of the apoptotic process.

Caspases are a family of cysteine proteases that play a key role in mediating apoptosis *(6)*. Caspases are synthesized as inactive proenzymes that are cleaved at specific peptide bonds to form a large subunit (17–21 kDa) and a small subunit (10–13 kDa) that interact to form heterotetramers that function as the active enzyme. In addition to autocatalytic cleavage, caspases can activate other members of the family in a cascade that induces programmed cell death. There are two main pathways that signal apoptosis: the cell death receptor pathway and the mitochondrial pathway (*see* **Fig. 2**).

Table 1
Examples of Techniques Available to Detect Apoptosis

1. Morphology
   Transmission electron microscopy
   Scanning electron microscopy
   DNA intercalating dyes (acridine orange, ethidium bromide, propidium iodide [PI])
2. Change in surface constituents and cell membrane permeability
   Permeable dyes (DAPI, Hoescht)
   Phospholipid externalization (Annexin V assays)
3. DNA fragmentation
   TUNEL assay
   In situ nick translation
   DNA gel electrophoresis
   Comet assay
   Oligonucleosome histone complex detection
4. Dissolution of mitochondrial function
   Release of cytochrome-c
   Disruption of mitochondrial permeability transition
5. Biochemical changes
   Caspase activation and detection of cleavage products
   Transglutaminase activity
   Alteration in proapoptotic or antiapoptotic molecules

Caspases can be activated following signaling through an extracellular death receptor such as Fas/CD95 or tumor necrosis factor (TNF) receptor 1 *(6)*. Receptor binding triggers the recruitment of caspases into a multiprotein complex known as the death-inducing signaling complex (DISC), which is essential for caspase activation. The mitochondrial death pathway is activated by proapoptotic members of the Bcl-2 family *(7)*. The prodeath Bcl-2 homologs mediate the release of cytochrome-c and other other proteins from the mitochondria. Subsequently, caspase activation is induced through the formation of the multiprotein complex known as the apoptosome.

Thus, proteins in the Bcl-2 family are additional key regulators of the programmed cell death cascade. Bcl-2 proteins consist of both proapoptotic and antiapoptotic homologs *(7)*. The mechanisms by which the Bcl-2 family regulates cell death are just beginning to be characterized. The ability of the proteins to form heterodimers and homodimers suggests that, through competitive dimerization, proapoptotic and antiapoptotic members regulate cell death. Additional factors involved in Bcl-2 family mediated cell death, such as disruption of mitochondrial function, also have been considered *(8)*. Bcl-2 and Bax can form channels leading to alteration of the permeability transition pore, thereby regulating the release of cytochrome-c and other proteins from mito-

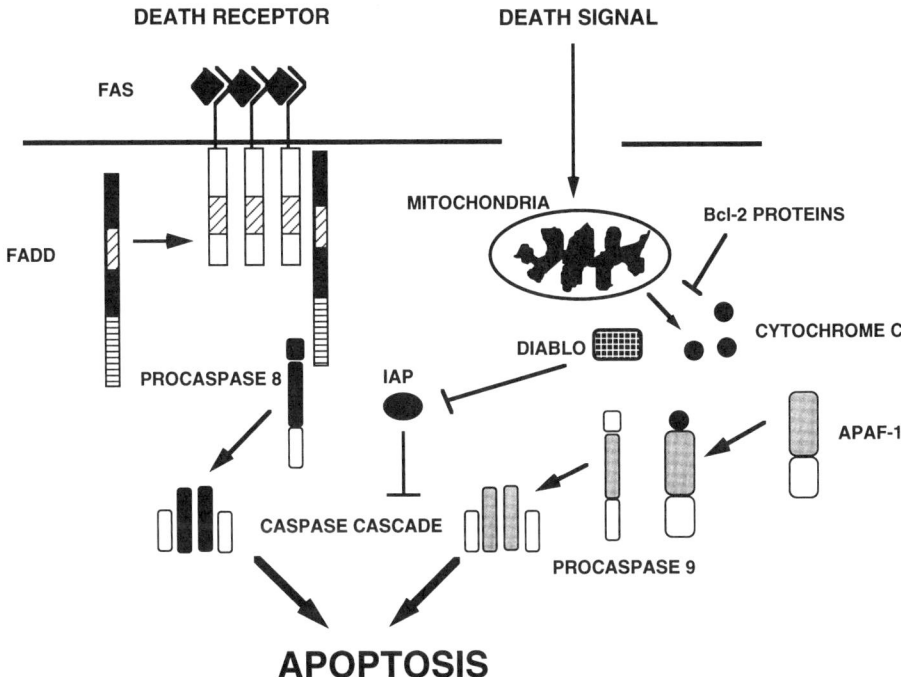

Fig. 2. Schematic representation of the apoptotic signal cascade triggered by two pathways, ligation of a death receptor or the mitochondrial pathway. Ligation of the Fas death receptor by the Fas ligand causes trimerization and mediates binding of the adaptor protein FADD through death domains. The death effector domain of FADD then causes oligomerization of procaspase 8. The formation of this multiprotein complex, known as the death-inducing signaling complex, autocatalytically activates caspase 8 triggering the caspase cascade. Cleavage of key substrates by the caspases results in apoptosis. Alternatively, the cell receives a death signal that results in the release of mitochondrial cytochrome-*c* and Diablo (also termed Smac). Cytochrome-*c* associates with Apaf-1, likely resulting in a conformational change causing exposure of the CARD domain, allowing binding to procaspase 9 and formation of the apoptosome. Activation of the caspase cascade then mediates programmed cell death. The two death signaling pathways can be disrupted by inhibitors of apoptosis proteins (IAP), which block caspase activation. Release of Diablo/Smac binds to IAPs and prevent them from inhibiting apoptosis.

chondria. Proapoptotic Bcl-2 homolog-mediated release of these mitochondrial molecules activates the caspase signaling cascade *(7)*.

Recently, additional levels of complexity have been identified that regulate these death-signaling pathways *(9)*. For example, inhibitors of apoptosis proteins (IAPs) modulate the caspase cascade by binding to caspases and inhibiting their activation. Furthermore, IAP molecules can promote ubiquitin-mediated degradation of caspases. Smac/Diablo, a recently identified mitochondrial protein, is released during apoptosis along with cytochrome-*c* *(10,11)*. Smac/Diablo promotes caspase activation by inhibiting IAPs. It is likely that an even higher degree of complexity of apoptotic cell death regulation will be identified in the future.

The detection of enhanced expression of particular proapoptotic or antiapoptotic proteins such as members of the Bcl-2 family during toxin-mediated cell death suggests their potential involvement in the cell death cascade. In order to demonstrate the requirement for a particular cell death gene in the induction of apoptosis, several strategies may be utilized. Disruption of the cell death-signaling pathway using specific inhibitors of gene products, such as the caspase cascade, have been employed to successfully block apoptosis.

Gene inactivation can also determine if a putative gene is essential to induce cell death. Techniques that are effective for gene inactivation include targeted mutagenesis and the use of antisense oligonucleotides *(12)*. These methodologies are time-consuming and expensive. An alternative approach that has proven efficacious is transient transfection assays utilizing either antisense constructs to inhibit particular genes or constructs to allow overexpression of candidate genes *(12)*. One limitation to this technique is the generally low transient transfection rate in tissue culture cells. This obstacle can be overcome by cotransfection of marker genes, such as green fluorescent protein *(13)* or a cell surface marker *(12)*, which can be detected by either fluorescence microscopy or flow cytometry. The concentration of the two plasmids for transfection is adjusted such that it is probable that a cell transfected with the gene of interest will also receive the marker gene.

Dysregulation of apoptosis is involved in different pathologic conditions, including cancer and autoimmunity. In addition, infectious agents, including viruses and bacteria, have developed mechanisms to exploit the apoptotic machinery of the eukaryotic cell, which can result in the induction of programmed cell death. Virus-encoded proteins can interact with several apoptotic pathways to either inhibit or induce apoptosis to maximize viral replication. A growing list of bacterial pathogens also has been recognized as mediators of the cell death cascade *(14)*. For example, *Shigella flexneri* infection induces apoptosis of infected macrophages both in vitro and in vivo *(14)*. Recent evidence indicates that Shiga toxins produced by *Escherichia coli* are also capable of inducing apoptosis in epithelial cells *(13,15)*, lymphocytes *(16)* and renal cortical cells *(17)*.

The aim of this chapter is to outline techniques that can be used to detect *E. coli* Shiga toxin-mediated apoptosis, including both morphologic and biochemical techniques. In addition, methodology to identify signaling pathways involved in regulating programmed cell death will be described.

2. Materials

2.1. Determination of Epithelial Morphology and Apoptosis

2.1.1. Transmission Electron Microscopy

1. Glutaraldehyde, 2%, in 0.1 M phosphate buffer (pH 7.4).
2. Osmium tetroxide, 2%.
3. Acetone.
4. Epoxy resin.
5. Copper grids.
6. Uranyl actetate.
7. Lead salts for staining.
8. Purified Shiga toxins Stx1 and Stx2 (1.25–25 ng/mL) using standard methods *(18)* (*see also* Chapters 3–7), can be utilized to induce apoptosis. Alternatively, recent studies have described the induction of apoptosis in cells that are transfected with constructs encoding the binding (B) subunit of the toxin *(19)*.
9. HEp-2 (ATCC CCL 23) epithelial cells.
10. Minimal essential media (MEM) culture medium supplemented with 10% fetal calf serum (FCS), antibiotics (2% penicillin–streptomycin), and 0.1% sodium bicarbonate *(13)*. Alternative media may be required for other cell lines.
11. EDTA-trypsin (0.05%) is employed to remove cells from the tissue culture flasks. Alternatively, cells can be scraped off the bottom of tissue culture flasks with a rubber policeman.
12. Phosphate-buffered saline (PBS) is used for washing and suspending cells.
13. Assorted disposable plasticware that is required include culture flasks, 12-well plates, pipet tips, and 15 mL and 50 mL tubes.

2.1.2. Fluorescent Dye Staining

1. Acridine orange (100 µg/mL) and ethidium bromide (100 µg/mL) diluted in phosphate-buffered saline.
2. Microscope slides and coverslips.

2.2. Detection of Apoptosis Biochemically

A commercially available kit that can be used to examine apoptosis biochemically is the Cell Death Elisa[plus] kit (Boerhinger-Mannheim, Indianapolis, IN).

2.3. Delineation of Apoptotic Signaling Pathways

2.3.1. Western Blotting to Detect Bcl-2 Family Members

1. Lysis buffer: 150 mM NaCl, 1% Triton X 100, 10 mM Tris-HCl (pH 7.4), 5 mM EDTA, 1 mM phenylmethylsulfonyl fluoride (PMSF), apoprotinin (2 µg/mL), leupeptin (50 µg/mL), 1 mM benzamidine, pepstatin (1 µg/mL), 1 mM sodium vanadate, 50 mM sodium fluoride, and 2 mM sodium pyrophosphate.
2. Blocking buffer: 5% skim milk in PBS plus 0.5% Tween-20.
3. Sodium dodecyl sulfate (SDS)–polyacrylamide gel.

4. Immobilon-P transfer membrane (Millipore, Bedford, MA).
5. Anti-Bcl-2 (Santa Cruz Biotechnology, Santa Cruz, CA), anti-Bak (Oncogene Research Products, Cambridge, MA), and anti-Bax (Santa Cruz Biotechnology, Santa Cruz, CA) antibodies.
6. Peroxidase conjugated secondary antibodies.
7. ECL Western blotting detection system (Amersham Life Science; Buckinghamshire, UK).
8. Kodak film.

2.3.2. Caspase Inhibition

Caspase signaling cascade inhibitors: the broad-spectrum inhibitor benzyloxycarbonyl-Asp-Glu-Val-Asp-fluoromethyl ketone (Z-VAD-fmk) and the CPP-32 family-specific inhibitor benzyloxycarbonyl-Val-Ala-Asp-fluoromethyl ketone (Z-DEVD-fmk) (Calbiochem, San Diego, CA) at concentrations of 1–200 μM diluted in DMSO. The diluent is stable for 2 yr when stored at –20°C.

2.3.3. Transient Transfection of Tissue Culture Cells

1. FuGENE transfection reagent (Roche Biochemicals).
2. Bcl-2 plasmid construct pcDNA3.Bcl-2 *(20)*.
3. Control empty vector pCDNA3 (Invitrogen, San Diego, CA).
4. Green fluorescent protein (GFP) marker plasmid (pEGFP C2) (Clontech Laboratories, Palo Alto, CA).
5. Hoescht dye (5 µg/mL).

3. Methods
3.1. Morphologic Assessment of Apoptosis
3.1.1. Transmission Electron Microscopy

1. Tissue culture cells should be grown to confluence (*see* **Note 1**) and then incubated for differing lengths of time with purified Stx1 or Stx2 at various concentrations (1.25–25 ng/mL) to determine dose and time-response curves. Depending on the cell line utilized, Stx2 generally induces a less potent apoptotic response in comparison with Stx1.
2. Centrifuge both cells in suspension and trypsinized cells (*see* **Note 2**) at 1000*g* for 5 min.
3. Gently overlay pelleted cells with 2% glutaraldehyde in 0.1 *M* phosphate buffer (v/v) to allow fixation and store at 4°C. Postfix in 2% osmium tetroxide and dehydrate through a series of graded acetone washes. Embed samples in epoxy resin, cut ultrathin sections (50–60 nm), and place onto 300-mesh copper grids. Stain grids with uranyl acetate and lead salts.
4. Apoptotic cells can be detected by using a transmission electron microscope at an accelerating voltage of 60 kV (*see* **Note 3**).

5. **Figure 1** is an example of the characteristic transmission electron microscopy (TEM) findings of apoptotic and necrotic cells.

3.1.2. Fluorescent Dye Staining

1. Grow tissue culture cells to confluence (*see* **Note 1**) and incubate with purified Stx1 or Stx2 (dose and time studies can be performed by using varying concentrations and incubation times).
2. Centrifuge both cells in suspension and trypsinized cells (*see* **Note 2**) and resuspend in 1 mL of PBS. Add acridine orange/ethidium bromide (under dark conditions) in PBS (100 µg/mL) to the suspension and agitate gently (*see* **Note 4**).
3. Add a drop of the suspension to a microscope slide, apply a coverslip and observe under fluorescent microscopy (*see* **Note 5**). Enumerate apoptotic cells by counting 500 cells at multiple randomly selected fields (*see* **Note 6**).
4. **Figure 3** is an example of a fluorescent micrograph of Stx-treated cells incubated with acridine orange and ethidium bromide.

3.2. Biochemical Detection of Apoptosis

3.2.1. Cell Death Detection by Immunoassay

1. Cells are grown to confluence and incubated with the toxin as described in **Subheading 3.1.**. Cells incubated with 1 *M* sorbitol for 2 h at 37°C, which has previously been demonstrated to induce apoptosis *(21)*, serve as a positive control for apoptosis (*see* **Note 7**). Cells undergoing 10 min of freezing at –20°C followed by 10 min of thawing for a total of 1 h serve as a control for necrosis.
2. Adherent control and Stx1-treated cells are trypsinized. Lyse 5×10^4 cells by incubating in the supplied lysis buffer and centrifuge at 14,000g for 10 min.
3. Transfer supernatants of cell lysates to streptavidin-coated microtiter wells (the samples may need to be diluted to be in the detection range of the assay).
4. As described by the manufacturer, quantitate cytoplasmic nucleosomes using biotinylated antihistone antibody and peroxidase conjugated anti-DNA antibody following incubation with the peroxidase substrate 2,2'-azino-di[3-ethyl-benzthiazolin-sulfonate]. The histone–DNA complex provided by the manufacturer serves as an additional positive control for the assay.
5. Measure absorbance of samples spectrophotometrically at 405 nm.
6. An enrichment factor, to quantitate the relative increase in nucleosomes, is calculated as the ratio of specific absorbances in lysates from Stx1-treated cells compared with untreated cells.

3.3. Delineation of Signaling Pathways Mediating Cell Death

3.3.1. Western Blotting to Detect Bcl-2 Family Members

1. Lyse Stx-treated cells and control cells in 0.15 mL of ice-cold lysis buffer for 30 min on ice.
2. Centrifuge the cell lysate at 14,000 rpm for 15 min at 4°C. Obtain supernatants and determine protein concentrations by the Bradford method. Dilute samples with Laemmli buffer.

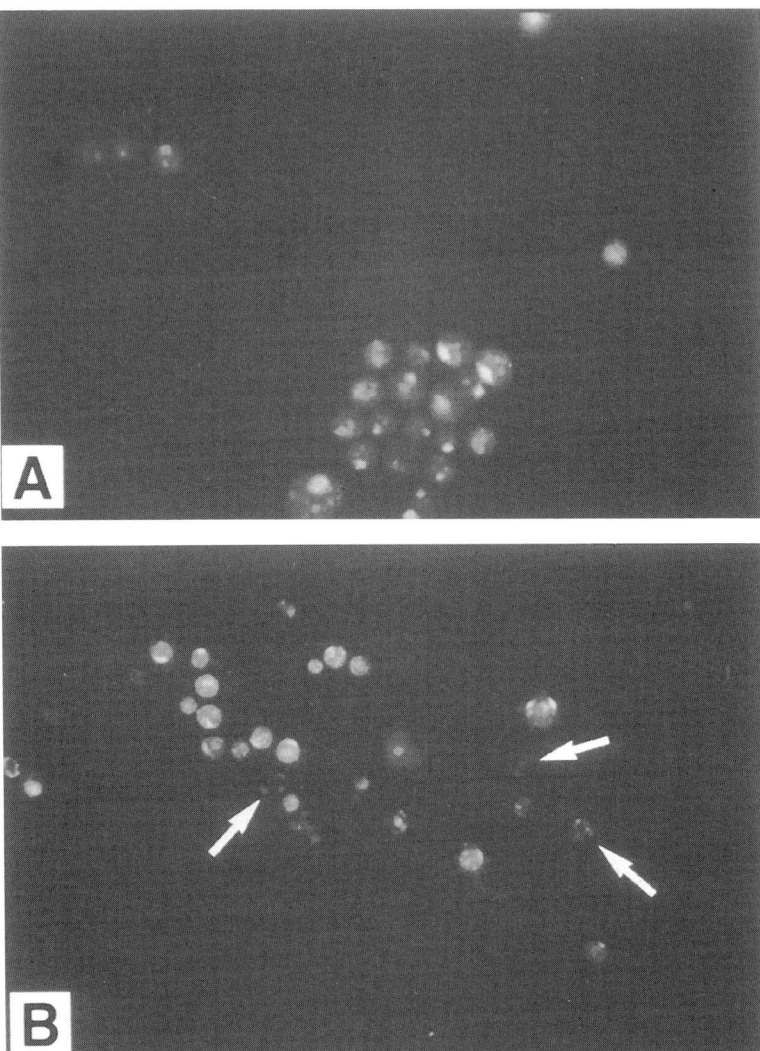

Fig. 3. Identification of HEp-2 cells with normal morphology (**A**) and toxin-medi-ated apoptotic HEp-2 cells (**B**) using acridine orange/ethidium bromide staining and fluorescence microscopy. Apoptotic cells demonstrate condensed and marginated chromatin with enhanced fluorescence (arrows). Nonviable apoptotic cells fluoresce orange but show nuclear features of apoptosis. Approximate magnification ×400.

3. Boil samples with equivalent concentrations of protein for 2 min and subject to 14% sodium dodecyl sulfate–polyacrylamide gel electrophoresis (SDS-PAGE).
4. Electroblot SDS-PAGE gels onto an Immobilon-P transfer membrane.
5. Block blots overnight at 4°C with blocking buffer.

6. Incubate separate blots with primary antibodies to Bcl-2, Bak, and Bax for 1 h at 37°C.
7. Wash blots in PBS plus 0.5% Tween-20 (3X 5 min) and then incubate with peroxidase conjugated secondary antibodies.
8. Using the ECL chemiluminescence Western blotting detection system (*13*), develop blots.
9. Perform densitometry to compare protein expression between experimental groups.

3.3.2. Caspase Inhibition

1. Preincubate HEp-2 cells with caspase inhibitor for 1 h at 37°C prior to Stx treatment. Varying concentrations can be utilized to determine a dose-response curve.
2. Determine ability of caspase inhibition to prevent cell death using one of the methods described above for determining apoptosis. Fluorescence microscopy of cells stained with acridine orange and ethidium bromide provides a reliable method of semiquantitating the degree of apoptosis.

3.3.3. Transient Transfection of Cultured Cells to Determine the Involvement of Cell Death Genes

1. Transiently transfect subconfluent HEp-2 cells in six-well Costar cell culture dishes using FuGENE transfection reagent with both pcDNA3.Bcl-2 and a GFP marker plasmid (pEGFP C2) at a molar ratio of 4:1. Cells transfected with the vector pCDNA3 and pEGFP C2 serve as controls. Culture cells for 48 h and then wash and add fresh medium.
2. Incubate cells in the presence or absence of Stx1 for up to 24 h at 37°C.
3. Collect floating and trypsinized adherent cells (*see* **Note 2**) and stain with Hoescht (5 µg/mL).
4. Enumerate the percentage of apoptotic nuclei among the GFP-positive cells by fluorescence microscopy.

4. Notes

1. Overnight serum starvation prior to incubation with purified Shiga toxins can enhance toxin-mediated apoptosis in certain tissue culture cell lines.
2. Cells grown in a monolayer tend to detach and float in suspension when undergoing apoptosis. Therefore, the detached cells should be included in the sample, otherwise, the degree of apoptosis is underestimated.
3. The earliest detectable morphologic features of apoptotic cells are nuclear changes (*1*). Apoptotic cells have uniformly condensed and sharply marginated chromatin at the edge of the inner surface of the nuclear envelope. Budding of the nucleus into apoptotic bodies also can be observed. Cytoplasmic condensation and vacuolation occurs in association with blebbing of the cell membrane. The cytoplasmic organelles remain intact. In contrast to in vivo studies where apoptotic cells are engulfed and degraded, apoptotic tissue culture cells eventually undergo degradation with breakdown of the cell membrane and disruption of organelles resembling necrosis. However, degenerating apoptotic cells can be

distinguished based on the presence of apoptotic-like nuclear fragments with even condensation and margination of the chromatin around the nuclear envelope.

4. Additional permeable and impermeable dyes can be used to distinguish apoptotic and necrotic cells *(22)*. In general, low-molecular-weight DNA-binding dyes such as DAPI, Hoescht, and acridine orange label apoptotic and normal cells. Higher-molecular-weight dyes such as propidium iodide or ethidium bromide only enter cells once they become permeable. Thus, necrotic cells and cells in the late phases of apoptosis will take up these dyes. As outlined in **Note 5**, the altered morphology of the nucleus and staining with specific dyes distinguishes these cell types.

5. Apoptotic cells can be distinguished by green fluorescence, reduced cell size and condensed brightly fluorescent nucleus *(23)*. Fragmentation of the nucleus also may be detected. In contrast, necrotic cells fluoresce orange and appear swollen with flocculation of the chromatin in the nucleus. Apoptotic tissue culture cells that have undergone secondary necrosis will also fluoresce orange. However, condensation and fragmentation of the nucleus of cells undergoing secondary necrosis distinguishes these cells from those undergoing primary necrosis.

6. As an alternative to enumeration by fluorescent microscopy, the degree of apoptosis can be determined using flow cytometry of cells stained with DNA intercalating dyes.

7. Depending on the cell line employed, sorbitol treatment may not be an effective inducer of apoptosis. Therefore, alternative positive controls (i.e., chemotherapeutic agents such as vinblastine *[24]* and glucocorticoid treatment *[25]*) can be employed.

References

1. Kerr, J. F., Gobe, G. C., Winterford, C. M., and Harmon, B. V. (1995) Anatomical methods in cell death. *Methods Cell. Biol.* **46,** 1–27.
2. Nagata, S. (2000) Apoptotic DNA fragmentation. *Exp. Cell Res.* **256,** 12–18.
3. Ben-Sasson, S. A., Sherman, Y., and Gavrieli, Y. (1995) Identification of dying cells—*in situ* staining. *Methods Cell. Biol.* **46,** 29–39.
4. Aragane, Y., Kulms, D., Metze, D., Wilkes, G., Poppelmann, B., Luger, T. A., et al. (1998) Ultraviolet light induces apoptosis via direct activation of CD95 (FAS/APO-1) independently of its ligand CD95L. *J. Cell Biol.* **140,** 171–182.
5. Rich, T., Watson, C. J., and Wyllie, A. (1999) Apoptosis: the germs of death. *Nat. Cell Biol.* **1,** E69–E71.
6. Stennicke, H. R. and Salvesen, G. S (2000) Caspases—controlling intracellular signals by protease zymogen activation. *Biochem. Biophys. Acta* **1477,** 299–306.
7. Tsujimoto, Y. and Shimuzu, S. (2000) Bcl-2 family: life or death switch. *FEBS Lett.* **466,** 6–10.
8. Kroemer, G. and Reed, J. C. (2000) Mitochondrial control of cell death. *Nat. Med.* **6,** 513–519.
9. Green, D. R. (2000) Apoptotic pathways: paper wraps stone blunts scissors. *Cell* **102,** 1–4.
10. Verhagen, A. M., Ekert, P. G., Pakusch, M., Silke, J., Connolly, L. M., Reid, G. E., et al. (2000) Identification of DIABLO, a mammalian protein that promotes apoptosis by binding to and antogonizing IAP proteins. *Cell* **102,** 43–53.

11. Du, C., Fang, M., Li, Y., Li, L., and Wang. X. (2000) Smac, a mitochondrial protein that promotes cytochrome c-dependent caspase activation by eliminating IAP inhibition. *Cell* **102,** 33–42.

12. Osborne, B. A., Smith, S. W., Liu, Z. G., McLaughlin, K. A., and Schwartz, L. M. (1995) Transient transfection assays to examine the requirement of putative cell death genes. *Methods Cell. Biol.* **46,** 99–106.

13. Jones, N. L., Islur, A., Haq, R., Mascarenhas, M., Karmali, M. A., Perdue, M., et al. (2000) *Escherichia coli* Shiga toxins induce apoptosis in epithelial cells that is regulated by the Bcl-2 family. *Am. J. Physiol.* **278,** G811–G819.

14. Weinrauch, Y. and Zychlinsky, A. (1999) The induction of apoptosis by bacterial pathogens. *Annu. Rev. Microbiol.* **53,** 155–187.

15. Inward, C. D., Williams, J., Chant, I., Crocker, J., Milford, D. V., Rose, P. E., et al. (1995) Verocytotoxin-1 induces apoptosis in Vero cells. *J. Infect.* **30,** 213–218.

16. Mangeney, M., Lingwood, C. A., Taga, S., Caillou, B., Tursz, T., and Wiels, J. (1993) Apoptosis induced in Burkitt's lymphoma cells via Gb3/ CD77, a glycolipid antigen. *Cancer Res.* **53,** 5314–5319.

17. Karpman, D., Hakansson, A., Perez, M. T., Isaksson, C., Carlemalm, E., Caprioli, A. et al. (1998) Apoptosis of renal cortical cells in the hemolytic uremic syndrome: in vivo and in vitro studies. *Infect. Immun.* **66,** 636–644.

18. Petric, M., Karmali, M. A., Richardson, S., and Cheung, R. (1987) Purification and biological properties of *Escherichia coli* verocytotoxin. *FEMS Microbiol. Lett.* **41,** 63–68.

19. Nakagawa, I., Nakata, M., Kawabata, S., and Hamada, S. (1999) Regulated expression of the Shiga toxin B gene induces apoptosis of mammalian fibroblasts. *Mol. Microbiol.* **33,** 1190–1199.

20. Zanke, B. W., Lee, C., Arab, S., and Tannock, I. F. (1998) Death of tumor cells after intracellular acidification is dependent on stress-activated protein kinases (SAPK/JNK) pathway activation and cannot be inhibited by Bcl-2 expression or interleukin 1beta-converting enzyme inhibition. *Cancer Res.* **58,** 2801–2808.

21. Koyama, A. H. and Miwa, Y. (1997) Suppression of apoptotic DNA fragmentation in herpes simplex virus type 1-infected cells. *J. Virol.* **71,** 2567–2571.

22. Sherwood, S. W. and Schimke, R. T. (1995) Cell cycle analysis of apoptosis using flow cytometry. *Methods Cell. Biol.* **46,** 77–97.

23. Jones, N. L., Day, A. S., Jennings, H. A., and Sherman, P. M. (1999) *Helicobacter pylori* induces gastric epithelial cell apoptosis in association with increased Fas receptor expression *Infect. Immun.* **67,** 4237–4242.

24. Lala, P., Ito, S., and Lingwood, C. (2000) Retroviral transfection of Madin Darby canine kidney cells with human MDR1 results in a major increase in globotriaosylceramide and 10(5)–10(6) fold increased cell sensitivity to verocytotoxin. Role of *p*-glycoprotein in glycolipid synthesis. *J. Biol. Chem.* **275,** 6246–6251.

25. Ashwell, J. D., Lu, F. W., and Vacchio, M. S. (2000) Glucocorticoids in T cell development and function. *Annu. Rev. Immunol.* **18,** 309–345.

20

Interaction of Shiga Toxin with Endothelial Cells

Martin Bitzan and D. Maroeska W. M. te Loo

1. Introduction

The vascular endothelium was initially viewed as a "semipermeable" barrier between circulating blood and the interstitium and organ parenchyma. It is now evident that the endothelium plays a much more complex and active physiological and pathophysiological role. For example, endothelial cells regulate vascular tone by the release of vasoactive mediators, such as nitric oxide and endothelin, and orchestrate the response to inflammatory stimuli by the expression of vascular adhesion molecules and increased "leakiness" to circulating cells and proteins.

The successful isolation and propagation of endothelial cells in vitro allowed the detailed investigation of many of the biological properties of the endothelium. Primarily the result of its easy availability and relative robustness, endothelial cells from human umbilical veins (HUVEC) were the first to be propagated in pure culture *(1,2)* and continue to be the "workhorse" of endothelial studies today. The cumulative work of various groups led to the characterization of various biological and molecular features peculiar to endothelium and distinct from other tissues. Saphenous vein and bovine aorta are other sources of robust endothelial cells amenable to cell biological and molecular studies *(3)*. However, phenotypic and molecular differences between endothelial cells from various vascular beds and species become increasingly evident, thus limiting the generalizability of results obtained with cells from a particular type of blood vessel or from different animal species. Inferences from in vitro studies with cultured endothelial cells toward physiological and pathological events in vivo, many of which result from changes in smaller vascular beds, have to be drawn with caution. Standard endothelial cell cultures are further limited by the lack of important physiological parameters, such as

From: *Methods in Molecular Medicine, vol. 73: E. coli: Shiga Toxin Methods and Protocols*
Edited by: D. Philpott and F. Ebel © Humana Press Inc., Totowa, NJ

mechanical stress (flow and shear) and cellular interaction ("crosstalk") with neighboring tissue or activated blood cells. Another concern is the potential loss of in vivo gene expression in cultured endothelial cells.

Evidence for the role of endothelial cells in Shiga toxin-producing *Escherichia coli*-mediated disease in humans or experimental animals is mainly circumstantial. Shiga toxin (Stx) binds to the vascular endothelium in vivo. For example, Richardson et al. injected rabbits with purified toxin and showed, using indirect immunofluorescence or direct detection of isotope-labeled toxin, that Stx primarily localize to the endothelium of small and medium-sized blood vessels, predominantly of the gut (cecum and colon) and central nervous system *(4)*. Boyd et al., injecting pigs with wild-type and mutated Shiga toxins, again showed a target organ- and receptor-specific distribution of the administered toxin in vivo *(5)*. Finally, Uchida et al. reported the detection of Stx in glomerular capillaries studying the renal biopsies of a child with *E. coli*-associated hemolytic uremic syndrome (HUS) *(6)*. Yet, the precise mechanisms(s) leading to the pathologic lesions and events in hemorrhagic colitis and HUS are still elusive. Recent studies by te Loo et al. revealed that granulocytes bind Stx. They postulated that these cells serve as a vehicle for the transport and delivery of Stx from the gut to their target organs, especially kidney. The potential relevance of this novel paradigm is supported by the observation that generic Gb_3 present on endothelial cells, but not on granulocytes, binds Stx with approx 100-fold greater affinity than the yet to be characterized Stx receptor on neutrophils *(7)*. Subsequently, these authors showed that the granulocytes of children with acute enteropathic HUS were indeed loaded with Stx *(8)*.

Early work addressing the biological effects of Stx on endothelial cells established that the toxin blocks protein synthesis and induces morphological changes and ultimately causes the demise of cultured endothelial cells. Overall, however, HUVEC emerged as relatively resistant to the protein synthesis-inhibiting and cytotoxic effects of Stx. Most authors agree that Stx concentrations in excess of 10 n*M* are necessary to kill confluent HUVEC monolayers, whereas subconfluent cultures proved susceptible to lower toxin concentrations. Effects of Stx on (large vessel) saphenous vein endothelial cells and on microvascular endothelial cells, primarily of dermal (foreskin), renal cortical *(9)*, or glomerular origin *(10)* vary considerably. In general, toxin concentrations required to induce cytotoxic effects in large-vessel endothelial cells and endothelial cells from select capillary beds (e.g., glomerular and gut mucosal capillaries) are several orders of magnitude greater than those leading to Vero cells (African green monkey epithelial cells), human proximal tubule epithelial, and HeLa S3 cell toxicity *(11–14)*.

Stx1 and Stx2 bind primarily to Gb_3, when endothelial cell extracts are studied in vitro *(11,15)*. Gb_3 facilitates endocytosis in a clathrin-dependent process and regulates the intracellular movement of the toxin to the endoplasmic reticulum (ER) *(16)*. Endothelial cell susceptibility to Stx increases about 10-fold when the cells are presensitized with lipopolysaccharide (LPS) or cytokines, such as tumor necrosis factor (TNF-α) or interleukin (IL-1α). This is partly the result of the increased expression of Gb_3 on cellular membranes *(10,15,17)* and not to the modulation of the apparent affinity of Stx to endothelial cells *(11,15)*. However, TNF-α fails to augment Stx-mediated cytotoxicity of dermal microvascular and renal cortical endothelial cells *(18,19)* and of bovine aortic endothelial cells (Bitzan, unpublished observation).

Various laboratories have now shown that Stx can activate cells, especially those with modest susceptibility to the cytotoxic effect of Stx (e.g., monocytes, gut epithelial and renal mesangial cells). In endothelial cells, "activation" refers to the induction of adhesion molecules, such as E-selectin, ICAM-1, and VCAM-1 upon stimulation with cytokines, which facilitates the adherence to and translocation of leukocytes through the vascular wall and involves a complex system of signal transduction events *(20)*. Endothelial cell activation following cytokine stimulation also results in the expression of a procoagulant or pro-oxidant phenotype. The latter has been linked to the generation of microthrombi, for example, in renal glomeruli in models of sepsis. Evidence for "classic" endothelial activation, originally defined by the response to inflammatory cytokines, has yet to be shown to occur in response to Stx in endothelial cells. However, there is no doubt that perturbation of the vascular endothelium plays an important role in human HUS and in animal models of verotoxemia.

It is now increasingly recognized that apoptosis represents the preferred mechanism of Stx-mediated cytotoxicity in epithelial cells, germinal center B cells, and cancer tissue, among others *(12,21–23)*. Apoptosis, a term borrowed from Greek, describes the shedding of leaves from a tree and was introduced in 1972 by Kerr et al. to describe a process that differs from necrotic cell death *(24)*. Whereas necrosis typically affects groups of contiguous cells and invokes an inflammatory reaction in the adjacent, viable tissue in response to the released cellular debris *(25)*, apoptosis is a physiological process that occurs throughout the life-span of an organism to regulate tissue development and adaptation. However, apoptosis has been increasingly implicated in a variety of renal and vascular diseases *(20,26)*. More recently, several laboratories demonstrated that Stx could indeed cause endothelial cells to commit suicide. Dermal microvascular endothelial cells are as sensitive to apoptosis as Vero cells and renal tubule epithelial cells *(19,27)*. Yoshida et al. showed that Stx-induced apoptosis in HUVEC is limited to freshly isolated cells, but that their susceptibility is lost after the first passage *(28)*. Although TNF-α alone failed to induce apoptosis in endothelial cells, Pijpers et al. observed that TNF-α sensitized normally "resistant" endothelium to Stx-mediated apoptosis and thus provided an explanation for the previously noted increase in Stx toxicity by inflammatory cytokines *(19)*, a situation reminiscent of the combined effect of TNF-α and cycloheximide *(20)*. However, Pijpers et al. also showed that in the absence of TNF-α, Stx, but not cycloheximide, induces apoptosis, at concentrations that induced comparable peptide synthesis inhibition *(19)*.

This chapter will detail methods for the isolation and propagation of HUVEC, glomerular endothelial cells, and bovine aortic endothelial cells (BAEC), followed by a brief description of and references to the isolation of endothelial cells from additional microvascular beds (foreskin, gut, brain, and lung) and potential sources and peculiarities of commercially available endothelial cells. Subsequently, classic methods to assess cytotoxicity, glycosphingolipid extraction, and Stx receptor binding are detailed for their use in vascular endothelial cells.

2. Materials

2.1. General Equipment

1. Tissue culture equipment: phase-contrast (inverted) microscope; water-conditioned tissue culture incubator (37°C; 5% CO_2); laminar-flow (safety) cabinet, cooled centrifuge; hemocytometer or automated cell counter.
2. Disposable plasticware: cell culture dishes (100 and 35 mm in diameter) or tissue culture flasks (25, 75, 175 cm²); multiwell dishes (6, 24, and 96 wells; Corning Costar Corp., Cambridge, MA). 15- and 50-mL sterile polypropylene tubes; disposable, sterile pipets (Falcon/Becton Dickinson); sterile filter 0.2 μm (Luer Lok; e.g., Corning).

2.2. Equipment and Tools for Endothelial Cell Isolation and Propagation

1. General items: sterile tweezers, scissor, scalpel, Kelly clamps, angiocaths (16 gage/2 cm); surgical silk (2-0) (e.g., Ethicon, Inc.) or nylon ties (SST-2, Extracorporeal Medical Specialties, Inc.), syringes (1, 10, 60 mL; Luer Lok), and needles (18G; e.g., Becton Dickinson).
2. Stainless-steel sieves (opening sizes 2000 to 38 μm; Endecotts, London, UK; also available from Bellco, Vineland, NJ).
3. Magnetic particle collector (Dynal, A.S., Oslo, Norway; cat. no. M1536).
4. Cloning glass cylinders, outer diameter 8 mm (Bellco Glass, Inc.).

2.3. Buffers and Solutions for Tissue Culture Work (see Note 1)

1. "HUVEC isolation buffer": 4 m*M* KCl, 140 m*M* NaC, 10 m*M* HEPES, 11 m*M* D-glucose, pH 7.3. Add 100 IU/mL penicillin and 100 μg/mL streptomycin (Pen/Strep) (Life Technologies/Gibco-BRL, Rockville, MD) after autoclaving.

2. Hanks balanced salt solution (HBSS) with phenol red and Ca^{2+}/Mg^{2+} (Gibco-BRL cat. no. 24020); HBSS without divalent cations ($HBSS^{--}$; Gibco-BRL cat no. 14170).
3. Dulbecco's phosphate-buffered saline (D-PBS) with (Gibco-BRL cat. no. 14040) and without Ca^{2+}/Mg^{2+} ($D-PBS^{--}$; Gibco-BRL cat. no. 14190)
4. 0.9% Saline, sterile (e.g., infusion bags with tubing).
5. Ringer's lactate solution.

2.4. Tissue Culture Media and Supplements

1. Heat-inactivated (56°C, 30 min) fetal bovine (FBS) and bovine calf serum (BCS), from Gibco-BRL, Boeringer-Mannheim, or Hyclone Laboratories (Logan, UT). Pooled human serum (HS), heat inactivated (local blood bank).
2. Antibiotics (Pen/Strep): penicillin (100 U/mL), streptomycin (100 μg/mL), final concentrations.
3. Heparin.
4. Endothelial cell growth supplement (Biomedical Technologies, Inc, Stoughton, MA; cat. no. BT-203; 50 mg/vial, lyophilized). Reconstitute with 10 mL tissue-culture-grade distilled water (dH_2O) and sterile-filtrate (0.2 μm). May be stored up to several weeks after reconstitution at 4°C.
5. Alternatively, crude endothelial cell growth factor is extracted from bovine brain as described by Maciag *(29)*.
6. Human (HSA) and bovine serum albumin (BSA; fraction 5).
7. Medium 199 (M199) for HUVEC. Basal medium: Earle's salts, 2.2 g/L sodium bicarbonate, 2 m*M* L-glutamine, and 25 m*M* HEPES buffer (Gibco-BRL; BioWhittaker, Walkersville, MD). Complete M199: 20% (v/v) FBS, 0.05 mg/mL (1:100 of reconstituted) endothelial cell growth supplement, Pen/Strep, and 17 U/mL heparin.
8. Medium 199 (M199) for glomerular microvascular endothelial cells (GMVEC) and HUVEC. Complete medium (alternative to 5.7): 10% FBS plus 10% pooled human serum, crude endothelial cell growth factor preparation 0.15 mg/mL, Pen/Strep, and 5 U/mL heparin.
9. RPMI 1640 for BAEC. Basal medium containing 25 m*M* HEPES, Phenol Red, and 200 m*M* L-glutamine (Gibco-BRL, cat. no. 11875). Glutamine decays when medium stored at 4°C and should be replaced after 2 wk.
10. Complete medium: 15% (v/v) FBS or BCS and Pen/Strep (*see* **Note 2**).

2.5. Isolation and Propagation of Endothelial Cells

1. Type 4 collagenase (for isolation of HUVEC and BAEC); type 2 collagenase (for GMVEC) (both from Worthington Biochemical Corp, Lakewood, NJ). Reconstitute crystalline enzyme before use with $D-PBS^{++}$ or (basal) tissue culture medium (1 mg/dL) and filter sterilize.
2. Gelatin (denatured tropocollagen; e.g., from Sigma or Fluka BioChemika, Buchs, Switzerland). To coat tissue culture dishes, dilute gelatin in tissue-culture-grade

water to 0.2% (HUVEC and BAEC) or 1% (GMVEC). Store at 4°C, keep sterile; may add Pen/Strep to avoid contamination.

3. Trypsin-EDTA: 0.5 g/L trypsin and 0.2 g/L sodium EDTA (0.05% trypsin/0.02% EDTA) in HBSS^{--} (without Ca^{2+}/Mg^{2+}). Store at –5°C to –20°C. Dilute further in HBSS^{--} if necessary. Avoid repeated freezing and thawing.

4. Antibodies and reagents for cell epitope characterization: monoclonal antibodies to PECAM-1 (XD31) (XLB, Amsterdam, The Netherlands) and von Willebrand factor (Dako, Oslo, Norway) (endothelial cells); cytokeratin 1–19, FITC-conjugated (Boehringer-Mannheim) (epithelial cells); alpha-smooth-muscle actin (Sigma) (mesangial cells); CD77 (Gb$_3$) (Biodesign International, Kennebunk, ME). If ascites are used, remove sodium azide by dialysis or column purification and filtersterilize. Acetylated low-density lipoprotein labeled with 1,1' dioctadecyl-1-3,3,3'3'-tetramethyl indocarbocyanine perchlorate (Dil-Ac-LDL) *(30)* (Biomedical Technologies, Inc.).

5. Dynabeads coated with goat anti-mouse IgG (Dynal). Beads are not sterile and have to be washed five times in sterile M199 before use.

6. Dimethyl sulfoxide (DMSO).

2.6. Glycolipid Extraction and Thin-Layer Chromatography

1. Chloroform, methanol, 2-propanol, acetone, plastic-backed Polygram silica gel plates (Machery-Nagel, Dueren, Germany) or plastic-coated silica gel F1500 thin-layered chromatography (TLC) plates (Schleicher and Schuell, Dassel, Germany); orcinol ferric chloride spray reagent [0.9% ferric chloride and 0.55% Orcinol (5-methylresorcinol) in acidified ethanol] for the detection of sugars, glycosides, and sulfolipids, including neutral glycosphingolipids (GSL) (Sigma); Tween-20 (polyoxyethylenesorbitan; Sigma); 4-chloro-1-naphthol and hydrogen peroxide (Sigma). Tris-buffered saline 0.05 *M*, pH 7.4 (TBS).

2. Other reagents: crystal violet dye (0.13% crystal violet in 5% ethanol–2% formalin in PBS); formaldehyde (formalin) 37%; Trypan blue (0.4% stock solution); MTT (methyl thiazolyl tetrazolium bromide; Sigma).

3. Methods

3.1. Human Umbilical Vein Endothelial Cells (HUVEC)

3.1.1. Isolation of HUVEC

1. Obtain umbilical cord and immediately store in antibiotic-containing ice-cold HUVEC isolation buffer or HBSS (200-mL bottle accommodates two cords). Start cell isolation procedure within 3 h. Autoclave tools and glassware ahead of time.

2. Coat tissue culture dishes with 0.2% (w/v) gelatin.

3. Inspect umbilical cord for clamped or otherwise damaged areas. Crop cord ends with a clean cut. Remove at least 1 cm from each end. Exclude damaged areas (clamp marks, nicks, bruises) and gelatinous mass. Umbilical cords have typically three large blood vessels: one vein (large patent vessel) and two arteries (narrow, constricted vessels).

4. Cannulate umbilical vein at both ends with plastic cannula of a 16-gage angiocath. Secure angiocaths with Kelly clamps, attach a 10-mL syringe and gently flush with isolation buffer or 0.9% saline. All solutions used during the procedure are of room temperature unless indicated otherwise. Observe for leaks. Angiocaths are fastened tightly with silk or nylon ties.

5. Clear vessel completely from blood with isolation buffer using 25- or 60-mL syringes or an infusion system (need 300–500 mL/cord).

6. Fill cord vessel with approx 10 mL of 0.1% (1 mg/mL in D-PBS or M199) sterile filtered, prewarmed (37°C) collagenase. Seal ends with 10-mL syringes. Immerse secured cord in a beaker. Beaker can be filled with prewarmed 0.9% NaCl. Incubate for 20 min at 37°C in water bath or incubator.

7. Flush collagenase solution back and forth and collect dislodged cells in a sterile 50-mL polypropylene tube. Repeat flushing the cord with 10 mL D-PBS or M199 at a time and collect cells (*see* **Note 3**).

8. Centrifuge at 200g to 500g for 5 min, resuspend pellet in complete M199, distribute cells into two or more wells of a gelatin-coated six-well-plate or one 10-cm dish and incubate overnight at 37°C/5% CO_2.

9. Aspirate culture medium, rinse repeatedly with D-PBS to remove red blood cells and add fresh medium. Replace medium every 2–3 d. Cells become confluent within 5–10 d and can be passaged at a split ratio of 1:4 to 1:6.

10. HUVEC are characterized with monoclonal antibodies to von Willebrand factor, PECAM-1 (CD31) or V,E-cadherin *(31)*, using immunofluorescence. Alternatively, Dil-Ac-LDL can be used (*see* **Notes 4** and **5**).

3.1.2. Propagation of HUVEC

1. To split monolayers, gently rinse cells with HBBS⁻ and detach cells by adding 0.5–1.0 mL of 0.05% trypsin/0.02% EDTA. Observe detachment under inverted microscope. Tap dish gently to dislodge adherent cells.

2. Add prewarmed, complete M199 to block trypsin activity. Transfer cells to 15-mL tube, spin at 200g to 500g for 5 min, resuspend cells in medium, and count cells.

3. Seed cells into new 0.2% gelatin-coated dishes. The optimal seeding density is $(2.5–5) \times 10^3/cm^2$ (or $[2–4] \times 10^4/mL$).

4. HUVEC are best used at passages 2 to 3. Following passage 5, the percentage of senescent cells increases and affects doubling time and cell phenotype. Senescent cells may appear flattened, with a large rim of cytoplasm.

3.2. Human Renal Glomerular Microvascular Endothelial Cells

Suitable tissue for the isolation of renal glomerular endothelial cells includes discarded donor kidneys (within 24 h to maximal 72 h postexplantation) or healthy portions of kidneys excised because of a renal tumor. Use only tissue that appears morphologically intact. The whole procedure is performed under sterile conditions.

3.2.1. Isolation of GMVEC

1. Remove and discard the capsula fibrosa of the kidney and separate the cortex from the medulla using a scalpel. Dissect the cortex into small pieces with a sterile pair of scissors.
2. Isolate the glomeruli by a gradual sieving procedure using sterile stainless-steel screens with decreasing opening sizes of 180, 125, 108, and 90 μm and a glass pestle. Rinse generously with isolation buffer. For pediatric kidneys, add a 53-μm-opening screen.
3. Collect retained glomeruli from each sieve into separate 50-mL tubes and spin at 200g for 5 min.
4. Remove supernatant and digest glomeruli for 1.5 h at 37°C with 1.5 mL filter sterilized, 0.1% (1 mg/mL) collagenase type 2 CLS in (incomplete) M199. Shake pellet repeatedly to loosen up the glomeruli.
5. Coat small flasks or six-well plates with 1% gelatin for at least 30 min at room temperature.
6. Inhibit collagenase by adding 5 mL complete M199 and collect the glomerular remnants by centrifugation at 200g to 500g for 5 min. Resuspend pellet in complete medium.
7. Seed glomerular remnants onto gelatin-coated wells or flasks. Use 25-cm^2 flasks for 1 mm^3 (1 g) of tissue; use 175-cm^2 flasks for whole kidney. Incubate at 37°C and 5% CO_2.
8. *Do not move dishes during the first 2 d* to allow attachment of glomerular remnants. Subsequently, replace culture medium every other day and monitor outgrowth of cells under a phase-contrast microscope.

3.2.2. Purification of GMVEC

1. Determine the ratio of epithelial/endothelial cells and start purification before epithelial cells overgrow endothelial cells (usually within 3–8 d).
2. Rinse flask with D-PBS⁻ or basal M199 and detach cells with 0.05% trypsin/ 0.02% EDTA at 37°C. To selectively detach endothelial cells, trypsinize only briefly, about 1 min, at room temperature and inspect cells through the phase-contrast microscope.
3. Neutralize trypsin by adding complete M199 immediately after endothelial cells become detached.
4. Filter cells through 38-μm sieve to reduce contamination by nonendothelial cells from the glomerular remnants during subculture (glomerular remnants are retained by the sieve).
5. Collect filtrate and centrifuge at 200g for 5 min. Wash pellet twice in HBSS containing 10% FBS at 200g for 5 min.
6. Resuspend pellet in 150 μL PBS/10% FBS supplemented with 10 μg/mL of monoclonal anti-PECAM antibody and incubate mixture 30 min on ice. Wash three times with HBSS/10% FBS to remove unbound antibody.
7. Resuspend pellet in 150 μL Hank's/10% FBS and add Dynabeads coated with goat-anti-mouse antibody. Calculate three beads/endothelial cell:

(Total number of cells) × (estimated fraction of endothelial cells present) × 3

   Excess beads interfere with the subsequent attachment of endothelial cells to the tissue culture dish. Incubate 15 min on ice.

8. Wash immunomagnetic positive fraction with HBSS/10% FBS. Collect magnetic particles as detailed by the manufacturer using the magnetic particle collector.
9. Resuspend isolated fraction in complete M199 and seed onto a new gelatin-coated dish. Regrow cells and exchange medium every 2–3 d.
10. Trypsinize monolayer, usually after 3–5 d, as described under **step 2**, count cells and repeat magnetic separation as above. The procedure can be combined with a negative selection using an epithelial-cell-specific antibody.
11. Seed purified cells at a split ratio of 1:2 to 1:3 using trypsin/EDTA as above. Continue feeding every 2–3 d. Repeat the immunomagnetic separation procedure one to two more times.
12. The obtained GMVEC culture is characterized with monoclonal antibodies to von Willebrand factor, PECAM-1 (CD31), or V,E-cadherin *(31)* and immunofluorescence methodology. The absence of contamination with nonendothelial cells is monitored by staining with cytokeratin 8 (epithelial cells) and α-smooth-muscle actin antibodies (mesangial cells). Alternatively, stain with Dil-Ac-LDL.
13. GMVEC can be used for experiments until passage 10.

3.3. Bovine Aortic Endothelial Cells

3.3.1. Isolation of BAEC

1. Resect a 15- to 20-cm-long portion of the aortic arch and thoracic aorta with a clean scalpel. Remove the adherent fat and collect the aorta segment in a wide-mouthed transport bottle containing ice-cold HBSS or D-PBS with antibiotics (Pen/Strep) (*see* **Note 6**).
2. Cut the aorta open lengthwise and position it with the edges slightly raised so that the luminal surface forms a shallow bowl and leakage from branching arteries is avoided. Rinse inner surface with sterile HBSS^{--} or PBS^{--} until clear of blood.
3. Add 2–3 mL of 0.05% trypsin/0.02% EDTA to cover the endothelial surface. After 2 min, collect detached endothelial cells with a Pasteur pipet. Rinse several times with serum-free RPMI and collect remaining cells. Pellet cells by centrifugation at 200g to 500g and 4°C, wash twice in serum-free RPMI, and, finally, resuspend in RPMI containing 15% BCS or FBS.
4. Seed cells into two to three 0.2% gelatin-coated 10-cm dishes with prewarmed medium. If cells are to be cloned, distribute pellet at three different densities.
5. Inspect cells daily using an inverted microscope for confluence and purity. Replace medium the next morning and then every 2–3 d.
6. Cells are characterized as endothelial cells by labeling with Dil-Ac-LDL, using fluorescence microscopy or FACS. Endothelial-specific antibodies can be used if they react with the corresponding bovine antigens.

3.3.2. Cloning and Propagation of BAEC

1. At appropriate plating density, BAECs form individual colonies or cell islands originating from a single, attached endothelial cell. Encircle separate endothelial

colonies on the reverse side of the culture dish with a fine marker under microscopic view. Rinse dish briefly with HBSS⁻⁻.

2. Using a pair of sterile tweezers, dip the cloning cylinder at one end in autoclaved Vaseline and place the cylinder greased side down firmly onto the plate so that the circled endothelial colony is contained within the cylinder. Use the microscope for the placement.

3. Add a few drops of 0.05% trypsin/0.02% EDTA into each cylinder and incubate for 2–3 min. Detach and collect cells by gentle, repeated aspirations using a lab pipet with thin plastic tips or a Pasteur pipet.

4. Seed cells onto separate, gelatin-coated, medium-filled, prewarmed culture dishes.

5. In the absence of contaminating nonendothelial cells and if cloned cells are not needed, let cells grow to monolayer and split them directly with 0.05% trypsine/ 0.02% EDTA at a ratio of 1:6.

6. Preferably, cells are used for four or five passages.

3.4. Endothelial Cells from Commercial Sources

1. Endothelial cells can be purchased commercially (e.g., from Clonetics/ BioWhittaker). This company offers human macrovascular and microvascular endothelial cells from various vascular beds, including umbilical veins, coronary and pulmonary arteries, aorta, and microvessels of the lung and foreskin (*see* **Note 7**).

2. The cells are shipped at passages 3 or 4 in flasks (at ambient temperature) or cryopreserved with 10% DMSO (on dry ice). One cryovial can be seeded into at least four T-25 flasks or two 10-cm plates or T-75 flasks.

3. The shipment includes optimized basal medium and supplements that can be mixed at the time when the cultures are started, along with trypsin/EDTA and balanced salt solutions. Detailed descriptions of the shipped items and instructions for the initiation and maintenance of the cells, including cell counting methods and suggestions for seeding densities and so forth are supplied by the distributor ("Endothelial Cell Systems Instructions," Clonetics BioWhittacker, brochure 04/1998).

4. The manufacturer assures experimental use for 10–15 population doublings. Continued passaging beyond 15 doublings (4–5 passages after receipt) may result in decreased growth rates and loss of biological responsiveness.

5. **Figure 1A** depicts a just confluent monolayer of human (dermal) microvascular endothelial cells.

3.5. Endothelial Cell Storage/Freezing and Thawing

1. Grow cells to 70–80% density.

2. Trypsinize cells with 0.05% trypsin/0.02% EDTA. Immediately after cell detachment, add 5 vol ice-cold complete growth medium. Combine cells from several plates. Centrifuge at 200*g* to 500*g* for 5 min at 4°C.

3. Resuspend pellet in complete medium containing 8% (v/v) DMSO. Use 1 mL DMSO containing medium per 100-mm plate. Aliquot into labeled cryovials. Keep cells and medium on ice.

A **B**

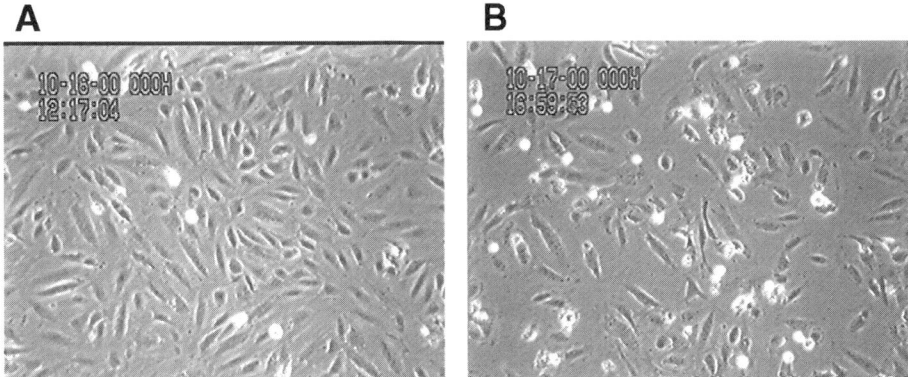

Fig. 1. Human dermal (foreskin) microvascular endothelial cells. The monolayer was exposed to Stx2 at a concentration of 0.1 nM and monitored using time-lapse video microscopy. (**A**) Cell morphology at time 0, (**B**) apoptotic changes evident 28 h after the addition of Stx.

4. Freeze vials for exactly 1 h at –20°C. Transfer vials to –80°C overnight. Store indefinitely in liquid nitrogen.
5. To thaw the cells, warm cryovial briefly in a 37°C water bath, disinfect vial by immersion into 70% ethanol, wipe dry, and immediately transfer into a ninefold volume of culture medium.

3.6. Cytotoxicity Assay

1. Cytotoxicity assays allow one to quantitate differences in the susceptibility of endothelial cultures to the cytotoxic effect of Stx and to analyze additive and synergistic effects of additional agents with Stx by determining the 50% cytotoxic dose ($CD_{50\%}$).
2. Trypsinize and count cells. Seed cells in 96-well microtiter plate; 5×10^4 cells/well is a good starting point. Grow to the desired density (just confluent or days postconfluence) under daily inspection. Change medium every 2–3 d.
3. When the monolayer has reached the desired density, add logarithmic (10-fold) toxin dilutions (in tissue culture medium) to triplicate wells. Omit the outermost wells. Inspect cells daily and record any damage to monolayers.
4. Assay cytotoxic effect. The simplest is a colorimetric method (e.g., staining with crystal violet). The integrity of cell function can be assessed with the MTT test. Alternatively, cells can be trypsinized and counted or stained with Trypan blue to determine the number and percentage of viable and dead cells.
5. Crystal violet staining (*32*). Rinse wells with PBS, aspirate all liquid. Fix cells with 70 µL of 2% formalin per well for 1 min. Remove formalin (dump into bleach with vigorous shaking). Add 70 µL of crystal violet stain for at least 20 min. Rinse microtiter plates generously with tap water until no more dye is flowing off; air-dry. Elute bound stain from cells with 50% (v/v) ethanol in water (100 µL).

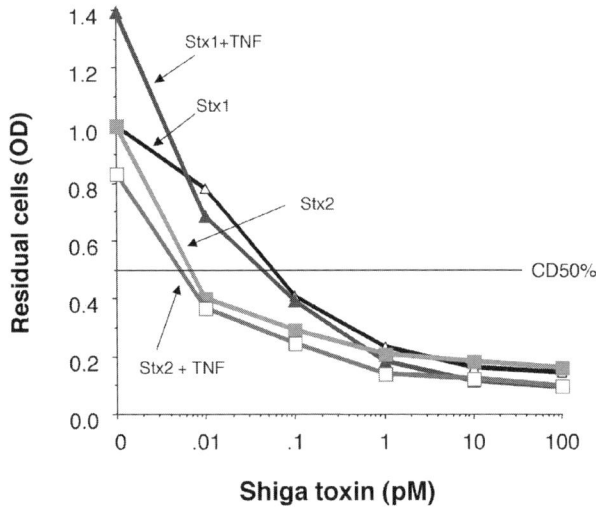

Fig. 2. Cytotoxicity assay (crystal violet stain). Confluent human dermal microvascular endothelial cells (96-well plate) were exposed for 2 d to Stx1 and Stx2 at the indicated dilutions, with and without TNF-α (10 ng/mL). The optical density obtained with vehicle-treated cells was set as 1. The 50% cytotoxic dose can be estimated graphically or derived mathematically as detailed elsewhere *(32,33)*. The Stx2 preparation is approximately one order of magnitude more potent than Stx1 in these microvascular endothelial cells. TNF-α, when added to the monolayer together with Stx, failed to modulate the effect of Stx1 or Stx2.

Tap plate or use orbital shaker. Read absorbance at 490 nm in the microplate reader. The optical density (OD) directly correlates with the number of attached cells. If the OD exceeds the linear range, because of high cell density, transfer 50-μL aliquots from each well to a new 96-well plate and add 150 μL PBS, or stain with less concentrated dye. Plot toxin concentration against optical density or express OD_{Stx} as a fraction of the OD_0 (Control or vehicle treated). Determine 50% cytotoxic dose ($CD_{50\%}$). *See* **Fig. 2** as an example.

6. Trypan blue exclusion test. Dilute cell suspension (v/v) with Trypan blue stock solution, incubate for 5 min, and count with hemocytometer. Dead cells appear blue.

7. MTT assay (*see* **Note 8**): Remove tissue culture medium from wells by aspiration or inversion. Rinse with PBS; blot dry gently. Add 50 μL MTT solution per well (1 mg/mL). Incubate for 2 h at 37°C. Add 30 μL SDS 3% and 150 μL of DMSO. The presence of metabolically active cells is reflected by dark purple color. Seal plate and shake 10 min on an orbital shaker. Read plate at 560 nm.

3.7. Stx Endothelial Cell-Binding Studies and Scatchard Analysis

1. Grow endothelial cells in 24-well plates to the desired density. One well of confluent endothelial cells (surface area 2 cm^2) corresponds to approx 5.6×10^4 cells (**Table 1**).

Table 1
Surface Areas of Typical Cell Culture Dishes

Culture dish	Dish size	Effective growth area[a]
Flasks	T-25	25 cm^2
	T-75	75 cm^2
Round dishes	35-mm diameter	9.6 cm^2
	100-mm diameter	78.5 cm^2
Multiwell dishes	6-well	9.6 cm^2/well
	12-well	3.8 cm^2/well
	24-well	2.0 cm^2/well
	96-well	0.32 cm^2/well

[a]Data from "Endothelial Cell Systems Instructions," Clonetics BioWhittacker, brochure 04/1998.

2. Radiolabeling of Stx is typically carried out with Na-^{125}I using the Iodogen procedure of Salacinski et al. *(34)*. The expected yield is 10–12 μCi/μg of protein.

3. Incubate cells with experimental agent(s) or vehicle. Wash wells three times with ice-cold serum-free medium containing 0.1% (1 mg/mL) human or bovine serum albumin for human and bovine cells, respectively.

4. Add various concentrations of ^{125}I-Stx in 0.1% albumin-containing medium and incubate on ice for 3 h. In a typical GMVEC experiment, a 100-fold range is used with final concentrations of ^{125}I-Stx between 0.3 n*M* (minimal) and 30 n*M* (maximal expected binding) *(10)*. Use duplicate or triplicate determinations.

5. Separate the free fraction of ^{125}I-Stx (free or "F") from the fraction of cell-bound VT (bound or "B"). Wash cells five times with 0.1% albumin-containing basal medium or HBSS and solubilize pellet cells in 400 μL of 1 *M* NaOH at room temperature.

6. Remove aliquots and assay for cell-associated ^{125}I-Stx ("B") and free radiolabeled toxin ("F") using a γ-counter.

7. Nonspecific binding is determined by incubating the cells with ^{125}I-Stx in the presence of a 25-fold to 100-fold excess of unlabeled Stx. Determine specific binding by subtracting the nonspecific fraction (binding in the presence of "cold" toxin) from the counts in the absence of unlabeled toxin. If the binding is specific, unlabeled Stx will outcompete the bound labeled fraction. Substantial counts in the presence of excess unlabeled toxin indicate nonspecific binding. Perform all determinations in duplicates.

8. To establish binding curves, the concentrations of cell-bound ^{125}I-Stx are plotted on the *y*-axis and concentrations of the added ^{125}I-Stx on the *x*-axis (*see* **Fig. 3A**).

9. Data can be analyzed by the method of Scatchard *(35)* to determine the binding affinity and the number of occupied receptors (*see* **Fig. 3B** and **Note 9**).

10. Specificity of binding and of receptors can be further studied by incubating the endothelial cells with an antibody to Gb$_3$ for 1 h prior to and during the Stx-binding step.

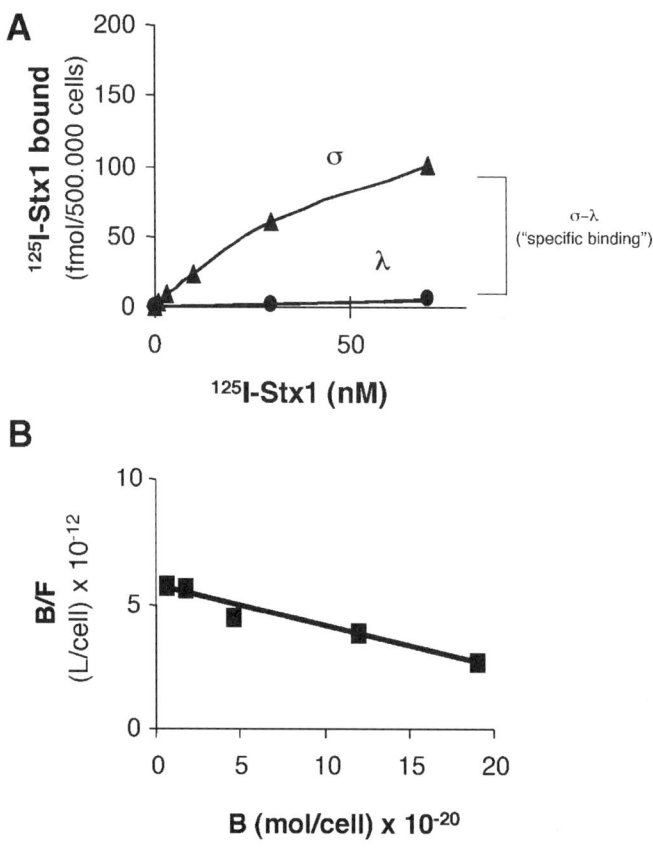

Fig. 3. Example for Stx-binding studies (human granulocytes). (**A**) Cells were incubated with increasing concentrations of ^{125}I-Stx$_1$ (0.3–70 n*M*) at 4°C for 3 h (σ). Nonspecific binding (λ) was demonstrated by incubating ^{125}I-Stx1 in the presence of 25-fold excess of unlabeled Stx1. Cells were washed and cell-attached radiolabel determined in a γ-counter. The difference (σ–λ) represents "specific binding." (**B**) Scatchard plot analysis. Calculation of *F* (unbound fraction) and *B* (cell-bound fraction) are detailed in **Subheading 3.7.** (Reproduced from te Loo, D. M. et al. (2000) Binding and transfer of verocytotoxin by polymorphonuclear leukocytes in hemolytic uremic syndrome. *Blood* **95,** 3396–3402, by copyright permission of the American Society of Hematology *[7].*)

3.8. Extraction and Characterization of Stx Receptor Glycosphingolipids by TLC

3.8.1. Extraction of Neutral Glycosphingolipids

1. Rinse cells with D-PBS^{--} and detach with 0.05% trypsine/0.02% EDTA.
2. Resuspend cells in ice-cold D-PBS and wash three times in D-PBS at 1000 g for 3 min at 4°C.

3. Resuspend pellet in D-PBS and add 20 vol of chloroform/methanol (2:1, v/v).
4. Filter suspension through glass wool.
5. Partition residue against 10 mL of 1 M NaCl (*see* **Note 10**), vortex, and separate phases by centrifugation for 5 min at 1800g.
6. Remove lower phase and dry under a stream of nitrogen.
7. Dissolve dried material in 1 mL of 0.4 M KOH in ethanol; keep for 2 h at 37°C.
8. Add 2 vol of chloroform and 2 vol of 1 M NaCl (*see* **Note 10**), centrifuge as above.
9. Remove lower phase and store in tightly sealed tube at –80°C.
10. Alternatively, dry lower phase under nitrogen, weigh dried extract, resuspend in chloroform:methanol (2:1, v/v), seal and store at –80°C.

3.8.2. Separation of Neutral Glycosphingolipids (TLC)

1. Separate desired amount of glycolipid extract (approx 100 µg/lane, cell-type dependent) by TLC in chloroform:methanol:water (60:25:4, v/v/v) (*36*). For reference, add 0.5–2 µg of standard neutral GSLs, at least lactosylceramide, Gb$_3$ and Gb$_4$.
2. To visualize glycosphingolipid bands, spray TLC plate with Orcinol and heat at 80°C until GSLs appear as purple bands.
3. To investigate the binding of Stx to endothelial GSLs, block replicate TLC plate with 1% gelatin in dH$_2$O at 37°C overnight, rinse with 0.05 M Tris-buffered saline, pH 7.4, and incubate with 0.1–0.5 µg/mL Stx in TBS–Tween for 1 h at room temperature.
4. For immunodetection of bound Stx, incubate plate with toxin-specific primary antibody, horseradish peroxidase conjugated, secondary antibody, both for 1 h at room temperature followed by generous washings with TBS–Tween. Develop signal with 4-chloro-1-naphthol in the presence of H$_2$O$_2$ (*37*). For quantitative analyses, incubate TLC plates with ^{125}I-labeled toxin. Visualize and analyze Stx binding by PhosphorImager autoradiography (*10*).

3.9. Molecular Studies to Investigate Cell Biological and Molecular Effects of Stx on Endothelial Cells

1. Standard molecular protocols are used for protein, DNA and RNA extraction and molecular analyses, signal transduction experiments, or apoptosis studies as detailed elsewhere (*19,38–41*).
2. **Figure 1B** depicts the effect of Stx treatment on human (dermal) microvascular endothelial cells. As confirmed by additional techniques using parallel dishes, almost all cells showed evidence of apoptotic cell death under these conditions.
3. A standardized cell culture approach is essential to obtain reproducible results (*see* **Notes 11–13**). A particular problem is the use of cells after multiple passages. The yield of primary endothelial cells is generally insufficient for most experimental purposes and several passages are required, especially for microvascular endothelial cells.
4. Rapidly dividing endothelial cell cultures differ from confluent monolayers with respect to their phenotype. An example is the expression of vasoactive mediators

and related molecules, such as endothelin-1, endothelial nitric oxide synthase, growth factors and their receptors *(42)*. Inferences from endothelial culture experiments therefore have to be interpreted with caution.

4. Notes

1. All solutions and material used for tissue culture must be sterile (autoclaved or sterile filtered) and should be endotoxin (pyrogen) free.
2. Bovine calf and human sera may contain Stx-neutralizing activity *(10,33,41)*. For Stx experiments, the use of fetal bovine serum is recommended. Confluent BAEC monolayers tolerate serum-free medium for up to 48 h without excessive rates of apoptosis, in contrast to HUVEC and microvascular endothelial cells.
3. HUVEC medium: Some laboratories add 10% FBS and 10% heat-inactivated (56°C, 30 min) human serum (e.g., from the local blood bank) instead of 20% FBS. There is no documented advantage for either modality with respect to cell growth.
4. For the characterization of endothelial cells by immunofluorescence, cells are grown on glass cover slips (plastic cover slips can give background fluorescence). Washings and incubations can be performed in 12- or 24-well plates. For microscopy the cover slip is mounted on a glass slide. Alternatively, cells can be grown in 22-mm dishes, stained *in situ* and monitored directly under the fluorescence microscope. The size of the 22-mm dishes suits the objective of most microscopes.
5. Electron microscopy can be used to identify Weibel–Palade bodies (storage granules containing vWF and P-selectin) that are specific for endothelial cells.
6. Bovine aortas can be obtained at the local abattoir. Avoid gross contamination and place aorta immediately in antibiotic containing, ice-cold HBSS^{++} or d-PBS.
7. Additional microvascular endothelial cell isolation and purification protocols are found in **refs. *43–48***.
8. The MTT assay is based on the reduction of the tetrazolium moiety of MTT to formazan. The reduction is mediated in part by the succinate dehydrogenase complex at the inner leaf of the mitochondrial membrane and by subsequent oxidative phosphorylation. The assay assumes that the capacity of cells to reduce MTT is proportional to cell number. Cells with low metabolic activity show reduced sensitivity *(49,50)*.
9. Scatchard analysis suffers from various statistical drawbacks and is only valid where the ligand binds to a single receptor population. A curved Scatchard plot will result where there are more than one type of binding sites, cooperativity in the binding, or experimental artifacts. It is therefore increasingly replaced by computer modeling to fit a mathematical model *(51)*. Peter and Lingwood, indeed, presented binding data that indicate the presence of at least two discrete Gb$_3$-binding sites resulting in two distinct rate constants and positive cooperativity. They imply that the concentration of Gb$_3$ on the surface of target cells has a nonlinear effect on Stx binding and, subsequently, biological activity *(52)*.
10. This extraction protocol follows the method published by Lingwood et al. *(36)*, but uses 1 *M* NaCl instead of water to increase the yield of neutral GSLs.
11. Prevent inadvertent activation of endothelial culture: Prewarm culture medium to 37°C before feeding, remove dishes only briefly from incubator, and so forth.

12. Lot-to-lot variations in the biological response of cultured endothelial cells may occur *(18)*, but are usually not a problem when testing for Stx cytotoxicity. Donor-specific, genetic differences of the response to Stx have yet to be defined.
13. Test Stx preparations for biologically relevant amounts of LPS (endotoxin). It may be useful to perform parallel tests with various dilutions of defined LPS. Heating Stx or incubation with polymyxin B has also proved effective to differentiate between Stx and LPS effects.

References

1. Jaffe, E. A., Nachman, R. L., Becker, C. G., and Minick, C. R. (1973) Culture of human endothelial cells derived from umbilical veins. Identification by morphologic and immunologic criteria. *J. Clin. Invest.* **52,** 2745–2756.
2. Gimbrone, M. A. (1976) Culture of vascular endothelium. *Prog. Hemost. Thromb.* **3,** 1–28.
3. Schwartz, S. M. (1978) Selection and characterization of bovine aortic endothelial cells. *In Vitro* **14,** 966–980.
4. Richardson, S. E., Rotman, T. A., Jay, V., Smith, C. R., Becker, L. E., Petric, M., et al. (1992) Experimental verocytotoxemia in rabbits. *Infect. Immun.* **60,** 4154–4167.
5. Boyd, B., Tyrrell, G., Maloney, M., Gyles, C., Brunton, J., and Lingwood, C. (1993) Alteration of the glycolipid binding specificity of the pig edema toxin from globotetraosyl to globotriaosyl ceramide alters in vivo tissue targetting and results in a verotoxin 1-like disease in pigs. *J. Exp. Med.* **177,** 1745–1753.
6. Uchida, H., Kiyokawa, N., Horie, H., Fujimoto, J., and Takeda, T. (1999) The detection of Shiga toxins in the kidney of a patient with hemolytic uremic syndrome. *Pediatr. Res.* **45,** 133–137.
7. te Loo, D. M., Monnens, L. A., van Der Velden, T. J., Vermeer, M. A., Preyers, F., Demacker, P. N., et al. (2000) Binding and transfer of verocytotoxin by polymorphonuclear leukocytes in hemolytic uremic syndrome. *Blood* **95,** 3396–3402.
8. te Loo, D. M., Hinsbergh, V. W., Heuvel, L. P., and Monnens, L. A. (2001) Detection of verocytotoxin bound to circulating polymorphonuclear leukocytes of patients with hemolytic uremic syndrome. *J. Am. Soc. Nephrol.* **12,** 800–806.
9. Louise, C. B. and Obrig, T. G. (1994) Human renal microvascular endothelial cells as a potential target in the development of the hemolytic uremic syndrome as related to fibrinolysis factor expression, in vitro. *Microvasc. Res.* **47,** 377–387.
10. van Setten, P. A., van Hinsbergh, V. W., van der Velden, T. J., van de Kar, N. C., Vermeer, M., Mahan, J. D., et al. (1997) Effects of TNF alpha on verocytotoxin cytotoxicity in purified human glomerular microvascular endothelial cells. *Kidney Int.* **51,** 1245–1256.
11. Van Setten, P. A., van Hinsbergh, V. W., Van den Heuvel, L. P., van der Velden, T. J., van de Kar, N. C., et al. (1997) Verocytotoxin inhibits mitogenesis and protein synthesis in purified human glomerular mesangial cells without affecting cell viability: evidence for two distinct mechanisms. *J. Am. Soc. Nephrol.* **8,** 1877–1888.
12. Williams, J. M., Boyd, B., Nutikka, A., Lingwood, C. A., Barnett Foster, D. E., Milford, D. V., et al. (1999) A comparison of the effects of verocytotoxin-1 on primary human renal cell cultures. *Toxicol. Lett.* **105,** 47–57.

13. Jacewicz, M. S., Acheson, D. W., Binion, D. G., West, G. A., Lincicome, L. L., Fiocchi, C., et al. (1999) Responses of human intestinal microvascular endothelial cells to Shiga toxins 1 and 2 and pathogenesis of hemorrhagic colitis. *Infect. Immun.* **67,** 1439–1444.

14. Thorpe, C. M., Hurley, B. P., Lincicome, L. L., Jacewicz, M. S., Keusch, G. T., and Acheson, D. W. (1999) Shiga toxins stimulate secretion of interleukin-8 from intestinal epithelial cells. *Infect. Immun.* **67,** 5985–5993.

15. van de Kar, N. C., Monnens, L. A., Karmali, M. A., and van Hinsbergh, V. W. (1992) Tumor necrosis factor and interleukin-1 induce expression of the verocytotoxin receptor globotriaosylceramide on human endothelial cells: implications for the pathogenesis of the hemolytic uremic syndrome. *Blood* **80,** 2755–2764.

16. Lingwood, C. A. (1999) Glycolipid receptors for verotoxin and Helicobacter pylori: role in pathology. *Biochim. Biophys. Acta* **1455,** 375–386.

17. Louise, C. B. and Obrig, T. G. (1991) Shiga toxin-associated hemolytic-uremic syndrome: combined cytotoxic effects of Shiga toxin, interleukin-1 beta, and tumor necrosis factor alpha on human vascular endothelial cells in vitro. *Infect. Immun.* **59,** 4173–4179.

18. Obrig, T. G., Louise, C. B., Lingwood, C. A., Boyd, B., Barley-Maloney, L., and Daniel, T. O. (1993) Endothelial heterogeneity in Shiga toxin receptors and responses. *J. Biol. Chem.* **268,** 15,484–15,488.

19. Pijpers, A. H., Setten, P. A., Heuvel, L. P., Assmann, K. J., Dijkman, H. B., Pennings, A. H., et al. (2001) Verocytotoxin-induced apoptosis of human microvascular endothelial cells. *J. Am. Soc. Nephrol.* **12,** 767–778.

20. Pober, J. S. (1998) Activation and injury of endothelial cells by cytokines. *Pathol. Biol. (Paris)* **46,** 159–163.

21. Inward, C. D., Williams, J., Chant, I., Crocker, J., Milford, D. V., Rose, P. E., et al. (1995) Verocytotoxin-1 induces apoptosis in vero cells. *J. Infect.* **30,** 213–218.

22. Mangeney, M., Lingwood, C. A., Taga, S., Caillou, B., Tursz, T., and Wiels, J. (1993) Apoptosis induced in Burkitt's lymphoma cells via Gb3/CD77, a glycolipid antigen. *Cancer Res.* **53,** 5314–5319.

23. Arab, S., Murakami, M., Dirks, P., Boyd, B., Hubbard, S. L., Lingwood, C. A., et al. (1998) Verotoxins inhibit the growth of and induce apoptosis in human astrocytoma cells. *J. Neurooncol.* **40,** 137–150.

24. Kerr, J. F., Wyllie, A. H., and Currie, A. R. (1972) Apoptosis: a basic biological phenomenon with wide-ranging implications in tissue kinetics. *Br. J. Cancer* **26,** 239–257.

25. Granville, D. J., Carthy, C. M., Hunt, D. W., and McManus, B. M. (1998) Apoptosis: molecular aspects of cell death and disease. *Lab. Invest.* **78,** 893–913.

26. Mitra, D., Kim, J., MacLow, C., Karsan, A., and Laurence, J. (1998) Role of caspases 1 and 3 and Bcl-2-related molecules in endothelial cell apoptosis associated with thrombotic microangiopathies. *Am. J. Hematol.* **59,** 279–287.

27. Bitzan, M. (2000) Verotoxin-induced endothelial cell activation and death∇molecular and pathogenetic implications. 4th International Symposium and Workshop on "Shiga toxin (Verocytotoxin)-producing *Escherichia coli* infections," Kyoto.

28. Yoshida, T., Fukada, M., Koide, N., Ikeda, H., Sugiyama, T., Kato, Y., et al. (1999) Primary cultures of human endothelial cells are susceptible to low doses of Shiga toxins and undergo apoptosis. *J. Infect. Dis.* **180,** 2048–2052.

29. Maciag, T., Cerundolo, J., Ilsley, S., Kelley, P. R., and Forand, R. (1979) An endothelial cell growth factor from bovine hypothalamus: identification and partial characterization. *Proc. Natl. Acad. Sci. USA* **76,** 5674–5678.

30. Voyta, J. C., Via, D. P., Butterfield, C. E., and Zetter, B. R. (1984) Identification and isolation of endothelial cells based on their increased uptake of acetylated-low density lipoprotein. *J. Cell. Biol.* **99,** 2034–2040.

31. Lampugnani, M. G., Resnati, M., Raiteri, M., Pigott, R., Pisacane, A., Houen, G., et al. (1992) A novel endothelial-specific membrane protein is a marker of cell–cell contacts. *J. Cell Biol.* **118,** 1511–1522.

32. Gentry, M. K. and Dalrymple, J. M. (1980) Quantitative microtiter cytotoxicity assay for Shigella toxin. *J. Clin. Microbiol.* **12,** 361–366.

33. Bitzan, M., Klemt, M., Steffens, R., and Muller-Wiefel, D. E. (1993) Differences in verotoxin neutralizing activity of therapeutic immunoglobulins and sera from healthy controls. *Infection* **21,** 140–145.

34. Salacinski, P. R., McLean, C., Sykes, J. E., Clement-Jones, V. V., and Lowry, P. J. (1981) Iodination of proteins, glycoproteins, and peptides using a solid-phase oxidizing agent, 1,3,4,6-tetrachloro-3 alpha,6 alpha-diphenyl glycoluril (Iodogen). *Anal. Biochem.* **117,** 136–146.

35. Scatchard, G. (1949) The attractions of proteins for small molecules and ions. *Ann. NY Acad. Sci.* **51,** 660–672.

36. Lingwood, C. A., Law, H., Richardson, S., Petric, M., Brunton, J. L., De Grandis, S., and Karmali, M. (1987) Glycolipid binding of purified and recombinant *Escherichia coli* produced verotoxin in vitro. *J. Biol. Chem.* **262,** 8834–8839.

37. Bitzan, M., Richardson, S., Huang, C., Boyd, B., Petric, M., and Karmali, M. A. (1994) Evidence that verotoxins (Shiga-like toxins) from *Escherichia coli* bind to P blood group antigens of human erythrocytes in vitro. *Infect. Immun.* **62,** 3337–3347.

38. Sambrook, J., MacCallum, P. and Russell, D. (2001) *Molecular Cloning: A Laboratory Manual.* Cold Spring Harbor Laboratory Press, Cold Spring Harbor, NY.

39. Darzynkiewicz, Z. and Traganos, F. (1998) Measurement of apoptosis. *Adv. Biochem. Eng. Biotechnol.* **62,** 33–73.

40. Willingham, M. C. (1999) Cytochemical methods for the detection of apoptosis. *J. Histochem. Cytochem.* **47,** 1101–1110.

41. Bitzan, M. M., Wang, Y., Lin, J., and Marsden, P. A. (1998) Verotoxin and ricin have novel effects on preproendothelin-1 expression but fail to modify nitric oxide synthase (ecNOS) expression and NO production in vascular endothelium. *J. Clin. Invest.* **101,** 372–382.

42. Flowers, M. A. and Marsden, P. A. (1994) Expression of endothelin-1 and nitric oxide is coupled to endothelial phenotype, especially growth state. *Exp. Nephrol.* **2,** 115–126.

43. Manconi, F., Markham, R., and Fraser, I. S. (2000) Culturing endothelial cells of microvascular origin. *Methods Cell Sci.* **22,** 89–99.

44. Marelli-Berg, F. M., Peek, E., Lidington, E. A., Stauss, H. J., and Lechler, R. I. (2000) Isolation of endothelial cells from murine tissue. *J. Immunol. Methods* **244,** 205–215.

45. Lamszus, K., Schmidt, N. O., Ergun, S., and Westphal, M. (1999) Isolation and culture of human neuromicrovascular endothelial cells for the study of angiogenesis in vitro. *J. Neurosci. Res.* **55,** 370–381.

46. Lou, J. N., Mili, N., Decrind, C., Donati, Y., Kossodo, S., Spiliopoulos, A., et al. (1998) An improved method for isolation of microvascular endothelial cells from normal and inflamed human lung. *In Vitro Cell. Dev. Biol. Anim.* **34,** 529–536.

47. Richard, L., Velasco, P., and Detmar, M. (1998) A simple immunomagnetic protocol for the selective isolation and long-term culture of human dermal microvascular endothelial cells. *Exp. Cell, Res.* **240,** 1–6.

48. Chung-Welch, N., Patton, W. F., Shepro, D., and Cambria, R. P. (1997) Two-stage isolation procedure for obtaining homogenous populations of microvascular endothelial and mesothelial cells from human omentum. *Microvasc. Res.* **54,** 121–134.

49. Isobe, I., Yanagisawa, K., and Michikawa, M. (2001) 3-(4,5-Dimethylthiazol-2-yl)-2,5-diphenyltetrazolium bromide (MTT) causes Akt phosphorylation and morphological changes in intracellular organellae in cultured rat astrocytes. *J. Neurochem.* **77,** 274–280.

50. van de Loosdrecht, A. A., Ossenkoppele, G. J., Beelen, R. H., Broekhoven, M. G., Drager, A. M., and Langenhuijsen, M. M. (1993) Apoptosis in tumor necrosis factor-alpha-dependent, monocyte-mediated leukemic cell death: a functional, morphologic, and flow-cytometric analysis. *Exp. Hematol.* **21,** 1628–1639.

51. Anonymous (2001) *Pharmacology Guide.* Glaxo Wellcome.

52. Peter, M. G. and Lingwood, C. A. (2000) Apparent cooperativity in multivalent verotoxin–globotriaosyl ceramide binding: kinetic and saturation binding studies with [(125)I]verotoxin. *Biochim. Biophys. Acta* **1501,** 116–124.

21

Shiga Toxin Interactions with the Intestinal Epithelium

Cheleste M. Thorpe, Bryan P. Hurley, and David W. K. Acheson

1. Introduction

Many different types of *Escherichia coli* are known to cause gastrointestinal disease, including enteroaggregative, enteropathogenic, enterotoxigenic, and enteroinvasive *E. coli*. However, none cause the severe morbidity that is sometimes seen in patients infected with Shiga toxin-producing *E. coli* (STEC). Although the majority of STEC infected patients have non-life-threatening diarrhea, STEC infection can result in serious gastrointestinal and systemic complications such as hemorrhagic colitis and hemolytic uremic syndrome (HUS) *(1)*. STEC produces a potent toxin that is biochemically, biologically, and genetically very similar to the Shiga toxin produced by *Shigella dysenteriae* type 1. Shiga toxins are considered to be the principal virulence factors involved in the systemic complications that follow STEC infection; however, the precise mechanisms by which Shiga toxins cause such disorders remain unclear.

Individuals are exposed to STEC following ingestion of the bacteria in contaminated foods, often undercooked ground beef, but also in milk, water, fruit, and vegetables. Once STEC organisms have been ingested, they are able to efficiently colonize the lower gastrointestinal (GI) tract. It is believed that STEC make Shiga toxins in the lower GI tract once colonization has occurred. This belief is supported by the observation that STEC-infected individuals have detectable amounts of free Shiga toxin(s) in their stools.

Usually, people infected with STEC do not develop the life-threatening complications of hemorrhagic colitis or HUS. However, in some infected individuals, serious systemic complications can result. These complications are primarily related to the development of thrombotic microangiopathy (TMA) in one or more organs, especially the kidney and central nervous system. Thrombotic microangiopathy is thought to be the result, at least in part, of the direct

From: *Methods in Molecular Medicine, vol. 73: E. coli: Shiga Toxin Methods and Protocols*
Edited by: D. Philpott and F. Ebel © Humana Press Inc., Totowa, NJ

action of Shiga toxins on endothelial cells in the microvasculature of affected organs. Because STEC is noninvasive, it has been hypothesized that Shiga toxins must be capable of gaining access to the systemic vasculature from the gastrointestinal tract in order for TMA to develop. If this conjecture is true, preventing Shiga toxin transport across the host intestinal epithelium and subsequent uptake into the systemic circulation should effectively prevent the systemic microvascular complications of STEC infection. Understanding how Shiga toxins interact with and cross the host gastrointestinal epithelium may ultimately prove to be critical in designing strategies to prevent such systemic complications. For this reason, developing a better understanding of Shiga toxin movement across intestinal epithelial cells (IECs) has been a key goal of our laboratory.

Shiga toxins (Stxs) belong to a family of bacterial protein toxins that share the same enzymatic activity and recognize structurally related host neutral glycolipid receptors *(2)*. There are various members of the Stx family including Stx1 and several different types of Stx2 *(2)*. Stxs are AB5 heterodimers consisting of a single enzymatically active A-subunit and a complex of five identical B-subunits that mediates binding to host cells. The cellular receptor for Stx1 and Stx2 is a blood-group-active glycolipid called globotriaosyl ceramide (Gb$_3$) *(3)*. The Stx1A-subunit is homologous with the A-subunit of the plant toxin ricin. In 1987, ricin was reported to be a single-site RNA *N*-glycohydrolase for the 28S rRNA of the mammalian ribosome, and soon thereafter, Stx1 and Stx2 were shown to have the same enzyme specificity *(4)*. Hydrolysis of this specific adenine residue irreversibly inhibits protein synthesis by blocking chain elongation. The precise way in which Stxs inactivate the 28S rRNA may prove to be very important in understanding exactly how Stxs affect host cells. Toxins that act by causing site-specific damage to the α-sarcin loop of the 28S rRNA have been shown to alter host signal transduction pathways involved in responses to stress and mitogens *(5)*. Our work has shown that Stxs may act in a similar way in intestinal epithelial cells to alter host gene regulation *(6)*.

In order for Stx to gain access to the systemic vasculature, it has to cross the IEC barrier. Theoretically, Stx could accomplish this in at least three ways. First, Stx could simply destroy the IECs, thereby creating a breach in the barrier. Second, Stx could cross an intact barrier via a transcellular route. Third, the toxin could damage tight junctions and cross via a paracellular route. Although any or all of these mechanisms may operate in vivo, they do not take into account a potential active role of the Stx–IEC interaction itself. As discussed below, we have developed models to examine Stx–IEC interactions and we have used the methods outlined in this chapter to investigate how Stx crosses the intestinal epithelial epithelium and how Stx may be affecting its own movement via inducing polymorphonuclear transmigration through the epithelial barrier.

In order to gain a better understanding of Stx–IEC interactions, we have focused our efforts on three areas using in vitro models described in this chapter. First, we have investigated Stx movement across polarized IECs grown on permeable polycarbonate filters. Second, we are assessing the effects of Stxs on IECs themselves, specifically relating to the regulation of cytokines and other inflammatory mediators that may alter host gastrointestinal epithelium permeability. Third, we are investigating the effect of acute inflammatory cells on Stx movement across polarized IECs in vitro.

In this chapter we describe the methods and models we have used to study Stx movement across, and interaction with, intestinal epithelial cells in vitro. Our goal is to provide a detailed description of the methods we have used to study these phenomena. The reader is then referred to individual published articles if further details are desired regarding our results using these methods.

2. Materials

2.1. Assessing Shiga Toxin Translocation

2.1.1. Collagen-Coating Transwell Filters (see **Note 1**)

1. Polycarbonate filter inserts: 0.33 cm^2, 5-μm pore Transwell membranes (Costar, Cambridge, MA) (*see* **Note 2**).
2. Tail from large rat.
3. 70% Ethanol.
4. 1% Glacial acetic acid.
5. Dialysis tubing (Spectrophor #1, molecular weight [MW] cutoff of 6000–8000).

2.1.2. Intestinal Epithelial Cells and Reagents (see **Note 3**)

1. CaCo-2A cells *(7)* and tissue culture medium: Dulbecco's modified Eagle medium (DMEM) with 100 mg/L sodium pyruvate, 10% fetal calf serum (FCS), 100 U/mL penicillin G, 100 μg/mL streptomycin sulfate, and 25 mM HEPES.
2. T84 cells (ATCC) and tissue culture medium: DMEM/F12 with L-glutamine and 15 mM HEPES, 10% FCS, 100 U/mL penicillin G, and 100 μg/mL streptomycin sulfate.
3. HCT-8 (ATCC) and tissue culture medium: RPMI 1640 with L-glutamine, 10% FCS, 100 U/mL penicillin G, 100 μg/mL streptomycin sulfate, 25 mM HEPES, and 1 mM sodium pyruvate.
4. Millicell-ERS apparatus (Millipore Corporation, Bedford, MA).

2.1.3. Movement of Stx Across the IEC Monolayer

1. IECs grown for 8–10 d on collagen-coated Transwell filters.
2. [^{3}H]-inulin (1.25 μCi/mL).
3. Purified Stx1 or Stx2 (*see* elsewhere in this volume for purification procedure).
4. Scintillation fluid and counter.

2.2. Effect of Polymorphonuclear Leukocyte (PMN) Transmigration on Stx Movement In Vitro

2.2.1. Growing IECs on Inverted Filters

1. IECs and appropriate medium (*see* **Subheading 2.1.2.**).
2. Collagen-coated Transwell filters (*see* **Subheading 2.1.1.**).
3. Millicell-ERS apparatus.

2.2.2. Isolation of Human PMNs

1. Syringes containing 10 U of heparin per milliliter.
2. Sterile saline.
3. Ficoll solution: 50 g Ficoll (Sigma) to 557 mL endotoxin-free bottled sterile water. Do not mix; add powder to the surface of the water and cover with aluminum foil. Let sit overnight, and the next day, filter through a 0.2-μm filter. Store at room temperature.
4. Hypaque solution: four 50-mL bottles of Hypaque-76 (NycoMed) to 248 mL sterile water. Filter though a 0.45-μm filter and store at room temperature.
5. Ficoll–Hypaque solution: Combine Ficoll and Hypaque solutions to a ratio of 2.4:1 Ficoll:Hypaque under sterile conditions on the day of experiment.
6. 3% Dextran (approx 500,000 kDa).
7. Hank's balanced salt solution (HBSS) without Ca^{2+} or Mg^{2+}.
8. 0.2% NaCl (w/v).
9. 1.6% NaCl (w/v).
10. Trypan blue stain.
11. Hemocytometer and microscope.

2.2.3. PMN Transmigration and Stx Movement

1. IECs grown inverted on collagen-coated filters.
2. Isolated neutrophils.
3. [^{3}H]-Inulin (1.25 μCi/mL).
4. Purified Stx1 or Stx2 (*see* elsewhere in this volume for purification procedure).
5. Formyl-Met-Leu-Phe (fMLP) (stock concentration 10^{-2} M in dimethyl sulfoxide (DMSO), stored at –70°C).

2.2.4. Myeloperoxidase Assay

1. 1% Triton X-100 in HBSS.
2. ABTS solution: 1 mM ABTS [2,2-azino-bis(3-ethylbenz-thiazoline)-6-sulfonic acid; Sigma]), 100mM citric acid, pH 4.2, and 0.03% H_2O_2. For example, to make up 10 mL, add 9 mL distilled water (dH$_2$O), 1 mL of 1 M citric acid, pH 4.2, 10 μL of 30% H_2O_2, and 5.5 mg ABTS.
3. Spectrophotometer.

3. Methods

3.1. Assessing Stx Translocation

3.1.1. Collagen-Coating Transwell Filters

1. Remove tail from a large rat and remove skin, exposing the shiny white tendon, which is mainly collagen.
2. Pull tendons off, mince with a razor blade, and weigh. Tendons may be stored in 0.75-g aliquots at –80°C.
3. Soak a 0.75-g aliquot of tendon in 70% ethanol for 10–20 min.
4. Solubilize in 100 mL prechilled 1% glacial acetic acid, then stir at 4°C overnight.
5. Centrifuge the solution at 25,000g for 30 min, after which the supernatant is removed and the pellet discarded.
6. Place the supernatant in dialysis tubing presoaked in distilled water for 1 h. The dialysis tubing with collagen preparation is then placed in 2 L of distilled water and dialyzed for 24 h at 4°C with two to three changes of distilled water. Use approx 10 vol of water in the bath for each volume of collagen in the dialysis tube.
7. Remove collagen from the bag. The material should be somewhat viscous. Protein concentration of this solution is approx 0.5 mg/mL and should be stored at 4°C.
8. Prior to coating Transwells, dilute collagen solution 1:100 in 60% ethanol (can be stored for up to 1 yr).
9. Add 50 μL per Transwell filter and leave the filters to dry open under a sterile hood for 1 h minimum or overnight.
10. Rinse 1X with phosphate-buffered saline (PBS) and leave filters to dry for 2 h.

3.1.2. Growth of Tissue Culture Cells

1. Seed intestinal epithelial cell lines (Caco-2A, T84, or HCT-8) at a density of $(2–5) \times 10^5$ cells/mL; add 100 μL of cells in medium to the upper chamber of collagen-coated Transwell filters placed in a 24-well plate and 600 μL of medium to the bottom chamber (*see* **Note 4**; schematic shown in **Fig. 1**).
2. Grow the cells 8–10 d postseeding to allow the cells to polarize and form tight junctions. Change media in both top and bottom wells every 2 d (*see* **Note 5**).
3. The development of tight junctions is determined by measuring electrical resistance across the IEC monolayers *(8)*. This is performed by using the Millicell-ERS resistance meter, as described in the manual.
4. Before measuring resistance, remove Transwell filters from the incubator and allow them to reach room temperature, which takes about 10 min under a tissue culture hood with the cover off.
5. Place the electrode in the Transwell filter, one electrode per chamber, and note the reading in ohms. Test at least two times to ensure that measurements represent a stable reading (*see* **Note 6**).
6. Convert the measured value into an $\Omega \cdot cm^2$ value by subtracting the Ω measurement received with a cell-free Transwell filter (usually around 150 Ω) and multiplying by the surface area of the Transwell filter.

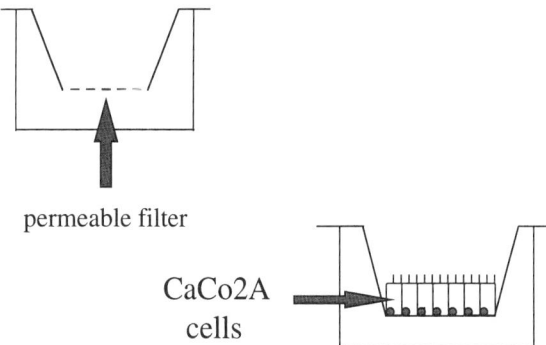

permeable filter

CaCo2A
cells

Fig. 1. Schematic model of intestinal epithelial cells grown on Transwell filters used to study Shiga toxin translocation. In the model shown, the upper surface of the filter is seeded with CaCo-2A cells, which then form polarized monolayers.

3.1.3. Movement of Stx and Inulin Across the IEC Monolayer (see **Note 7**)

1. Dilute Stx and [³H]-inulin in an appropriate culture medium to 1 µg/mL and 2.5 µCi/mL and add in a volume of 100 µL to the upper chamber of the Transwell filter.
2. Incubate the filters at 37°C in 5% CO_2 for 24 h.
3. Collect samples from the lower and upper chambers and freeze at –70°C prior to further processing.
4. Assay Stx either using Vero cells for determining biological activity *(9)* (described elsewhere in this volume) or enzyme immunoassay using Stx-specific monoclonal antibodies *(10,11)* (described elsewhere in this volume).
5. Determine the amount of [³H]-inulin in sample by adding 10 µL of sample to 4 mL scintillation fluid and measure counts per minute in a scintillation counter.
6. Record values as the amount of translocated toxin or [³H]-inulin as a percentage of the amount added. An example of these data is shown in **Table 1**.

3.2. Effects of PMN Transmigration on Stx Movement

3.2.1. Growing Inverted IEC Monolayers

1. Add 70 µL of cell suspension ($[2–5] \times 10^5$ cells/mL) to the bottom side of collagen-coated filters that are placed upside down in a 12-well plate.
2. Leave filters upside down overnight with lid on in the incubator, which is a sufficient amount of time to permit cell attachment.
3. The following day, flip the filter membranes upright in a 24-well plate and add 100 µL of media to the top well, with 600 µL of media added to the bottom well. This maneuver reverses the polarity of the cells with the basolateral surface facing up and the apical surface facing down *(13)*.
4. Measure electrical resistance of inverted monolayers in experiments as with the conventional monolayers 8–10 d postseeding (described in **Subheading 3.1.2.**).

Table 1
Comparison of Stx1 and Stx2 Translocation Across CaCo-2A Cells

	n	\multicolumn{3}{c}{Percent movement[a]: mean (SD)}		
		Toxin (72 kDa)	^{3}[H]-Inulin (5 kDa)	HRP (44 kDa)
Stx1	4	4.2(1.0)[b]	6.1(0.2)[c]	0.05(0.02)[d]
Stx2	6	0.1(0.1)[b]	6.2(0.5)[c]	0.08(0.09)[d]

[a]Percent movement represents the amount of toxin, [^{3}H]-inulin, and HRP measured in both the starting material as well as in the medium recovered from the opposite chamber following translocation (24 h). Values are reported as the amount translocated as a percentage of the amount added.
[b]$p < 0.0001$.
[c]p = not significant (ns).
[d]p = ns.
Source: Adapted from **ref. 12**.

3.2.2. Isolation of Human PMNs

1. Draw blood from healthy donors into syringes containing 10 U of heparin per milliliter of blood.
2. Dilute blood in saline 1:2.
3. Mix Ficoll and Hypaque 2.4:1 on the day of the experiment and use a 3:1 volume of blood/saline:Ficoll–Hypaque. For example, mix 10 mL of blood with 20 mL of saline, and then overlay onto 10 mL of Ficoll–Hypaque using a pipet. The rate of overlay should be 1 mL of blood/saline mix per minute or less to minimize mixing at the interface.
4. Centrifuge Ficoll–Hypaque overlays at 300g for 35–40 min at room temperature. Do not use the brake to slow the centrifuge, as this will disturb the gradient that is formed.
5. Discard the top layer of the gradient, which consists of saline and plasma.
6. Discard the interface of the plasma and Ficoll–Hypaque, which contains lymphocytes and monocytes, as well as the Ficoll–Hypaque layer, leaving the red blood cell (RBC) pellet.
7. Separate the neutrophils from the RBCs; these cells are present as a thin layer of cells above the RBC pellet. Dextran sedimentation followed by hypotonic lysis is used to isolate a cell suspension containing 95–98% PMNs. First, add an equal volume of 3% dextran (approx 500,000 kDa) to the neutrophil/RBC pellet, mix, and allow to sit for 20 min at room temperature.
8. Remove the supernatant carefully and place in a clean 50-mL conical centrifuge tube. Discard the RBC pellet.
9. Add enough ice-cold HBSS without Ca^{2+} or Mg^{2+} to the supernatant to fill the 50-mL tube.
10. Centrifuge at 300g for 5 min and carefully discard the supernatant while saving the neutrophil pellet with residual RBCs.
11. To lyse, add 25 mL ice-cold 0.2% NaCl to the pellet and mix gently by inversion for 20 s.

12. Add 25 mL cold 1.6% NaCl and centrifuge at 500*g* for 5 min.
13. Retain the pellet and repeat the hypotonic lysis (from **step 11**).
14. Resuspend the final pellet in 10 mL HBSS without Ca^{2+} and Mg^{2+}.
15. Determine the number of viable cells by staining with Trypan blue and counting with a microscope and hemocytometer.

3.2.3. PMN Transmigration and Stx Movement

1. Add 2×10^6 neutrophils to the top chamber of Transwell filters containing T-84 or CaCo-2A cells grown in the inverted fashion (and therefore corresponding to the basolateral IEC surface).
2. Add [^{3}H]-inulin and Stx1 or Stx2 to the bottom chamber (apical IEC surface).
3. Add the neutrophil chemoattractant formyl-Met-Leu-Phe (fMLP) to a final concentration of 1 μ*M* to the bottom chamber of some filters and the vehicle to others for a control.
4. Allow neutrophils to transmigrate for 2 h at 37°C in 5% CO_2.
5. Collect supernatants from the top chamber, filter through a 0.2-μm filter for measurement of [^{3}H]-inulin and Stx.
6. Count the number of neutrophils in the bottom chamber using the myeloperoxidase assay *(14)* (*see* **Subheading 3.2.4.**).

3.2.4. Myeloperoxidase Assay

1. Centrifuge the plates at 300*g* for 5 min to pellet the PMNs that have migrated.
2. Remove supernatant, leaving approx 300 μL that is suspected to contain neutrophils.
3. Add 300 μL of 1% Triton X-100 in HBSS to this supernatant and incubate on an orbital shaker for 20 min at room temperature.
4. Add 600 μL ABTS solution and allow the subsequent reaction to proceed at room temperature for about 20 min.
5. Read optical density (OD) values at 405 nm. Include a standard curve of neutrophils ranging from 5000 to 600,000 neutrophils in 300 μL, with the number of neutrophils determined by counting. Our data show that neutrophil transmigration does indeed increase Stx translocation (*see* **ref. 15**).

3.3. Concluding Remarks

We do not understand how and why certain individuals infected with STEC go on to develop severe systemic complications such as HUS. It is probable that there are both bacterial and host factors that contribute to the likelihood of getting severe disease following STEC infection. However, it seems very likely that the amount of Stx that moves from the intestine and into the systemic vasculature is a critical determinant of whether an STEC-infected individual develops systemic complications such as HUS. We have used the methods outlined in this chapter to ask basic questions about how Stxs get from the gut lumen into the bloodstream and what factors regulate this process. Data from our group and others suggest that Stxs can cross an intact intestinal epithelial cell monolayer in vitro and remain biologically active *(16,17)*. We have also shown that Stx1 and Stx2 likely traverse this monolayer through different pathways *(12)*. New data using PMNs indicate that the acute inflammatory response may also be playing a role in the movement of Stxs from the intestinal lumen and into underlying tissues *(15)*.

We are beginning to explore other questions related to understanding the Stx–host intestinal epithelium interaction. What factors alter translocation of Stxs across the intestinal epithelium? How critical is the host inflammatory response in altering Stx uptake into the systemic circulation? We are also trying to understand the specific ways in which Stxs alter host signal transduction to permit primary response mRNA and protein expression despite an overall inhibition of protein synthesis. If our in vitro observations are applicable in vivo, then one may speculate that Stx itself is contributing to a positive feedback loop. In this scenario, the toxin may lead to an increase in chemokines, resulting in increased PMN infiltration and transmigration that results in greater amounts of toxin crossing the IEC barrier. Hopefully, understanding Stx–intestinal epithelial cell interactions will allow us to find strategies to prevent the severe and sometimes deadly sequelae of STEC infection.

4. Notes

1. Alternatively, rat tail collagen type 1 can be purchased from Gibco-BRL. As the concentration of these solutions changes with the lot ordered, ensure that 5 µg/cm^2 of Transwell filter is used for coating.
2. At the time this chapter was written, the manufacturer of the Transwell filters had recently changed their manufacturing process. Filters are now usually heat bonded to the filter support, rather than solvent bonded. This change in manufacture had detrimental effects on our experimental system when Transwell filters were seeded with intestinal epithelial cells for inverted monolayer experiments. Using the heat-bonded filters, the intestinal epithelial cells would not attach. Solvent-bonded Transwell filters are still available by special order.
3. A variety of human intestinal epithelial cell lines such as CaCo-2A, T-84, and HCT-8 derived from colon carcinomas have been used to study Shiga toxin movement. CaCo-2A and HCT-8 cells have villous characteristics, whereas T-84 cells are more cryptlike. All three cell lines form polarized monolayers that generate high electrical resistance and develop good barrier function. Any of the cell lines described above are useful for assessing the effects of Shiga toxin translocation without the complications of cytotoxicity. T-84 cells are essentially devoid of the Stx receptor Gb$_3$ and are resistant to Stxs. CaCo-2A cells have low levels of Gb$_3$ and are naturally highly insensitive to Stx1 and Stx2, but both Gb$_3$ levels and sensitivity can be increased by treatment with sodium butyrate *(18)*. Concentrations of Stxs as high as 10 µg/mL do not inhibit protein synthesis in either cell line (unpublished observations). HCT-8 cells are sensitive to Stxs during their initial growth period; however, once they have formed polarized monolayers, in our hands they appear to be insensitive to Stxs (up to levels of 10 µg/mL).

4. These volumes are recommended by the manufacturer to minimize the effects of hydrostatic pressure.

5. It is critical to use extreme care when removing media from the top chamber during feeding to avoid touching the monolayer. Aspirating is not recommended; we prefer to tip the Transwell filter to remove spent media.

6. After 8 d, measurements are typically 800–2000 $\Omega \cdot cm^2$ depending on the cell line. For T84 cells at 9–13 d postseeding, the measurements are typically > 2000 Ω thus, it is necessary to use the 20,000-Ω scale on the Millicell-ERS meter.

7. To ensure that each monolayer remains intact and represents an epithelial barrier during the course of the experiment, the widely used paracellular marker [^{3}H]-inulin is added in each experiment *(14)*. Only a small amount of [^{3}H]-inulin (3–10% over 24 h) is able to move across polarized monolayers unless cell injury occurs or tight junctions are altered *(12,16)*. [^{3}H]-Inulin, Stx1, and Stx2 freely equilibrate across the filters in the absence of cells, demonstrating that the collagen-coated filter itself offers no significant barrier to the movement of molecules.

Acknowledgments

The work described in this chapter was supported by the National Institutes of Health (AI-39067) and by the Charles H. Hood Foundation (Boston MA).

References

1. Paton, J. C. and Paton, A. W. (1998) Pathogenesis and diagnosis of Shiga toxin-producing *E. coli* infections. *Clin. Microsc. Rev.* **11,** 450–479.

2. Acheson, D. W. K. and Keusch, G. T. (1999) The family of Shiga toxins, in *The Comprehensive Sourcebook of Bacterial Protein Toxins* (Alouf, S. E. and Freer, J. H., eds.), Academic, London, pp. 229–242.

3. Jacewicz, M., Feldman, H. A., Donohue-Rolfe, A., Balasubramanian, K. A., and Keusch, G. T. (1989) Pathogenesis of Shigella diarrhea. XIV. Analysis of Shiga toxin receptors on cloned HeLa cells. *J. Infect. Dis.* **159,** 881–889.

4. Endo, Y., Tsurugi, K., Yatsudo, T., Takeda, Y., Ogasawara, T., and Igarashi, K. (1988) Site of action of a verotoxin (VT2) from *E. coli* O157:H7 and of Shiga toxin on eukaryotic ribosomes: RNA *N*-glycosidase activity of the toxin. *Eur. J. Biochem.* **171,** 45–50.

5. Iordanov, M., Pribnow, M. D., Magun, J. L., Dinh, T.-H., Pearson, J. A., Chen, S. L-Y., et al. (1997) Ribotoxic stress response: activation of the stress-activated protein kinase JNK1 by inhibitors of the peptidyl transferase reaction and by sequence-specific damage to the α-sarcin loop in the 28S rRNA. *Mol. Cell. Biol.* **17,** 3373–3381.

6. Thorpe, C. M., Hurley, B. P., Lincicome, L. L., Jacewicz, M. S., Keusch, G. T., and Acheson, D. W. K. (1999) Shiga toxins stimulate secretion of interleukin-8 from intestinal epithelial cells. *Infect. Immun.* **67,** 5985–5993.

7. Van Beers, E. H., Al, R. H., Rings, E. H., Einerhand, A. W., Dekker, J., and Buller, H. A. (1995) Lactase and sucrase-isomaltase gene expression during Caco-2 cell differentiation. *Biochem. J.* **308,** 769–775.

8. Parkos, C. A., Delp, C., Arnaout, M. A., and Madara, J. L. (1991) Neutrophil migration across a cultured intestinal epithelium. Dependence on a CD11b/CD18 mediated event and enhanced efficiency in a physiological direction. *J. Clin. Invest.* **88,** 1605–1612.

9. Keusch, G. T., Donohue-Rolfe, A., Jacewicz, M., and Kane, A. V. (1988) Shiga toxin: production and purification. *Methods Enzymol.* **165,** 152–162, 399–401.

10. Donahue-Rolfe, A., Acheson, D. W. K., Kane, A. V., and Keusch, G. T. (1989) Purification of Shiga and Shiga-like toxins I and II by receptor analog affinity chromatography with immobilized P1 glycoprotein and the production of cross-reactive monoclonal antibodies. *Infect. Immun.* **57,** 3888–3893.

11. Acheson, D. W. K., Jacewicz, M., Kane, A. V., Donahue-Rolfe, A., and Keusch, G. T. (1993) One step high yield purification of Shiga like toxin II variants and quantitation using enzyme-linked immunosorbent assays. *Microb. Pathogen.* **14,** 57–66.

12. Hurley, B. P., Jacewicz, M., Thorpe, C. M., Lincicome, L. L., King, A. J., Keusch, G. T., et al. (1999) Shiga toxins 1 and 2 translocate differently across polarized intestinal epithelial cells. *Infect. Immun.* **67,** 6670–6677.

13. Madara, J. L. and Dharmsathaphorn, K. (1985) Occluding junction structure-function relationships in a cultured epithelial monolayer. *J. Cell Biol.* **101,** 2124–2133.

14. Madara, J. L. (1998) Regulation of the movement of solutes across tight junctions. *Annu. Rev. Physiol.* **60,** 143–159.

15. Hurley, B. P., Thorpe, C. M., and Acheson, D. W. K. (2001) Shiga toxin translocation across intestinal epithelial cells is enhanced by neutrophil transmigration. *Infect. Immun.* **69,** 6148–6155.

16. Acheson, D. W. K., Moore, R., De Bruecker, S., Lincicome, L. L., Jacewicz, M., Skutelsky, E., et al. (1996) Translocation of Shiga toxin across polarized intestinal cells in tissue culture. *Infect. Immun.* **64,** 3294–3300.

17. Philpott, D. J., Ackerley, C. A., Kiliaan, A. J., Karmali, M. A., Perdue, M. H., and Sherman, P. M. (1997) Translocation of verotoxin-1 across T-84 monolayers: mechanism of bacterial toxin penetration of epithelium. *Am. J. Physiol.* **273,** G1349–G1358.

18. Jacewicz, M. S., Acheson, D. W. K., Mobassaleh, M., Donahue-Rolfe, A., Balasubramanian, K. A., and Keusch, G. T. (1995) Maturational regulation of globotriaosylceramide, the Shiga-like toxin I receptor, in cultured human gut epithelial cells. *J. Clin. Invest.* **96,** 1328–1335.

22

Protocols to Study Effects of Shiga Toxin on Mononuclear Leukocytes

Christian Menge

1. Introduction

Endothelial cells are regarded as the main targets of the Shiga toxins (Stxs) during infections caused by Stx-producing *Escherichia coli* (STEC). However, several investigations also confirmed an effect of these toxins on immune cell functions in species naturally infected with STEC. Human B-cell lines *(1)* and tonsillar B-cells *(2)* are highly susceptible to the cytotoxic activity of Stx1, which also hampers activation and proliferation of bovine B- and T- cell subpopulations in vitro *(3)*. Although Stxs appear to be immunosuppressive, they do not prevent the development of a specific antibody response in STEC-infected individuals *(4–6)*. Thus, the question of an immunosuppressive effect of Stx in the pathogenesis of STEC-mediated diseases needs to be addressed. STEC infections lead to an immunocompromised condition in gnotobiotic pigs and calves *(7,8)*, which is assumed to contribute to the observed persistency of infection (e.g., in calves and humans) *(9,10)*.

The investigation of immunomodulation through products of the enteric flora is usually biased by the fact that a variety of those molecules is known to positively or negatively regulate inflammatory responses. Apart from the well-known biological effects of lipopolysaccharide, two factors first decribed for enteropathogenic *E. coli* (EPEC) and *Citrobacter rodentium*, but also present in STEC, have been shown to modulate the mucosal immune system. First, lymphostatin, a novel large toxin from EPEC, has been shown to specifically inhibit lymphocyte proliferation and interleukin-2 (IL-2), IL-4, and γ interferon production by murine mucosal lymphocytes *(11)*. Genes-encoding proteins that are homologous to lymphostatin are present on the large STEC

From: *Methods in Molecular Medicine, vol. 73: E. coli: Shiga Toxin Methods and Protocols*
Edited by: D. Philpott and F. Ebel © Humana Press Inc., Totowa, NJ

plasmid *(12,13)* and recent data implicate this STEC protein also in bacterial binding to target cells *(14)*. Second, the surface protein intimin of EPEC and *C. rodentium* induces a massive T-helper-cell type 1 immune response in the colonic mucosa of mice *(15)*. Hence, when experiments to study effects of Stx on mononuclear leukocytes are designed, it must be taken into account that STEC lysates most likely contain additional bacterial factors able to interfere with the immune system of the infected host. Thus, these studies greatly rely on the availability of pure toxin preparations. (For a detailed protocol describing the isolation of Stx, the reader is referred to Chapter 15). Only those effects that can be blocked by specific, neutralizing antibodies should be ascribed to Stx. Studies aimed at the detection of Stx receptors on the surface of cells can be performed without purified toxin using CD77-specific antibodies (*see* below).

Mononuclear cells (monocytes and lymphocytes) isolated from peripheral blood (PBMCs) can be obtained easily and repeatedly even from humans without ethical reservations. However, circulating lymphocytes represent just 1–2% of all body lymphocytes and differ significantly from tissue lymphocytes because they lack the interaction with neighboring tissue cells and should thus be regarded as quiescent. A variety of mitogenic stimuli can be included in the experimental design to simulate effects of Stx on activated lymphocytes in the tissue. When PBMCs are incubated in tissue culture plasticware, monocytes tend to adhere tightly within some hours, whereas lymphocytes remain in the medium and can thus easily be submitted to flow cytometry analysis. This analytical approach relies on expensive laboratory equipment, but offers the opportunity to quantitatively determine several parameters in parallel for a large number of single cells within a short time. Thus, most of the protocols presented in this chapter contain a flow cytometry step. However, a protocol for the determination of cellular metabolic activity avoiding flow cytometry is also included. The MTT reduction assay described is a reliable system to detect any decrease in cellular metabolic activity whether it is the result of inhibition of cells or cytotoxicity *(16)*.

To further discriminate between inhibitory and cytotoxic effects of Stx, loss of cellular membrane integrity can be measured after propidium iodide staining. However, cells dying from apoptosis lose their membrane integrity at a very late stage of the cell death process, and quantification of dead cells by propidium iodide uptake may be insufficient. To monitor apoptotic effects of Stx on lymphocytes, cellular DNA fragmentation can be determined on a daily base using a commercially available kit. Whether or not Stx cause apoptosis in lymphoid cells is a matter of discussion. Human and bovine B-cell lines have been reported to be highly sensitive to the apoptotic effect of Stx1, whereas the toxin inhibits activation and proliferation of primary cultures of bovine lymphocytes without inducing cellular death *(1,3)*. Analysis of blast cell transformation ratio and blast cell composition by the protocols described herein are highly sensitive methods for demonstrating effects of Stx on immune cells independently of the underlying mechanism, cytotoxicity, or inhibition.

Most of the diverse biological effects of Stx reported so far are mediated via its binding to the specific cell surface receptors globotriaosylceramide (Gb_3/ CD77) and globotetraosylceramide (Gb_4) *(17,18)*. Most of the Stx variants, except Stx2e, preferentially bind to and act via Gb_3/CD77. Because a monoclonal CD77-specific antibody *(19)* is commercially available, immunological detection of this antigen on cell surfaces is a feasible way to identify presumably Stx-sensitive cell populations. However, biochemically diverse Gb_3/CD77 isoforms with varying affinities for Stx have been reported and solely binding studies with Stx holotoxin or the receptor binding B-subunit will confirm whether detected CD77 antigens truly serve as Stx receptors.

2. Materials

1. Disposable plastics: V-shaped centrifugation tubes, 50 mL (Greiner); flat-bottom (Nunc) and V-shaped (Greiner) 96-well microtiter plates; reaction vials, 75×12 mm, round bottom (Renner).
2. Ficoll–Paque (Amersham Pharmacia Biotech).
3. Mitogens (e.g., concanavalin A [ConA], phytohemagglutinin P [PHA-P], pokeweed mitogen [PWM], or lipopolysaccharide [LPS]; Sigma).
4. In Situ Cell Death Detection Kit, Fluorescein (Roche Molecular Biochemicals).
5. Monoclonal antibody against CD77 (clone 38.13, rat IgM; Coulter Immunotech Diagnostics) and leukocyte antigen-specific antibodies suitable for the species of interest and matching anti-immunoglobulin fluorescein–isothiocyanate (FITC) or R-Phycoerythrin (R-PE) conjugates.
6. Purified Stx tested for the absence of endotoxin and neutralizing monoclonal antibody against the respective type of Stx; Stx B-subunit and a monoclonal B-subunit-specific antibody (e.g., mouse monoclonal anti-StxB1 13C4, ATCC cat. no. CRL 1794)
7. Na–citrate solution (3.8% w/v), pH 7.0.
8. Phosphate-buffered saline (PBS): 10.0 g NaCl, 0.25 g KCl, 0.25 g KH_2PO_4, and 1.8 g $Na_2HPO_4 \cdot 2 H_2O$ per liter of distilled water, pH 7.4.
9. PBS supplemented with EDTA (PBS-EDTA): 8.0 g NaCl, 0.2 g KCl, 0.2 g KH_2PO_4, 1.42 g Na_2HPO_4 2 H_2O, and 2.0 g Na-EDTA per liter of distilled water, pH 7.4.
10. PBS supplemented with 1% bovine serum albumin fraction V (PBS-BSA; Serva).
11. Lysis buffer: 8.26 g NH_4Cl, 1.09 g $NaHCO_3$, 0.037 g Na-EDTA per liter of distilled water.
12. Modified cell culture medium: RPMI 1640 (Biochrom) supplemented with 10% fetal calf serum (FCS) (Invitrogen) and 3 μM of 2-mercaptoethanol (Sigma).
13. MTT (3-[4,5-dimethyl-2-thiazolyl]-2,5-diphenyl tetrazolium bromide, Sigma) stock solution (5 mg/mL in PBS): Freshly dissolved and filter sterilized through

0.2-µm-pore filter. The stock solution should be immediately aliquoted and frozen (–20°C) and is stable for months.

14. Sodium dodecyl sulfate (SDS)-solution: 10% (w/v) SDS, 0.01 N HCl in distilled water.
15. Propidium iodide (PI, Sigma) stock solution (100 µg/mL in PBS): keep at 4°C in the dark, stable for years.
16. Formaldehyde solution (4% [w/v] in PBS, pH 7.4); prepare freshly for each assay (*see* **Note 1**).
17. Permeabilizing solution: 0.1% (w/v) Triton X-100 in 0.1% (w/v) Na–citrate; keep at 4°C, stable for days.
18. Multichannel pipets (10–100 µL).
19. Standard cell culture laboratory equipment including laminar-flow bench, CO_2 incubator, and inverse microscope; centrifuge suitable for 50-mL tubes and microtiter plates, with a cooling system and disengageable brake.
20. Incubator set at 37°C equipped with a rocking table adjustable to about 40 rpm.
21. Enzyme-linked immunosorbent assay (ELISA) reader equipped with a filter set to read the optical density at 540 nm and 680 nm.
22. Flow cytometer equipped with an argon laser and standard filter configuration for FITC, R-PE, and PI (525, 575, and 630 nm, respectively).

3. Methods

3.1. Analyzing the Effect of Stx on Lymphocyte Viability

3.1.1. Preparation of Mononuclear Cells

1. Dilute 20 mL of citrated blood (4 mL Na–citrate solution plus 16 mL venous blood) with 17 mL PBS-EDTA (*see* **Note 2**) and layer it carefully onto 12 mL Ficoll–Paque® in a separate centrifugation tube (*see* **Fig. 1**). Avoid pertubation of the gradient.
2. Centrifuge (800g, 40 min, 20°C) without break.
3. Carefully recover the pale gray cell layer containing monocytes and lymphocytes, referred to as peripheral blood mononuclear cells (PBMCs) from the Ficoll–buffer interface. Avoid aspirating the erythrocyte sediment. Dilute the suspension thoroughly with 40 mL PBS-EDTA (*see* **Note 3**).
4. Centrifuge: 250g, 7 min, 4°C.
5. Discard the supernatant until the soft pellet leaks out and dilute the remaining suspension with 3 vol of lysis buffer to lyse contaminating erythrocytes; incubate for 5 min at room temperature.
6. Wash cells once with PBS-EDTA and then with PBS and adjust to 5×10^6 cells/mL in modified cell culture medium. In stimulation assays, supplement medium with ConA (at a final concentration of 5 µg/mL), PHA-P (5 µg/mL), PWM (10 µg/mL), or LPS (25 µg/mL) (*see* **Note 4**).
7. Transfer 50 µL of the cell suspension to 96-well flat-bottomed microtiter plates prepared as described in **Subheading 3.1.2.**

3.1.2. Preparation of Shiga Toxin Dilution Series on Microtiter Plates

1. To prepare 50 µL of 10-fold dilution series of toxin preparations in microtiter plates dispense 90 µL of 0.15 M NaCl per well, leaving the first and the last columns empty.

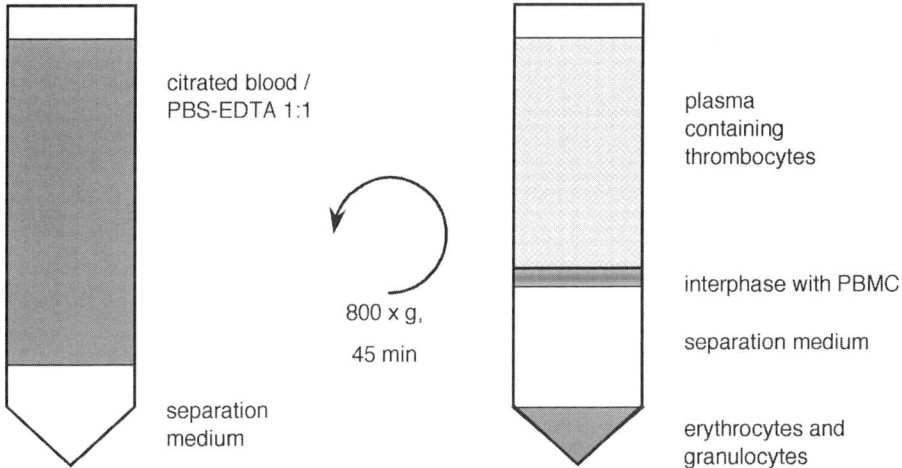

Fig. 1. Schematic drawing of the preparation of bovine PBMCs by density gradient centrifugation.

2. Add 60 μL of the toxin preparation to each well of the first column (*see* **Note 5**).
3. Transfer 10 μL from the first column to the second column and mix. Proceed in the same manner up to the second last column of the microtiter plate. Carefully blow out the pipet tips after each step to prevent excess carryover.
4. Remove 40 μL from each well-except the first and the last columns starting with the lowest dilution.
5. Add 50 μL of 0.15 *M* NaCl and 50 μL of 1% (w/v) SDS in 0.15 *M* NaCl as negative and positive controls, respectively, into separate wells of the last column.
6. Add 50 μL of modified cell culture medium to all wells. In neutralization studies, this medium component can be supplemented with neutralizing Stx-specific antibodies (*see* **Note 5**). In this case, the plates should be incubated for 30–60 min at room temperature before proceeding to **step 7**.
7. Apply 50 μL of cell suspension (5×10^6 cells/mL of cell culture medium) to each well; incubate at 37°C in 5% CO_2.

3.1.3. Measurement of Cellular Metabolic Activity

1. After 2–6 d, add 25 μL of MTT stock solution to each well of the microtiter plates. Place plates on a shaker and move gently for 4 h at 37°C.
2. Stop the reaction and dissolve dye crystals by adding 100 μL SDS solution to each well (*see* **Note 6**).
3. After overnight incubation, read optical density (OD) with an ELISA reader using a test wavelength of 540 nm and a reference wavelength of 690 nm.
4. Calculate percent cellular metabolic activity by the formula: [OD (sample) – OD (positive control)]/[OD (negative control) – OD (positive control)] × 100 (*see* **Fig. 2**).

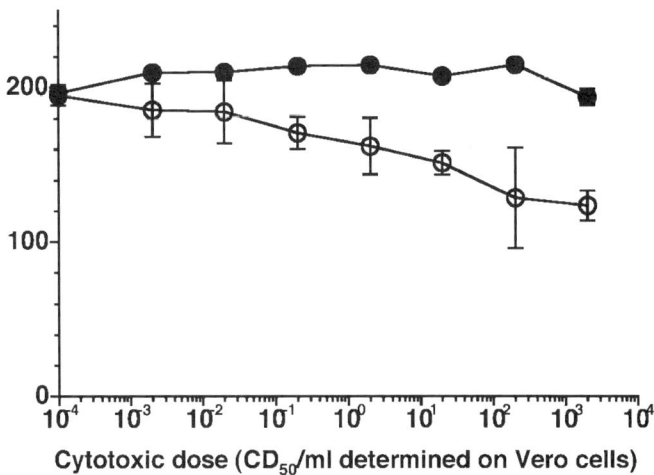

Fig. 2. Effect of purified Stx1 on the cellular metabolic activity of bovine PBMCs. Cells were incubated with 10-fold dilutions of purified Stx1 (0.002 to 2.000 CD_{50}/mL; quantified on Vero cells) for 96 h at 37°C. Culture medium was supplemented with 5 µg/mL PHA-P. Observed effects were assigned to Stx1 by comparison of the results obtained in the absence (open circles) or presence (filled circles) of 1.5 µg/mL monoclonal anti-StxB1 13C4 antibody. Cellular metabolic activity was determined by MTT reduction assay. Cells incubated with medium containing PHA-P alone were used as negative control, whereas cells treated with 1% (w/v) SDS served as a positive control to calculate percent activity. Data are means ± standard deviations of triplicate determinations. (Reprinted from **ref. 3** with permission.)

3.1.4. Determination of Propidium Iodide Uptake

1. Prepare and incubate the cells as described in **Subheadings 3.1.1.** and **3.1.2.**, but without SDS-treated controls. Daily monitor the cultures for the appearance of dead cells by quantifying the portion of cells able to take up PI.
2. Resuspend cells in the wells of a microtiter plate thoroughly (*see* **Note 7**), transfer them into reaction tubes suitable for the flow cytometer, and add 200 µL of PBS containing 2 µg/mL of PI. Include a blank control in each test series that is resuspended in PBS without PI (*see* **Note 8**).
3. Submit the samples to flow cytometry. Acquire 5000–10,000 events that are presumably leukocytes because of their light-scatter characteristics (*see* **Subheading 3.2.1.**). Plot the events in a one-parameter histogram depicting the PI fluorescence (630-nm filter) vs cell counts. Set an electronic threshold that defines less than 2% of the cells in the blank control to be PI positive.
4. Compare values of samples that received Stx with values of control samples (*see* **Note 9**).

3.1.5. Quantification of DNA Strand Breaks (TUNEL Method)

1. Prepare and incubate the cells as described in **Subheadings 3.1.1.** and **3.1.2.** and monitor them at different time-points for the appearence of DNA strand breaks (*see* **Note 10**). Resuspend cells thoroughly and transfer them to a V-shaped microtiter plate (*see* **Note 7**). To compensate for the loss of cells during the following manipulations, pool two identical samples for each determination.
2. Centrifuge plates (300*g*, 10 min, 4°C) and remove supernatants by inverted flicking of the plate (*see* **Note 11**).
3. Resuspend pellets in 150 µL of PBS-BSA (all volumes throughout this protocol are meant per well), centrifuge, remove supernatants by inverted flicking of the plate, and repeat this washing step once.
4. Add 50 µL of PBS-BSA and 50 µL of formaldehyde solution and resuspend the cells.
5. Following 30 min at room temperature, wash the cells once (*see* **step 3**).
6. Resuspend cells in 50 µL of permeabilizing solution (*see* **Note 11**).
7. Incubate for 2 min at 4°C and wash cells twice (*see* **step 3**).
8. Add 20 µL of the reaction mixture from the In Situ Cell Death Detection Kit, Fluorescein (containing FITC-labeled dUTP and terminal desoxynucleotidyl transferase, prepared according to the instructions of the provider) and mix well.
9. Add 20 µL of the nucleotide solution to a separate sample included as blank control and resuspend.
10. Incubate for 1 h at 37°C in 5% CO_2.
11. Centrifuge and wash cells twice (*see* **step 3**).
12. Optional: Proceed to **Subheading 3.2.2.** or **3.3.1.** for immunostaining (*see* **Note 12**).
13. Transfer the cells into reaction tubes suitable for the flow cytometer and add 200 µL of PBS.
14. Perform flow cytometry analysis and acquire 5000–10,000 events that are presumably leukocytes because of their light-scatter characteristics (*see* **Subheading 3.2.1.** and **Note 13**). Plot the events in a one-parameter histogram depicting the FITC fluorescence (525-nm filter) vs cell counts. Set an electronic threshold that defines less than 2% of the cells in the blank control to be FITC positive (*see* **Fig. 3**).
15. Compare values of samples that received Stx with values of control samples (*see* **Note 14**).

3.2. Analyzing the Effect of Stx on Lymphocyte Transformation and Proliferation

3.2.1. Analysis of Cell Morphology

1. Cells prepared as described in **Subheading 3.1.4.** can additionally be analyzed by flow cytometry recording detailed light-scatter characteristics of the cells.
2. To determine appropriate gates for the differentation of morphologically different PBMCs, analyze freshly isolated cells by flow cytometry (5000–10,000 events). In the forward versus sideward scatter histogram (scattergram), viable lymphocytes should appear as a population of medium size and little granularity. Define a gate surrounding this population and name it (e.g., "viable nonblast cells") (*see* **Fig. 4**).

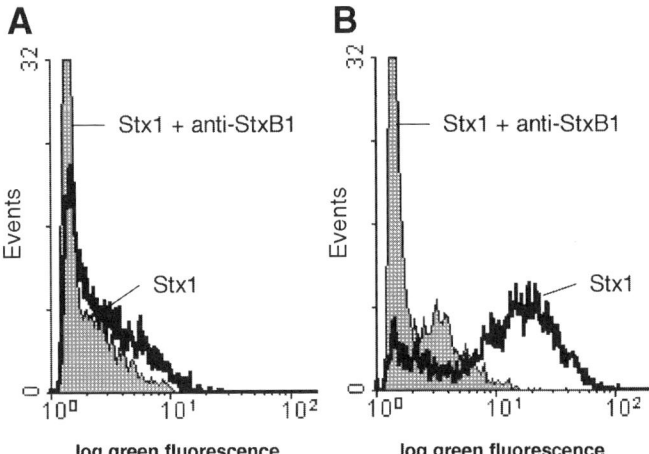

Fig. 3. Flow cytometric histograms illustrating the induction of DNA strand breaks in BL-3 cells (a bovine B lymphoma cell line) by Stx1. Cells were treated with 200 CD$_{50}$/mL (quantified on Vero cells) for 96 h at 37°C. Culture medium was free of mitogens (**A**) or supplemented with 25 µg/mL LPS (**B**). After incubation, DNA strand breaks were labeled by the TUNEL method. Observed effects were assigned to Stx1 by comparison of the results obtained in the absence or presence of 1.5 µg/mL monoclonal anti-StxB1 13C4 antibody as indicated. (Reprinted from **ref. 3** with permission.)

3. In the same way, analyze PBMCs incubated for more than 2 d in the absence of mitogens and Stx. As a result of necrosis and apoptosis in primary cultures of PBMCs, a second population of cells should appear in the scattergram character-ized by a smaller size and a somewhat increased granularity. Define a gate sur-rounding this population and name it (e.g., "subvital cells").

4. Analyze PBMCs incubated for more than 2 d in the presence of a potent mitogen (e.g., PHA-P). Because of the mitogen-induced transformation of quiescent lym-phocytes to enlarged and polygonal blast cells, a third population of cells should appear in the scattergram, characterized by a prominent increase in cell size and a small increase in granularity. Define a gate surrounding this population and name it (e.g., "viable blast cells") (*see* **Note 15**).

5. Incubate PBMCs in the presence or absence of Stx and analyze daily according to the protocol described above. Perform flow cytometry analysis acquiring 5000–10,000 events from each sample. Exclude cells from further analyis that are PI positive. Plot a scattergram for the viable cells only. Calculate the ratio of blast cell transformation by dividing the number of viable blast cells by the num-ber of viable nonblast cells.

6. Compare values obtained for cells that received Stx and those of control samples (*see* **Note 9**).

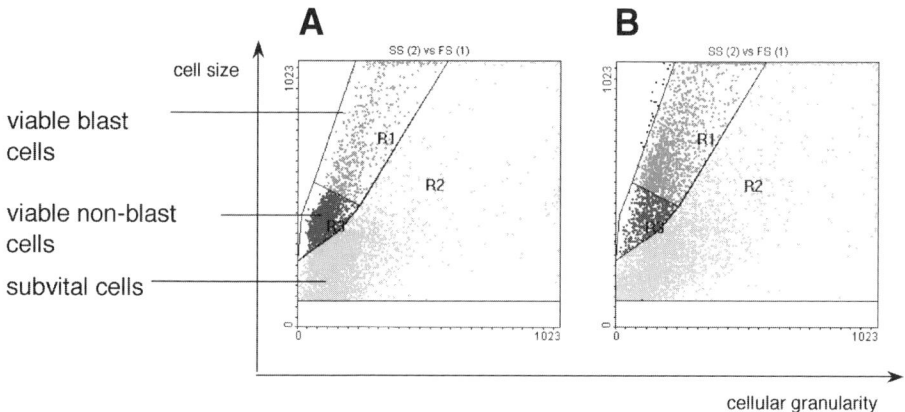

Fig. 4. Cellular morphology of cultured bovine PBMCs as assessed by flow cytometry. PBMCs of a 3-yr-old cow were incubated (96 h, 37°C) in the presence (**B**) or absence (**A**) of phytohemagglutinin P (PHA-P, 5 µg/mL).

3.2.2. Analysis of Blast Cell Composition

1. In order to improve the method described above, examine the effect of Stx on different leukocyte subtypes daily. Resuspend cells thoroughly and transfer them to a V-shaped microtiter plate (*see* **Note 7**). Keep the plate on ice and use pre-cooled (4°C) solutions throughout the entire protocol (*see* **Note 16**).
2. Centrifuge: 150*g*, 10 min, 4°C.
3. Remove supernatants by inverted flicking of the plate.
4. Resuspend pellets in 50 µL of buffer as a blank control or with buffer containing leukocyte subtype-specific antibody (all volumes throughout this protocol are meant per well). If these primary antibodies are already fluorochrome labeled, proceed to **step 8**.
5. Incubate the cells for 20 min on ice, centrifuge, and discard the supernatant.
6. Resuspend the cells in 150 µL of PBS and centrifuge again.
7. Resuspend the cells with 50 µL of a buffer containing a FITC-conjugated antibody recognizing the leukocyte-specific primary antibodies.
8. After 20 min on ice, wash cells twice (*see* **step 6**).
9. Transfer the cells into reaction tubes suitable for the flow cytometer and add 200 µL of PBS containing 2 µg/mL PI (*see* **Note 17**).
10. Perform flow cytometry analysis acquiring 5000–10,000 events from each sample. Define cells of the blast cell population by gating as described in **Subheading 3.2.1**. Exclude cells from further analyis that are PI positive. Create a histogram depicting the FITC fluorescence (525-nm filter) vs cell counts of the viable blast cells only. Set electronic gates according to the blank control included in each test series defining less than 2% of the control cells as positive.

11. Calculate the percentage of viable blast cells that are positive for a certain leukocyte marker among all cells in culture by dividing the absolute number of cells fulfilling all of these three criteria (viable + blast cells + marker positive) by the number of cells acquired in total.

12. Compare values obtained for cells that received Stx and those of control samples (*see* **Note 9**).

3.3. Detection of Stx Receptor Surface Expression

3.3.1. Detection of Gb₃/CD77

1. To examine the surface expression of Gb$_3$/CD77, freshly isolated PBMCs can be used as well as cells that have been stimulated as described in **Subheadings 3.1.1.** and **3.1.2.**, with the exception that Stx was not included. Using an anti-CD77 antibody and a matching secondary antibody conjugate, the procedure is essentially the same as described in **Subheading 3.2.** To determine the CD77 surface expression by different leukocyte subsets, the procedure described in **Subheading 3.2.2.** can be extended to detect CD77 and certain leukocyte markers simultaneously (*see* **Note 18**).

2. Transfer cells to a V-shaped microtiter plate (*see* **Note 7**). Keep the plate on ice and use precooled (4°C) solutions throughout the entire protocol. Incubate the cells with a leukocyte subtype-specific antibody as described. Spin the cells down once, discard the supernatant to remove the first antibody, and resuspend the pellet in 50 μL of a buffer as a blank control or with buffer containing the CD77-specific monoclonal antibody (all volumes throughout this protocol are meant per well).

3. Incubate on ice for 20 min.

4. Centrifuge (150g, 10 min, 4°C), resuspend the cells in 150 μL of PBS, and centrifuge again.

5. Resuspend the cells in 50 μL of a buffer containing a R-PE-labeled antibody recognizing the leukocyte-specific antibody; incubate for 20 min on ice. **Steps 5** and **6** are not necessary if R-PE-labeled leukocyte-specific antibodies are used.

6. Remove the conjugate by centrifugation and discard the supernatant.

7. Resuspend the cells in 50 μL of a buffer containing a rat IgM-specific antibody linked to FITC.

8. Incubate 20 min on ice; wash twice.

9. Transfer cells into reaction tubes suitable for the flow cytometer and add 200 μL of PBS containing 2 μg/mL PI (*see* **Note 17**).

10. Perform flow cytometry analysis acquiring 5000–10,000 events from each sample. Exclude PI-positive cells from further analysis. Create a two-parameter histogram for the viable cells only depicting the FITC (525 nm) vs the R-PE fluorescence (575 nm). Set electronic gates according to the blank controls included in each test series defining less than 2% of the control cells as positive for one or both colors (*see* **Note 19**).

3.3.2. Detection of Surface Toxin Binding

1. As with the detection of Gb₃/CD77 by a CD77-specific antibody, surface Stx-binding experiments can also be performed with both freshly isolated as well as cultivated PBMCs. At the end of the cultivation period, resuspend cells thoroughly (*see* **Note 7**) and transfer them to a V-shaped microtiter plate on ice, centrifuge ($150g$, 10 min, 4°C), and remove supernatants by inverted flicking of the plate. Keep the plate on ice unless otherwise indicated and use precooled (4°C) solutions throughout the entire protocol.
2. Resuspend pellets in 50 µL of buffer as a blank control or with buffer containing Stx or the Stx B-subunit, respectively (*see* **Note 20**).
3. Incubate the cells for 30 min and centrifuge.
4. Resuspend the cells in 150 µL of PBS and centrifuge again.
5. As an option, resuspend the cells in 100 µL PBS and incubate for 1 h at 37°C to induce internalization of the bound toxin. Afterwards centrifuge and discard the supernatant.
6. Resuspend with 50 µL of a buffer containing Stx- or Stx B-subunit-specific antibody, respectively.
7. Repeat **steps 3** and **4**.
8. Resuspend with 50 µL of a buffer containing a FITC-conjugated antibody recognizing the Stx-specific antibody.
9. Following another 30 min, wash the cells twice (*see* **step 4**).
10. Transfer the cells into reaction tubes suitable for the flow cytometer and add 200 µL of PBS containing 2 µg/mL PI (*see* **Note 17**).
11. Perform flow cytometry analysis acquiring 5000–10,000 events from each sample. Exclude PI-positive cells from further analysis. Create histograms containing only viable cells and depicting the FITC fluorescence (525 nm) vs cell counts. Set electronic gates according to the blank control included in each test series defining less than 2% of the control cells as positive.
12. To quantify toxin internalization, compare results of samples warmed up at **step 5** with those that were kept on ice throughout.

4. Notes

1. Formaldehyde is easier to dissolve than paraformaldehyde, which can be used instead, but much less stable; fresh preparation of the solution is thus recommended.
2. Supplementation of PBS with EDTA used as diluent and washing buffer throughout the preparation of bovine PBMCs efficiently prevents clumping of the leukocytes during gradient centrifugation and subsequent washing steps. However, this may not be necessary with leukocytes from other species. Carryover of EDTA into the cell culture medium should be avoided in any case. Thus, at least the last washing step (*see* **Subheading 3.1.1.**, **step 6**) should be carried out with PBS devoid of EDTA.
3. Density of Ficoll–Paque is designed for the isolation of human PBMCs. These cells accumulate on top of the Ficoll layer precisely. In contrast, bovine PBMCs

tend to enter the Ficoll layer upon centrifugation. Therefore, it is convenient to aspirate the pale gray interphase as well as the underlying cloudy Ficoll layer to achieve a good cell recovery.

4. In contrast to several lymphoma cell lines reported to be highly sensitive to the cytotoxic effect of Stx, primary cultures of lymphocytes are affected more gradually. Particularly with bovine PBMCs, an effect of Stx is best seen when the cells are stimulated by mitogens. Because mitogens stimulate PBMCs of different species to different extents, the stimulation protocol should be optimized for cells of the species of interest before including Stx in the experiments.

5. Before examining the effects of Stx on PBMCs in detail, a suitable concentration for both Stx and the neutralizing antibody should be determined by titrating both reagents. With regard to the limited stability of Stx, inclusion of a Vero cell assay in each set of experiments assuring the activity of the used toxin preparation is recommended.

6. Various methods exist to resolve the formazan crystals. In our hands, the method of Tada et al. *(16)* adding 10% (w/v) SDS/0.01 *N* HCl is the most reliable technique. The detergent solution is added to the wells without the need to thoroughly remove the medium, which interferes with the dye solubilization in most of the other methods. The addition of HCl changes the color of Phenol Red included in most culture media from red to yellow. A spectral overlap with the purple color of the formazan can thus be avoided. Nevertheless, the plates must be shaken very gently overnight to avoid foaming of the SDS-solution, which then would cause cross-contamination between wells.

7. Transfer of PBMCs cultured in flat-bottom microtiter plates to V-shaped plates and reaction tubes is a prerequisite for immunostaining that requires washing of the cells by centrifugation. Upon cultivation at 37°C, monocytes stick tightly to the surfaces of the flat-bottom wells, whereas lymphocytes do so much less. However, great care must be taken to thoroughly resuspend the cells before transfering them to the V-shaped plates because lymphocytes tend to be trapped in clusters including monocytes, particularly in the presence of mitogens. Microscopic evaluation of the successful transfer is strongly recommended to ensure that almost all lymphocytes are submitted to flow cytometry analysis.

8. In the protocol presented, the nucleic acids staining dye propidium iodide is used to detect cells that lost their membrane integrity, enabling the dye to enter the nucleus. No conclusion can be drawn from this type of experiment about whether stained cells died from necrosis or apoptosis. However, PI can also be used to specifically detect apoptotic cells using a modified protocol that stains the total DNA content of all cells *(20)*. Cells which have lost apoptotic DNA fragments can then be identified flow cytometrically by their reduced DNA content.

9. Controls should include cells incubated in the absence of Stx as well as cells incubated in the presence of Stx and a defined amount of neutralizing antibody.

10. Usually, apoptosis is a process that takes several hours. However, the best time-point to detect Stx induced apoptosis greatly relies on the time-point the cells experienced the lethal stimulus. This time-point is not necessarily when the cells come into contact with the toxin (i.e., the start of the cultivation period), but

when the cells are able to bind and internalize the toxin. This, in turn, may depend on the induction of the receptor expression. Because of host species variations, the appropriate time-point for analysis may be days after establishment of the culture.

11. This protocol will lead to higher losses of cells during the procedure as compared to the other procedures carried out in microtiter plates (*see* **Subheadings 3.2.** and **3.3.**). To reduce these losses, higher *g*-forces upon centrifugation of the plates (300*g* instead of 150*g*) and supplementation of PBS with 1% (w/v) BSA are strongly recommended. Additionally, keep the incubation time of 2 min when permeabilizing the cells with permeabilizing solution.

12. Performing the TUNEL method in a flow-cytometry-compatible format offers the opportunity to analyze DNA strand breaks in different lymphocyte subsets individually. However, prolonging the protocol by immunostaining will augment the above-mentioned cell losses. Furthermore, binding characteristics of antibodies to fixed and permeabilized cells may be drastically different from those of native cells.

13. In most of the cases, fixation alters the morphological features of cells recorded by flow cytometry. Establishment of cytometer settings suitable for fixed cells before measuring cells subjected to the TUNEL method is thus strongly recommended.

14. Control samples for the establishment of appropriate cytometer settings to quantify DNA strand breaks should include (1) a blank control with PBMCs incubated with labeled nucleotides but without terminal transferase, (2) a positive control with PBMCs incubated with an apoptosis inducing reagent (e.g., 0.15 μM of camptothecin, 1 μM dexamethason, or 0.5 μM of ionomycin), and, optional, (3) a control with cells of a cell line reported to be sensitive to the apoptosis inducing effect of Stx (e.g., Daudi) incubated in the presence of the toxin. When appropriate cytometer settings are found, it will be sufficient to include a blank control in each experiment. The effect of Stx can then be calculated from the comparison of samples as explained in **Note 9**.

15. Freshly isolated PBMCs consist of lymphocytes and monocytes. Because monocytes are larger in size than lymphocytes, these cells exhibit morphological features similar to viable blast cells that appear solely after mitogenic stimulation of PBMCs. On the other hand, monocytes adhere to the plastic surface of the microtiter plates upon cultivation and are barely recovered when the cells are resuspended and transfered to reaction tubes. PBMCs submitted to flow cytometry after incubation in plasticware thus mainly represent lymphocytes. Therefore, cells appearing larger than lymphocytes in the scattergram should be referred to as blast cells if cells had been cultivated previously. In contrast, cells with similar features in the scattergram of freshly isolated PBMCs should be referred to as monocytes.

16. Keeping cells at approx 4°C throughout the preparation process is crucial to avoid capping of surface molecules after binding of a ligand (antibody). Capping will reduce the signal for the detection of this particular antigen.

17. Dead cells tend to bind antibodies and fluorochromes unspecifically. Exclusion of dead cells from further analysis improves the distinction between antigen positive and negative cells.

18. As outlined in the protocol, double staining of the cells requires two primary antibodies and two Ig-specific conjugates, if labeled primary antibodies are not available. To avoid crossreactions between the Ig-specific conjugates, the use of primary antibodies of different isotypes and heavy-chain-specific conjugates is strongly recommended.

19. Because of a marked spectral overlap of FITC and R-PE, one has to put special emphasis on fluorescence compensation when setting up a protocol for the cytometer to simultaneously detect FITC and R-PE signals.

20. In functional assays, Stx exhibits biological activities in the nanogramm range. In contrast, the concentrations of Stx and Stx B-subunit needed to detect surface binding of the proteins to lymphocytes depend on the detection system used and may be in the microgram range.

References

1. Mangeney, M., Lingwood, C. A., Taga, S., Caillou, B., Tursz, T., and Wiels, J. (1993) Apoptosis induced in Burkitt's lymphoma cells via Gb_3/CD77, a glycolipid antigen. *Cancer Res.* **53,** 5314–5319.

2. Cohen, A., Madrid-Marina, V., Estrov, Z., Freedman, M. H., Lingwood, C. A., and Dosch, H. M. (1990) Expression of glycolipid receptors to Shiga-like toxin on human B lymphocytes: a mechanism for the failure of long-lived antibody response to dysenteric disease. *Int. Immunol.* **2,** 1–8.

3. Menge, C., Wieler, L. H., Schlapp, T., and Baljer, G. (1999) Shiga toxin 1 from *Escherichia coli* blocks activation and proliferation of bovine lymphocyte subpopulations in vitro. *Infect. Immun.* **67,** 2209–2217.

4. Pirro, F., Wieler, L. H., Failing, K., Bauerfeind, R., and Baljer, G. (1995) Neutralizing antibodies against Shiga-like toxins from *Escherichia coli* in colostra and sera of cattle. *Vet. Microbiol.* **43,** 131–141.

5. Wieler, L. H., Franke, S., Menge, C., Rose, M., Bauerfeind, R., Karch, H., et al. (1995) Investigations on the immunoresponse during edema disease of piglets after weaning by using a recombinant B subunit of Shiga-like-toxin IIe. *Dtsch. Tierarztl. Wochenschr.* **102,** 40–43.

6. Reymond, D., Johnson, R. P., Karmali, M. A., Petric, M., Winkler, M., Johnson, S., et al. (1996) Neutralizing antibodies to *Escherichia coli* Vero cytotoxin 1 and antibodies to O157 lipopolysacccharide in healthy farm family members and urban residents. *J. Clin. Microbiol.* **34,** 2053–2057.

7. Christopher-Hennings, J., Willgohs, J. A., Francis, D. H., Raman, U. A. K., Moxley, R. A., and Hurley, D. J. (1993) Immunocompromise in gnotobiotic pigs induced by Verotoxin-producing *Escherichia coli* (O111:NM). *Infect. Immun.* **61,** 2304–2308.

8. Hoffman, M., Casey, T., and Bosworth, B. (1997) Bovine immune response to *Escherichia coli* O157, Abstracts of the 3rd International Symposium and Workshop on Shiga Toxin (Verocytotoxin)-producing *Escherichia coli* infections, p. 117.

9. Cray, W. C. and Moon, H. W. (1995) Experimental infection of calves and adult cattle with *Escherichia coli* O157:H7. *Appl. Environ. Microbiol.* **61,** 1586–1590.

10. Karch, H., Rüssmann, H., Schmidt, H., Schwarzkopf, A., and Heesemann, J. (1995) Long-term shedding and clonal turnover of enterohemorrhagic *Escherichia coli* O157 in diarrheal diseases. *J. Clin. Microbiol.* **33,** 1602–1605.
11. Klapproth, J. M., Scaletsky, I. C. A., McNamara, B. P., Lai, L.-C., Malstrom, C., James, S. P., et al. (2000) A large toxin from pathogenic *Escherichia coli* strains that inhibits lymphocyte activation. *Infect. Immun.* **68,** 2148–2155.
12. Makino, K., Ishii, K., Yasunaga, T., Hattori, M., Yokoyama, K., Yutsudo, C. H., et al. (1998) Complete nucleotide sequences of 93-kb and 3.3-kb plasmids of an enterohemorrhagic *Escherichia coli* O157:H7 derived from Sakai outbreak. *DNA Res.* **5,** 1–9.
13. Burland, V., Shao, Y., Perna, N. T., Plunkett, G., Sofia, H. J., and Blattner, F. R. (1998) The complete DNA sequence and analysis of the large virulence plasmid of *Escherichia coli* O157:H7. *Nucleic Acids Res.* **26,** 4196–4204.
14. Nicholls, L., Grant, T. H., and Robins-Browne, R. M. (2000) Identification of a novel genetic locus that is required for in vitro adhesion of a clinical isolate of enterohaemorrhagic *Escherichia coli* to epithelial cells. *Mol. Microbiol.* **35,** 275–288.
15. Higgins, L. M., Frankel, G., Connerton, I., Goncalves, N. S., Dougan, G., and MacDonald, T. T. (1999) Role of bacterial intimin in colonic hyperplasia and inflammation. *Science* **285,** 588–591.
16. Tada, H., Shiho, O., Kuroshima, K., Koyama, M., and Tsukamoto, K. (1986) An improved colorimetric assay for Interleukin 2. *J. Immunol. Methods* **93,** 157–165.
17. Jacewicz, M., Clausen, H., Nudelman, E., Donohue-Rolfe, A., and Keusch, G. T. (1986) Pathogenesis of *Shigella* diarrhea XI: Isolation of a *Shigella* toxin-binding glycolipid from rabbit jejunum and HeLa-cells and its identification as globotriaosylceramide. *J. Exp. Med.* **163,** 1391–1404.
18. DeGrandis, S., Law, H., Brunton, J., Gyles, C., and Lingwood, C. A. (1989) Globotetraosylceramide is recognized by the pig edema disease toxin. *J. Biol. Chem.* **264,** 12,520–12,525.
19. Wiels, J., Fellous, M., and Tursz, T. (1981) Monoclonal antibody against a Burkitt lymphoma-associated antigen. *Proc. Natl. Acad. Sci. USA* **78,** 6485–6488.
20. Nicoletti, I., Migliorati, G., Pagliacci, M.C., Grignani, F., and Riccardi, C. (1991) A rapid and simple method for measuring thymocyte apoptosis by propidium iodide staining and flow cytometry. *J. Immunol. Methods* **139,** 271–279.

23

Animal Models for STEC-Mediated Disease

Angela R. Melton-Celsa and Alison D. O'Brien

1. Introduction

Escherichia coli O157:H7 is the most common infectious cause of bloody diarrhea or hemorrhagic colitis (HC) in the United States *(1)*. The potentially serious nature of infection with *E. coli* O157:H7 is illustrated by the fact that about 6% of those infected individuals (particularly children) develop a sequela called the hemolytic uremic syndrome or HUS *(2)*. The HUS is characterized by microangiopathic hemolytic anemia, thrombocytopenia, renal failure, central nervous system involvement, and, in 3–5% of children, death (reviewed in **ref. *3***). *E. coli* O157:H7 belongs to a subset of Shiga toxin-producing *E. coli* (STEC) named enterohemorrhagic *E. coli* (or EHEC) that not only make Shiga toxins (Stxs, formerly called Shiga-like toxins and alternatively known as verotoxins) but also attach to the bowel by a protein called intimin and evoke an attach and efface (A/E) lesion at the site of the bacterial–enterocyte interface *(4)*. The pathogenic process by which *E. coli* O157:H7 and some other STEC evoke blood in the stools and cause the HUS and the adult version of HUS, thrombotic thrombocytopenic purpura (TTP), remains incompletely understood. Nevertheless, Stx appears to play a pivotal role in these manifestations, and endothelial cells are a primary target of Stx action *(5,6)*. A number of animal species have been tried as models of STEC infection or Stx-mediated disease. However, in no single animal system is there replication of the entire spectrum of the infectious disease process as observed in humans (i.e., in humans, the steps in STEC pathogenesis include oral infection followed by diarrhea, sometimes HC, and occasionally the HUS). Nevertheless, there are animal models that mimic parts of the disease process. These models are described in the following subheadings, with the exception of pig and

From: *Methods in Molecular Medicine, vol. 73: E. coli: Shiga Toxin Methods and Protocols*
Edited by: D. Philpott and F. Ebel © Humana Press Inc., Totowa, NJ

Table 1
Summary of Animal Models, Treatments and Outcomes

Animal type	Animal strain	Treatment(s)	Route of inoculation	Outcome or symptoms
Mouse	CD-1, DBA/2J	str/most O157 strains	Oral feeding	Colonization
	CD-1	str/some Stx2 producers, Stx2d producers	Oral feeding	Renal damage, death
	ICR	str/mitomycin C/STEC strain E32511/HSC	Intragastric feeding	Encephalopathy
	CD-1	str/ciprofloxacin/ STEC strain 1:361R	Intragastric feeding	Death
	C3H/HeN	STEC strain 86-24	Intragastric feeding	Loose stool, mesangial matrix expansion
	C57BL/6	Low-protein diet/ STEC strain N-9	Intragastric feeding	Neurologic and systemic manifestations, cerebral hemorrhages, death
Rabbit	New Zealand white	STEC strain UC741	Intragastric feeding	Diarrhea, death
	New Zealand white	RDEC-H19A	Intragastric feeding	Diarrhea, colonic subserosal hemorrhages and submucosal edema and vascular changes
	New Zealand white, infant	Stx1	Intragastric	Diarrhea, mucosal damage in the colon, some death
	New Zealand white	Stx1	Injection into marginal ear vein	Some diarrhea, CNS symptoms, cecal mucosal edema and submucosal vascular changes in some animals
	Japanese white rabbits	Stx2	Injection into marginal ear vein	Hemorrhagic diarrhea, flaccid paresis, convulsions
	Not stated in chapter	Stx2	Continuous pump in peritoneum	Diarrhea and intestinal lesions similar to HC
Chicken		O157 STEC strains	Intragastric feeding	Colonization, A/E lesions
Dog	Greyhound	Stx1 or Stx2	Injection	Thrombocytopenia, anemia, HC
Baboon	*Papio c. cynocephalus* or *Papio c. anubis*	Stx1	Injection into cephalic vein	Damage to gastrointestinal mucusa, renal failure, thrombocytopenia, other HUS signs

292

bovine models, which are described elsewhere in this volume. A summary of the animal types, treatments, and outcomes is given in **Table 1**.

1.1. Mouse Models

Four mouse models have been developed for the evaluation of the pathogenesis of STEC strains. These models include the following: (1) orally fed streptomycin (str)-treated mouse; (2) str- and mitomycin C- or ciprofloxacin-treated, intragastrically inoculated mouse; (3) intragastrically-fed but not str-treated mouse, and; (4) malnourished mouse.

The outcomes in mice that are infected with an STEC strain vary somewhat with the STEC strain and the type of mice used, as well as how the mice are treated and infected. The str-treated mouse model was adapted from the model of Myhal et al. *(7)* by Wadolkowski et al. *(8)*. In that model, adult CD-1 mice are given 5 g/L str sulfate in their drinking water 1 d before oral infection with a str-resistant (Str^r) strain of STEC. Oral infection entails permitting the animals to eat the bacteria (suspended in a solution containing sucrose) from a weigh boat in their cages or by hand-feeding the animals the inocula in a dropper. After infection, animals are permitted food and water with 5 g/L str sulfate *ad libitum* throughout the duration of the experiment. The str pretreatment reduces the level of facultative anaerobic flora more than 6 logs in 24 h *(7)*, and this reduction permits any Str^r STEC to colonize the bowel at high levels (~10^7 colony-forming units [CFU] per g feces) for over 1 wk. Although CD-1 and inbred strains of mice (such as DBA/2J) become well colonized with O157:H7 by this protocol, only certain strains of *E. coli* O157:H7 (e.g., 86–24 Str^r) are virulent for these animals and only when challenged with a large inocula (10^{10} bacteria). Death of str-treated mice fed an *E. coli* O157:H7 strain that makes Stx1 and Stx2 is solely attributed to Stx2 and appears to be a consequence of toxin-mediated renal tubular necrosis *(9)*. When str-treated mice are fed a certain STEC that make the Stx2d variant [that toxin is activated by elastase in intestinal mucus *(10,11)*], the oral 50% lethal dose (LD$_{50}$) is less than 10 organisms. Animals infected with these highly pathogenic strains also die from renal tubular necrosis. Thus, the str-treated mouse is not a model for STEC-mediated diarrhea, HC, or HUS, but can be used to assess relative virulence of STEC for mice after oral challenge and to measure induction of Stx-dependent renal tubular necrosis. The str-treated mice can also be used to evaluate the efficacy of a potentially Stx-neutralizing compound, and, indeed, we have found that Stx2 antibody is protective in the str-treated, STEC-infected mouse *(9,12,13)*.

Two other groups have used a variation of the str-treated mouse model in which phage-inducing compounds are given to the mice at or after intragastric inoculation with an O157 STEC strain *(14,15)*. In one case, str-treated mice were given mitomycin C at the time of intragastric inoculation with O157:H⁻

strain E32511/HSC *(15)*. Death and acute encephalopathy in the mice during this experiment required str and mitomycin C treatment, as well as a high dose of bacteria (i.e., greater than 10^9 CFUs per mouse). In the second type of experiment, str-treated CD-1 mice were given ciprofloxacin at doses that induced a 1000-fold drop in the CFUs of the infecting strain shed into the feces *(14)*. Ciprofloxacin induces the phage that encodes Stx2 and thus increases the level of Stx2 produced by STEC strain 1:361R in vitro and in vivo.

Approximately two-thirds of mice given ciprofloxacin and 1:361R die, whereas those mice infected with 1:361R alone survive the infection. The finding that administration of ciprofloxacin to str-treated, 1:361R-infected mice promotes death of the mice and that death correlates with high levels of free fecal Stx suggests that this variation in the original str-treated mouse model may permit an otherwise mouse-attenuated STEC strain to be virulent.

The third mouse model involves intragastric inoculation of C3H/HeN (lipopolysaccaharide [LPS]-responder) or C3H/HeJ (LPS-nonresponder) mice with 10^7 to 10^8 CFUs of the Stx2-producing O157:H7 strain 86-24, or 86BL, a derivative of 86-24 that was recovered from the blood of mice infected with 86-24 *(16)*. As a control, Karpman et al. used strain 87-23, an Stx-negative O157:H7 strain that was isolated during the same outbreak as 86-24 *(17)*. Overall, the C3H/HeN mice develop more severe disease, a finding that indicates that the host response to LPS may play a role in the pathogenesis of STEC-mediated disease. Only mice that were infected with an Stx2-producing strain exhibit expansion of the mesangial matrix of the renal glomeruli, as is seen in HUS. Although there are other signs of O157-mediated damage to the mice, such as loose stools and vascular congestion, these same manifestations are observed in mice infected with the toxin-negative O157:H7 strain. Both strains also invade the bloodstream of some of the mice, a complication not seen in humans infected with STEC. In contrast, the non-O157 control strain causes only mild systemic symptoms that last less than 1 d.

The final mouse model of STEC infection necessitates induction of malnutrition in female C57BL/6 mice by maintaining the animals on a 5% protein diet for 2 wk prior to and at all times after intragastric infection with an O157:H7 strain, N-9 *(18)*. The malnourished mice, but not the control mice fed a 25% protein diet, are killed by an inoculum of greater than 10^6 CFUs of N-9. The malnourished mice develop systemic as well as neurological signs before death, as are seen in the other mouse models. However, the malnourished mice in these studies do not exhibit appreciable kidney damage but do develop cerebral hemorrhages that are thought to be the immediate cause of death. Why the malnourished mice do not exhibit kidney damage similar to that seen in the str-treated mouse is unclear but may be related to the nutritional status of the mice.

1.2. Rabbit Models

1.2.1. Oral Inoculation Models

An infant rabbit model of O157:H7 infection was first described in a letter to *Lancet* in 1983 *(19)*. That report described intragastric feeding of 5- to 10-d-old rabbits that subsequently develop watery diarrhea. Those same authors did not detect diarrhea in O157-infected older rabbits, guinea pigs, mice, or young rhesus monkeys. In 1986, Pai et al. described the infection of infant rabbits with an O157:H7 strain *(20)*. Rabbits that receive the highest inoculum exhibit severe diarrhea and death. A second intragastric feeding model of rabbits has also been described *(21)*. In that model, rabbits are given a natural rabbit diarrheal pathogen (RDEC) that has been lysogenized with an stx_1-converting phage *(21)*. The lysogenized RDEC strain, RDEC-H19A, was used to evaluate the role of Stx1 in the enteritis induced in the infected rabbits *(21)*. The Stx1-producing RDEC strain causes diarrhea faster and more often than the RDEC control. Additionally, the cecum and proximal colons of RDEC-H19A-infected rabbits exhibit subserosal hemorrhages as well as submucosal edema at d 7 postinoculation. Vascular changes in submucosal venules are also observed only in rabbits infected with the Stx1-producing strain. Another group of researchers intragastrically fed a number of different serotypes of STEC to postweanling rabbits and found that most of those strains are capable of inducing diarrhea in the rabbits *(22)*. However, this group's interest lay primarily in observing the type of attachment mediated by the STEC at 7 d postinfection. They found that the strains tested have the capacity to attach to the intestinal epithelium in an intimate fashion (A/E lesions). Because the primary focus of that study was on adherence, we will not discuss it further in this review.

1.2.2. Toxin Administration Models

Infant rabbits intragastrically administered a preparation of sodium bicarbonate and a preparation of Stx1 made from an O157:H7 strain (purified as described in **ref.** *23* and elsewhere in this volume), develop diarrhea (bloody in one animal) within 24 h and some animals die *(20)*. Histological sections of the colons of the rabbits given the Stx preparation show mucosal damage. A second group of researchers gave a smaller dose of purified Stx1 (0.2 µg/kg) to larger rabbits and found that diarrhea occurs in half of the animals *(6)*. A few animals have occult blood in their stools, but none show overtly bloody stools. Many animals also develop central nervous system (CNS) symptoms that include paresis. The Stx1-treated rabbits also develop mucosal edema in the cecum. Submucosal vascular changes are observed in about 50% of the animals. A third group of investigators injected rabbits with a large bolus of Stx2 (2 µg/kg) and found that the rabbits develop hemorrhagic diarrhea as well as

flaccid paresis and convulsions *(24)*. The authors of this study also found significant CNS involvement, but did not examine tissues other than brain. In a
fourth investigation, Stx2 was continuously infused into the peritoneum of rabbits who then developed diarrhea and intestinal lesions similar to those seen in
HC *(25)*. Finally, purified Stx or Stx1 injected into rabbit ileal loops causes
fluid accumulation and gross and microscopic lesions *(26)*. Most of the villus
absorptive cells are sloughed off and/or undergoing apoptosis by 24 h postinoculation. However, the rabbit ileal loop model will not be described in detail
because it is largely a model of enterotoxicity.

1.3. Baboon and Monkeys

The baboon model is established by intravenous infusion of Stx1 *(27)*. The
baboons that die have renal failure as well as damage to the gastrointestinal
mucosal. In the kidneys, there is damage in both the tubules and the glomeruli.
The renal histopathology is more severe in baboons that receive a lower dose
(0.05–0.2 μg/kg) of Stx1 than in animals that receive a higher dose (2 μg/kg).
The glomeruli of the baboons in the lower-dose group exhibit obliteration of
the capillary lumina, fragmented red cells, and some fibrin thrombi, findings
that are similar to those seen in histological sections of kidneys from patients
with the HUS. The low-dose animals also display thrombocytopenia. Damage
to the intestine, in contrast to the kidney damage, is dose dependent, with loss
of the villus tip epithelial cells in the baboons that receive a high dose of Stx1.

A Macaque model has been reported in which the animals display diarrhea
and exhibit A/E lesions after infection with an O157:H7 strain *(28)*, but the
description of that model has too few details to include here.

1.4. Chicken

A few investigators have tried oral inoculation of chicks with *E. coli*
O157:H7 *(29–31)*. The results suggest that chicks become colonized in the
cecum and colon *(29)* by the STEC and can shed those organisms for longer
than 5 mo *(30)*. In a separate study, A/E lesions were detected in 2/7 chicks fed
O157:H7 strain ATCC 43889 *(31)*. The materials and methods for these studies will not be described further because the chick appears to largely be a model
of colonization and other models of colonization have already been mentioned
here and are described elsewhere.

1.5. Greyhound Dog Model

The greyhound dog model of STEC disease is in the preliminary stages of
development. These animals were selected because of the similarity noted
between STEC disease and a naturally occurring condition in greyhounds
called idiopathic cutaneous and renal glomerular vasculopathy of greyhounds

(CRVG) (also called Alabama rot *[32]*). In contrast to STEC-mediated illness, greyhounds with CRVG generally have numerous cutaneous lesions on the abdomen and hind legs. However, the dogs with CRVG, like humans with HUS, exhibit thrombocytopenia and microangiopathic hemolytic anemia. Some of the animals also display progressive acute renal failure. The glomeruli of greyhounds with CRVG show necrosis and damage to the capillary endothelium. Fibrin may also accumulate. Later renal changes include endothelial cell hypertrophy and proliferation. STEC are suspected as the cause of CRVG, but an absolute causal link has not been made. Although the animals with CRVG do not have a prodrome of diarrhea, the clinical presentation shows a strong similarity to the HUS. Therefore, the development of an experimental dog model may prove useful to understanding the HUS disease process and for the evaluation of HUS therapies.

In a pilot study, administration of purified Stx1 or Stx2 (0.03 mg/kg) to experimental greyhounds resulted in thrombocytopenia and anemia, followed by HC at 48 h postinjection *(32)*. Why the dogs developed the HC following injection of toxin is not understood, but further pursuit of this model will likely answer many such questions.

2. Materials

In this section, model numbers are described in the respective subheadings.

2.1. Mouse Models

1. Male CD-1 mice (models 3.1.1 and 3.1.3); male ICR mice (equivalent to CD-1 mice, model 3.1.2); female or male 8- to 16-wk-old C3H/HeN or C3H/HeJ (model 3.1.4); specific-pathogen-free, 3-wk-old female C57BL/6 mice (model 3.1.5).
2. Streptomycin sulfate (models 3.1.1 and 3.1.3); mitomycin C (model 3.1.2) or ciprofloxacin (model 3.1.3).
3. STEC strain (models 3.1.4 and 3.1.5), Str^r STEC strain (models 3.1.1 and 3.1.3).
4. 20% Sucrose, w/v in distilled water (dH$_2$O) (models 3.1.1 and 3.1.3).
5. Catheters: 0.9-mm-diameter (model 3.1.2), 0.61 mm diameter (model 3.1.4), stainless-steel catheter with a blunt end (outer diameter 0.45 mm, model 3.1.5).
6. 22-Gage feeding needle (model 3.1.3).
7. Pipetman and pipets (all models).
8. Empty mouse cages and small weigh boats (model 3.1.1).
9. Saline or phosphate-buffered saline (PBS; all models).
10. Luria–Bertani broth (LB) and LB plates (models 3.1.1–3.1.4); tryptic soy broth (TSB) and plates (model 3.1.5).
11. Low-protein (5%) diet (Oriental Bioservice, Inc., Kyoto, Japan) and isocaloric diet with 25% protein (model 3.1.5).
12. Plasticware or glassware for making dilutions and solutions and growing bacteria (all models).

13. Centrifuge and centrifuge tubes (all models).
14. Ether (models 3.1.2 and 3.1.4).
15. Materials for necropsy, blood/urine collection, and histological analysis (all models, as necessary).

2.2. Rabbit Models

2.2.1. Oral Inoculation Models

1. New Zealand white rabbits, 2–3 d old (model 3.2.1); male New Zealand white rabbits, 1.0–1.7 kg (model 3.2.2).
2. STEC strain (UC741, model 3.2.1); RDEC-1A (nalidixic acid resistant) and RDEC-H19A (model 3.2.2).
3. TSB and TSB plates (model 3.2.1); MacConkey agar plates (model 3.2.2); Penassay broth (model 3.2.2).
4. PBS (both models); saline (model 3.2.2).
5. 10% Sodium bicarbonate (both models).
6. 17-Gage nylon catheter tube (model 3.2.1); orogastric tube (model 3.2.2).
7. Sterile swabs (both models).

2.1.2. Toxin Administration

1. Stx1 purified as in **ref. 23** (model 3.2.3); Stx1 purified as in **ref. 33** (model 3.2.4); Stx2 purified as in **ref. 34** (model 3.2.5) or in **ref. 35** (model 3.2.6).
2. Infant New Zealand white rabbits (model 3.2.3); 2-kg New Zealand white rabbits (model 3.2.4); 2-kg Japanese white rabbits (model 3.2.5).
3. 17-Gage nylon catheter (model 3.2.3).
4. Sodium bicarbonate (model 3.2.3).
5. Ketamine and xylazine (all models).
6. Necropsy tools (all models).
7. Materials for histological examination of tissues (all models).
8. Materials for urinalysis, hematological analysis, and serum biochemistry (models 3.2.4 and 3.2.6).

2.3. Baboon Model

1. *Papio c. cynocephalus* or *Papio c. anubis* baboons.
2. Stx1 purified as in **ref. 27**.
3. Saline.
4. Ketamine and sodium pentobarbital.
5. Percutaneous catheter.
6. Cannulas.
7. Catheters.
8. Heating pad.
9. Pressure transducer and recorder.
10. Telethermometer.
11. Heparin lock.

3. Methods

3.1. Mouse Models

3.1.1 Str-Treated, Orally Infected CD-1 Mice (8)

1. Remove food from 22-g male CD-1 mice 18 h prior to feeding STEC strain.
2. Give mice str (5 g/L) in the drinking water at the time food is removed.
3. Start overnight (o/n) culture of STEC strain in LB broth (no str) 18 h prior to feeding.
4. Shortly prior to feeding, o/n culture should be diluted in 20% sucrose, or, if a large dose is to be given, the o/n may be pelleted by centrifugation and resuspended in 20% sucrose to the appropriate concentration. Titer to determine actual dose.
5. The mice may be fed by pipetman or placed into empty mouse cages in which small weigh boats have been taped down and filled with 600 µL of the bacterial suspension.
6. Once mice have finished eating the bacterial suspension, they are returned to their cages and allowed food and str water *ad libitum*.
7. Mice should be monitored twice daily for clinical signs or death.
8. The colonization capacity of a strain may be monitored by collecting fecal pellets, adding 9 mL saline/g feces (first 10-fold dilution), breaking up the fecal pellets by a stomacher or by pipeting up and down, doing further 10-fold dilutions, and plating the dilutions to determine the number of CFUs per gram feces.
9. Organs may be removed from the mice for histological examination.

3.1.2. Intragastric Inoculation of str- and Mitomycin C-Treated Mice (15) (see *Notes 1* and *2*)

1. Three-week-old ICR mice are given drinking water with 5 g/L str for 3 d.
2. On d 2 after str treatment, a culture of strain E32511/HSC is inoculated into LB.
3. The o/n culture of E33511/HSC is centrifuged and the pellet washed with PBS and the concentration of the bacterial suspension adjusted to between 10^7 and 10^{11} CFUs/mL with PBS (d 3). Titer culture to determine inoculum.
4. Mice are starved for 6 h (d 3).
5. Mice are anesthetized with ether.
6. Mice are inoculated through a catheter tube (0.9-mm diameter, 70-mm length) attached to a syringe that contains the bacterial suspension.
7. Simultaneously, the mice are given 2.5 mg/kg mitomycin C intraperitoneally.
8. Mice are maintained on str water and allowed food *ad libitum* and monitored for clinical signs.
9. Blood is collected and mice are necropsied.

3.1.3. Intragastric Inoculation of CD-1 Mice Treated with Ciprofloxacin (14) (see *Notes 1* and *2*)

1. Five-week-old CD-1 male mice are given drinking water with 2 g/L str (d 1).
2. The d after str treatment, food is removed from the mice for 24 h (d 2).
3. A culture of 1:361R is inoculated into LB (d 2).

4. The o/n culture of 1:361R is centrifuged and washed in PBS and resuspended to the same volume in 20% sucrose (d 3). Titer culture to determine inoculum.
5. The mice are inoculated intragastrically with 0.15 mL (about 3×10^9 CFUs) of the bacterial suspension through a 22-gage feeding needle (d 3).
6. Ciprofloxacin 30 μg is injected intraperitoneally on d 4–6.
7. Mice are monitored daily for death.

3.1.4. Intragastric Inoculation of C3H Mice (16)

1. C3H/HeN or C3H/HeJ mice are fasted but given water for 20–24 h prior to inoculation (d 1).
2. Start culture of STEC strain (86-24 used in this model, d 1) in LB.
3. Bacterial culture is harvested, centrifuged, and resuspended in PBS to 109 CFUs/mL (d 2). Titer culture to determine inoculum dose.
4. Mice are anesthetized with ether (d 2).
5. One-tenth milliliter of the 10^8–10^9 CFU/mL bacterial suspension is inoculated into the mice through a soft polypropylene catheter (outer diameter 0.61 mm); controls receive 0.1 mL of PBS only (*see* **Note 1**).
6. Collect blood for hematological analysis and to check for systemic infection.
7. Necropsy and perform histological analysis on desired organs.

3.1.5. Malnourished Mouse Model (18)

1. Three-week-old female C57BL/6 mice are quarantined for 2 wk and fed a low-protein (5%) diet. Control mice are fed an isocaloric diet with 25% protein.
2. The d prior to inoculation, a culture of STEC strain (N-9) is started in TSB (d 1).
3. Harvest the bacteria by centrifugation and wash with PBS. Resuspend in PBS to 2×10^7 CFUs/mL. Titer culture to determine actual dose administered.
4. Infect mice intragastrically with 0.1 mL of the bacterial suspension through a stainless steel catheter with a blunt end (outer diameter 0.45 mm) (*see* **Note 1**).
5. Monitor mice for clinical signs.
6. Collect organs for histopathologic analysis.

3.2. Rabbit Models

3.2.1. Infection of Infant Rabbits (20)

1. New Zealand white rabbits, 2–3 d old, are observed for 1 d to determine that no prior diarrhea exists.
2. Start culture of STEC strain (UC741) in TSB (d 1).
3. Inoculate fresh TSB from the o/n STEC culture (1 : 10 ratio of inoculum to TSB) and grow for 4 h (d 2).
4. Harvest the bacteria by centrifugation and wash twice with PBS and resuspend in 10% sodium bicarbonate solution (d 2). Titer to determine actual dose.
5. Obtain a swab of the rectum of each rabbit for culture (d 2).
6. Inoculate rabbits with 1.0 mL bacterial suspension through a 17-gage nylon catheter tube (d 2) (*see* **Note 1**).

3.2.2. Infection with a Rabbit Diarrheal pathogen That Has Been Lysogenized with an stx₁-Converting Phage (21)

1. Rectal swabs are obtained from male New Zealand white rabbits the d prior to inoculation and plated onto MacConkey agar. Rabbits from which there was growth on the MacConkey agar at 48 h postinoculation are excluded from the study (d 1).
2. The rabbits are fasted overnight (d 1).
3. Inoculate Penassay broth with RDEC-1 or RDEC-H19A and incubate statically overnight at 37°C (d 1).
4. Collect the o/n bacterial growth by centrifugation and wash twice in sterile saline. Resuspend the bacterial pellet in PBS to 10^6 or 10^8 CFUs/mL (d 2). Titer to determine actual dose.
5. Place an orogastric tube into the fasted rabbits and instill 10 mL of 10% sodium bicarbonate followed by 1.0 mL of bacterial suspension (d 2).

3.2.3. Model of Stx1 inoculation into Rabbits (20)

1. The Stx preparation is diluted in sodium bicarbonate (10% final concentration). The rabbits are given 2.5×10^{10} or 7.5×10^{10} 50% cytotoxic doses (CD_{50}s) of the toxin preparation (*see* **Note 3**).
2. Three-day-old New Zealand white rabbits are inoculated with the toxin/sodium bicarbonate mixture directly into the stomach with a 17-gage nylon catheter tube.
3. The animals are observed for clinical signs and/or organs are removed for histological examination 1–2 d postinoculation.

3.2.4. Second Model of Stx1 Inoculation into Rabbits (6)

1. New Zealand white rabbits (2 kg) are given 1 mL of the pure Stx1 solution through the marginal ear vein (*see* **Note 3**).
2. Rabbits are observed for clinical signs and urine, blood, and tissues are collected from some animals for histological examination.

3.2.5. Large Bolus of Stx2 Administered to Rabbits (24)

1. Male 2-kg Japanese white rabbits are injected intravenously with 5 µg/kg of Stx2 in a volume of 2 mL through the marginal ear vein.
2. Rabbits are observed for clinical signs and some are used for histological examination.

3.2.6. Continuous Infusion of Stx2 (25)

1. The mini-osmotic pump is loaded with enough Stx2 to give 100 ng/kg each day for 2 wk.
2. Rabbits are anesthetized with 0.4 mL ketamine + 4 mg/mL xylazine per kilogram.
3. An abdominal midline incision is made in the rabbits and the pump is placed in the cavity. The incision is then sutured.
4. The rabbits are observed for clinical signs.
5. Blood and tissues are collected from the rabbits

3.3. Baboon Model (27)

1. The baboons are fasted overnight (d 1).
2. The baboons are immobilized with ketamine (14 mg/kg) intramuscularly (d 2).
3. Sodium pentobarbital (2 mg/kg) is infused every 20–40 min.
4. The baboons are intubated and allowed to breathe orally.
5. A femoral vein is exposed and cannulated and a catheter advanced to the inferior vena cava to sample blood and monitor central venous pressure.
6. The femoral artery is cannulated to monitor arterial pressure.
7. The baboons are placed on their sides on a temperature-controlled heating pad.
8. Blood pressure and rectal temperature are monitored.
9. Saline or Stx1 (0.053–0.2 µg/kg or 2 µg/kg) is infused via percutaneous catheter in the cephalic vein.

3.4. Concluding Remarks

Although none of the animal models described here fully reflects the pathogenesis of STEC in humans, a few of these systems show promise as models of the most severe manifestations of STEC infection, the HUS. Unfortunately, the models with the features that most closely resemble the clinical presentation in humans (i.e., the greyhound and baboon models) are expensive and not readily available to all researchers. However, certain features of the disease process can be studied in selected smaller animal model systems such as the mouse and then confirmed in the less accessible larger animal models. Another beneficial feature of the mouse models, although not perfect models of HUS, is that they readily demonstrate Stx-mediated damage, and, therefore, are useful for evaluating the efficacy of therapies directed against the toxin.

4. Notes

1. For all of the models that require intragastric feeding, care must be taken not to inadvertently introduce the bacteria into the bloodstream of the animals. If bacteremia were to result from the inoculation procedure, the results of the experiment would be skewed and the outcome could not necessarily be said to be the result of an intestinal infection.
2. For the models that use mitomycin C or ciprofloxacin, the infecting strain must contain an inducible prophage that encodes a gene for Stx.
3. For the models that involve the injection of toxin, the level of purity of the toxin preparation will have an impact on the results. Additionally, if a crude lysate of a strain that produces an Stx is used as the source of toxin, the presence of other bacterial factors, such as endotoxin, may have an influence on the outcome of the experiment.

References

1. Centers for Disease Control and Prevention (1994) Addressing emerging infectious disease threats: a prevention strategy for the United States. *MMWR* **43,** 1–18.

2. Griffin, P. M. (1998) Epidemiology of Shiga toxin-producing *Escherichia coli* infections in humans in the United States, in Escherichia coli *O157:H7 and Other Shiga Toxin-Producing* E. coli *Strains* (Kaper, J. B. and O'Brien, A. D., eds.), ASM Washington, DC, pp. 15–22.

3. Griffin, P. M. (1995) *Escherichia coli* O157:H7 and other enterohemorrhagic *Escherichia coli*, in Infections of the Gastrointestinal Tract (Blaser, M. J., Smith, P. D., Ravdin, J. I., Greenberg, H. B., and Guerrant, R. L., eds.), Raven, New York, pp 739–761.

4. Levine, M. M. (1987) *Escherichia coli* that cause diarrhea: enterotoxigenic, enteropathogenic, enteroinvasive, enterohemorrhagic, and enteroadherent. *J. Infect. Dis.* **155,** 377–389.

5. Obrig, T. G. (1998) Interaction of Shiga toxins with endothelial cells, in Escherichia coli *O157:H7 and Other Shiga Toxin-Producing* E. coli *Strains* (Kaper, J. B. and O'Brien, A. D., eds.), ASM, Washington, DC, pp. 303–311.

6. Richardson, S .E., Rotman, T. A., Jay, V., Smith, C. R., Becker, L. E., Petric, M., et al. (1992) Experimental verocytotoxemia in rabbits. *Infect. Immun.* **60,** 4154–4167.

7. Myhal, M. L., Laux, D. C., and Cohen, P. S. (1982) Relative colonizing abilities of human fecal and K-12 strains of *Escherichia coli* in the large intestines of streptomycin-treated mice. *Eur. J. Clin. Microbiol.* **1,** 186–192.

8. Wadolkowski, E. A., Burris, J. A., and O'Brien, A. D. (1990) Mouse model for colonization and disease caused by enterohemorrhagic *Escherichia coli* O157:H7. *Infect. Immun.* **58,** 2438–2445.

9. Wadolkowski, E. A., Sung, L. M., Burris, J. A., Samuel, J. E., and O'Brien, A. D. (1990) Acute renal tubular necrosis and death of mice orally infected with *Escherichia coli* strains that produce Shiga-like toxin type II. *Infect. Immun.* **58,** 3959–3965.

10. Kokai-Kun, J. F., Melton-Celsa, A. R., and O'Brien, A. D. (2000) Elastase in intestinal mucus enhances the cytotoxicity of Shiga toxin type 2d. *J. Biol. Chem.* **275,** 3713–3721.

11. Melton-Celsa, A. R. and O'Brien, A. D. (1996) Activation of Shiga-like toxins by mouse and human intestinal mucus correlates with virulence of enterohemorrhagic *Escherichia coli* O91:H21 isolates in orally infected, streptomycin-treated mice. *Infect. Immun.* **64,** 1569–1576.

12. Edwards, A. C., Melton-Celsa, A. R., Arbuthnott, K., Stinson, J. R., Schmitt, C. K., Wong, H. C., et al. (1998) Vero cell neutralization and mouse protective efficacy of humanized monoclonal antibodies against *Escherichia coli* toxins Stx1 and Stx2, in Escherichia coli *O157:H7 and Other Shiga Toxin-Producing* E. coli *Strains* (Kaper, J. B. and O'Brien, A. D., eds.), ASM, Washington DC, pp. 388–392.

13. Lindgren, S. W., Melton, A. R., and O'Brien, A. D. (1993) Virulence of enterohemorrhagic *Escherichia coli* O91:H21 clinical isolates in an orally infected mouse model. *Infect. Immun.* **61,** 3832–3842.

14. Zhang, X., McDaniel, A. D., Wolf, L. E., Keusch, G. T., Waldor, M. K., and Acheson, D. W. K. (2000) Quinolone antibiotics induce Shiga toxin encoding bacteriophages, toxin production, and death in mice. *J. Infect. Dis.* **181,** 664–670.

15. Fujii, J., Kita, T., Yoshida, S.-I., Takeda, T., Kobayashi, H., Tanaka, N., et al. (1994) Direct evidence of neuron impairment by oral infection with verotoxin-producing *Escherichia coli* O157:H⁻ in mitomycin-treated mice. *Infect. Immun.* **62,** 3447–3453.

16. Karpman, D., Connell, H., Svensson, M., Scheutz, F., Alm, P., and Svanborg, C. (1997) The role of lipopolysaccharide and Shiga-like toxin in a mouse model of *Escherichia coli* O157:H7 infection. *J. Infect. Dis.* **175,** 611–620.

17. Griffin, P. M., Ostroff, S. M., Tauxe, R. V., Greene, K. D., Wells, J. G., Lewis, J. H., et al. (1988) Illnesses associated with *Escherichia coli* O157:H7 infections. *Ann. Intern. Med.* **109,** 705–712.

18. Kurioka, T., Yunou, Y., and Kita, E. (1998) Enhancement of susceptibility to Shiga toxin-producing *Escherichia coli* O157:H7 by protein calorie malnutrition in mice. *Infect. Immun.* **66,** 1726–1734.

19. Farmer, J. J., III, Potter, M. E., Riley, L. W., Barrett, T. J., Blake, P. A., Cohen, M. L., et al. (1983) Animal models to study *Escherichia coli* O157:H7 isolated from patients with haemorrhagic colitis [Letter]. *Lancet* **1,** 702–703.

20. Pai, C. H., Kelley, J. K., and Meyers, G. L. (1986) Experimental infection of infant rabbits with verotoxin-producing *Escherichia coli. Infect. Immun.* **51,** 16–23.

21. Sjogren, R., Neill, R., Rachmilewitz, D., Fritz, D., Newland, J., Sharpnack, D., et al. (1994) Role of Shiga-like toxin I in bacterial enteritis: comparison between isogenic *Escherichia coli* strains induced in rabbits. *Gastroenterology* **106,** 306–317.

22. Sherman, P., Soni, R., and Karmali, M. (1988) Attaching and effacing adherence of vero cytotoxin-producing *Escherichia coli* to rabbit intestinal epithelium in vivo. *Infect. Immun.* **56,** 756–761.

23. O'Brien, A. D. and LaVeck, G. D. (1983) Purification and characterization of a *Shigella dysenteriae* 1-like toxin produced by *Escherichia coli. Infect. Immun.* **40,** 675–683.

24. Fujii, J., Kinoshita, Y., Kita, T., Higure, A., Takeda, T., Tanaka, N., et al. (1996) Magnetic resonance imaging and histopathological study of brain lesions in rabbits given intravenous verotoxin 2. *Infect. Immun.* **64,** 5053–5060.

25. Barrett, T. J., Potter, M. E., and Wachsmuth, I. K. (1989) Continuous peritoneal infusion of Shiga-like toxin II (SLT II) as a model for SLT II-induced diseases. *J. Infect. Dis.* **159,** 774–777.

26. Keenan, K. P., Sharpnack, D. D., Collins, H., Formal, S. B., and O'Brien, A. D. (1986) Morphologic evaluation of the effects of Shiga toxin and *E. coli* Shiga-like toxin on the rabbit intestine. *Am. J. Pathol.* **125,** 69–80.

27. Taylor, F. B., Jr., Tesh, V. L., DeBault, L., Li, A., Chang, A. C. K., Kosanke, S. D., et al. (1999) Characterization of the baboon responses to Shiga-like toxin: descriptive study of a new primate model of toxic responses to Stx-1. *Am. J. Pathol.* **154,** 1285–1299.

28. Kang, G., Pulimood, A., Mathan, M., and Mathan, V. I. (1997) A primate model of enterohemorrhagic *Escherichia coli* infection. 3rd International Symposium and Workshop on Shiga Toxin (Verocytotoxin)-Producing *Escherichia coli* Infections (VTEC '97), abstract V104/IV.

29. Beery, J. T., Doyle, M. P., and Schoeni, J. L. (1985) Colonization of chicken cecae by *Escherichia coli* associated with hemorrhagic colitis. *Appl. Environ. Microbiol.* **49,** 310–315.

30. Schoeni, J. L. and Doyle, M. P. (1994) Variable colonization of chickens perorally inoculated with *Escherichia coli* O157:H7 and subsequent contamination of eggs. *Appl. Environ. Microbiol.* **60,** 2958–2962.

31. Sueyoshi, M. and Nakazawa, M. (1994) Experimental infection of young chicks with attaching and effacing *Escherichia coli*. *Infect. Immun.* **62,** 4066–4071.

32. Fenwick, B. W. and Cowan, L. A. (1998) Canine model of the hemolytic uremic syndrome, in Escherichia coli *O157:H7 and Other Shiga Toxin-Producing* E. coli *Strains.* (Kaper, J. B. and O'Brien, A. D., eds.), ASM, Washington, DC, pp. 268–277.

33. Petric, M., Karmali, M. A., Richardson, S., and Cheung, R. (1987) Purification and biological properties of *Escherichia coli* verocytotoxin. *FEMS Microbiol. Lett.* **41,** 63–68.

34. Yutsudo, T., Nakabayashi, N., Hirayama, T., and Takeda, Y. (1987) Purification and some properties of a verotoxin from *Escherichia coli* O157:H7 that is immunologically unrelated to Shiga toxin. *Microb. Pathol.* **3,** 21–30.

35. Downes, F. P., Barrett, T. J., Green, J. H., Aloisio, C. H., Spika, J. S., Strockbine, N. A., et al. (1988) Affinity purification and characterization of Shiga-like toxin II and production of toxin-specific monoclonal antibodies. *Infect. Immun.* **56,** 1926–1933.

24

Gnotobiotic Piglets as an Animal Model for Oral Infection with O157 and Non-O157 Serotypes of STEC

Florian Gunzer, Isabel Hennig-Pauka, Karl-Heinz Waldmann, and Michael Mengel

1. Introduction

Over the last few decades, the use of swine as an animal model for human diseases in biomedical research has been steadily increasing because of similarities between the two species. The gnotobiotechnique, on the other hand, has been developed further since the beginning of the 20th century *(1–3)*, stimulated by the need for an experimental model to study bacteria–host interactions in sterile laboratory animals, during the course of an infection with a defined pathogen. The combination of both aspects led to the development of a complex isolator system that made the delivery of piglets by cesarean section and their rearing in a self-contained unit possible, shielded from undesirable contaminating germs *(4,5)*. In such a microbiologically well-defined environment, pathogen–host interactions can be studied without the influence of accompanying bacterial flora.

Naturally, *Escherichia coli* causes various diseases in pigs. These are often associated with several risk factors. For example, in *E. coli*-induced neonatal diarrhea, the immune status of the sow, the quality of colostrum, as well as the viability of the piglets after birth play the most important roles in pathogenesis. *E. coli* strains producing Shiga toxin 2e (Stx2e), a variant of Shiga toxin 2 (Stx2), are the infectious agent of edema disease *(6)*, an illness leading to widespread morbidity and mortality in weaner pigs. This disease is associated not only with certain serotypes of Stx2e producing *E. coli* strains but also with

From: *Methods in Molecular Medicine, vol. 73: E. coli: Shiga Toxin Methods and Protocols*
Edited by: D. Philpott and F. Ebel © Humana Press Inc., Totowa, NJ

failure in management and feeding, such as sudden changes in the protein and mineral contents of the feed or low room temperature and inadequate water supply after weaning. The toxin of the edema disease strains belongs to a family of toxins made by Shiga toxin-producing *E. coli* (STEC). It is weakly enterotoxigenic but cytotoxic on porcine vascular endothelial cells *(7)*. Viable counts of 10^9 colony-forming units (CFUs)/g *E. coli* are typically found in the small intestine of pigs early in the disease. Although Stx2e could clearly be shown to be the causative agent of edema disease in pigs *(8)*, the role of Shiga toxins in STEC mediated illness in humans is still not well understood.

Oral infection of gnotobiotic piglets with O157 or non-O157 STEC isolates from patients with hemorrhagic colitis (HC) or hemolytic uremic syndrome (HUS) leads to intestinal and extraintestinal manifestations of disease. Animals exhibit profound watery diarrhea 3–4 d after challenge as a result of severe mucosal damage associated with attaching and effacing lesions of colonocytes *(9,10)*. Up to 90% of them develop toxin-mediated neurological symptoms 2–3 d later, manifested clinically by ataxia, lateral recumbency, and death and histologically by foci of microhemorrhage and necrosis in the cerebellum *(11)*. Like humans, gnotobiotic piglets develop complications more often when the infecting STEC strain produces Stx2. A recently published study provides experimental proof for a higher pathogenic potential of Stx2 over Stx1 in gnotobiotic piglets *(12)*.

Virulent O157 STEC wild-type strains may be given in doses as little as 10^4 organisms and are still able to induce lethal disease in the animals. Animals can be inoculated orally, via the natural route, and do not need any antibiotical pretreatment. However, in addition to the development of diarrhea and neurological complications as well as the histological signs in the large intestine and the cerebellum, it is not entirely clear how far gnotobiotic piglets develop further symptoms of human STEC disease and HUS. Therefore, at present, this animal model can be well used to study the effects of STEC pathogenicity factors using mutant strains or to assess gene regulation in the bacteria as well as in the host. To further enlighten the mechanisms of human STEC disease and to develop intervention strategies based on animal data, the model has to be characterized more thoroughly as to which extent it really resembles HUS in humans.

2. Materials

Performing animal experiments with gnotobiotic piglets needs an animal facility with a well-organized infrastructure in general technical equipment (e.g., operating room, diagnostical laboratory for clinical chemistry, hematology, and microbiology, experimental laboratory complying to Biosafety Level 2 regulations), in special technical equipment (e.g., cages, isolators, automatic feeding system, waste disposal), and in personnel (veterinarians, technicians).

Every single experiment is preceded by laborious weeks of assembling and sterilizing the cages, preparing the diet, and carefully choosing and conditioning the sow. The combination of a sophisticated technical system with the unpredictability of living animals requires experience and a wide scope in timing and finances.

2.1. General Technical Needs

2.1.1. Sow Pen and Operating Room

1. Farrowing pen or crate (at least 200 cm × 70 cm).
2. Progestational hormone for prevention of spontaneous farrowing (e.g., chlormadinonacetate, Gestafortin®, Bayer, Leverkusen, Germany).
3. Operating room with sufficient space for the operating table.
4. Mobile and tiltable operating table (at least 200 cm wide, 90 cm deep, 80 cm high).
5. Venous permanent catheter, diameter: 0.9 × 1.4 mm, length: 35 cm (Vygoflex pur®; Vygon, Aachen, Germany).
6. General anesthesia: azaperon (Stresnil®; Janssen, Neuss, Germany), ketamine hydrochloride (Ursotamin®; Serumwerk Bernburg, Bernburg, Germany).
7. Spinal anesthesia: lidocaine hydrochloride (Lidocain 2% N®; Albrecht, Aulendorf, Germany).
8. Euthanasia: pentobarbital–Na 40% (Eutha 77®; Essex, Munich, Germany).
9. Shearing machine, shaving utensils, and depilatory cream.
10. Surgical soap and brush to clean the skin of the sow.
11. Diethylether to remove the grease of the skin.
12. Alcoholic iodine solution as skin disinfectant.
13. Infrared radiator for drying the skin of the sow immediately before surgery.
14. Contact glue based on polyurethane to seal the polyvinyl chloride (PVC) film to the skin of the sow (Ruderer U56; Ruderer Klebetechnik, Zorneding, Germany).

2.1.2. Preparation and Storage of Isolators

1. Room located close to the laboratory, containing all equipment needed for preparation of the isolators.
2. Storeroom for high-grade-steel cages, peristaltic pumps, electric low-pressure fans, PVC film rolls, and adhesive tapes.
3. Welding area equipped with a high-frequency film welding torch for heat sealing PVC tunnels of different diameters and a set of electrodes for different types of welds.

2.1.3. Biosafety Level 2 Laboratories

In the United States and many other countries, EHEC/STEC is considered a Biosafety Level 2 organism. Animal Biosafety Level 2 facilities and practices are recommended for activities with experimentally or naturally infected animals (for more detailed information on this topic see the 4th edition of the CDC/NIH guide *Biosafety in Microbiological and Biomedical Laboratories*, published in 1999 by the US Department of Health and Human Services). In

Germany, however, the wild-type organisms are classified Biosafety Level L3*, whereas genetically manipulated EHEC/STEC strains usually receive Biosafety Level S2 classification.

2.1.3.1. AIR CONDITIONED EXPERIMENTAL LABORATORY (ABOUT 24 M^2) THAT CAN BE HEATED UP TO 32°C

1. Beam scale at the ceiling with precise (up to 5 kg) and gross (up to 25 kg) scales.
2. High-pressure air supply.
3. High-pressure steam cleaner.
4. Repair materials for isolator covers.

2.1.3.2. AIR CONDITIONED LABORATORY TO PERFORM NECROPSIES AND BLOOD, SERUM, AND URINE TESTS

Basic equipment: laminar airflow workbench complying with Biosafety Level 2 guidelines, dissection instruments, dissection trays, chopping boards, bone saw, incubator, culture media, Bunsen burner, fridge, centrifuges, light microscope, container with liquid nitrogen, sample tubes with 4% formalin buffered at pH 7 or 2.5% glutaraldehyde to preserve tissue specimens for histology and electron microscopy, autoclave inside the laboratory that is designed to dispose of S2-level pathogenic materials.

2.2. Special Technical Needs for the Preparation of Isolators

2.2.1. Materials

1. Conventional clear PVC film (0.4 mm × 135 cm).
2. PVC adhesive tape (Scotch 470 soft PVC galvanic adhesive tape, 25 mm and 50 mm). Specifications: glue, gum resin; carrier, soft PVC; thickness, 0.18 mm; breaking burden, 352 N/100 mm; breaking stretch, 225%; temperature stability, up to 80°C.
3. Fiberglass adhesive tape (3M filament adhesive tape, 6 mm, 25 mm, and 50 mm). Specifications: glue, gum resin; carrier, polyester film; thickness, 0.13 mm; breaking burden, 5500 N/100 mm; breaking stretch, 5%.
4. Seamless vulcanized rubber rings.
5. Chrome nickel–steel screw clamps.
6. Flexible silicone tubes.
7. Plastic tube couplings.
8. Puncture safe neoprene gauntlets.
9. Chrome nickel steel rings without juts fitting the diameter of the neoprene gauntlets.
10. Yellow household gauntlets with an inside tissue layer giving a good grip to the surgeon and the assistant at the surgical isolator.

2.2.2. Forced Ventilation System to Build up a Slight Overpressure Inside the Isolators

1. Rigid plastic air tubes (size about 4 cm × 110 cm).
2. Electric low pressure fans.

3. Metallic round air filters for incoming air and outgoing air each with a 4 layer inorganic fiberglass filtration media as a barrier to microbial breakthrough (Fiber Glass Filtration Media AFS-4 1/2, Johns Manville, DN).

2.3. Isolators

2.3.1. Surgical Isolator

1. PVC film tunnel 200 cm × 80 cm with one pair of household gauntlets at each side.
2. Outlet for blood and secretions at the bottom (5 cm diameter) leading into a collection bag (35 cm × 40 cm).
3. Electrical outlet in the isolator wall for connecting an electrical cauterization knife to the transformer outside the sterile unit.
4. Air ducts for connection to the forced ventilation system.
5. Contents of the surgical isolator: towels, gauze swabs, electric cable with adaptors for the electroknife, about 12 serum tubes, 5 sample swabs, clamps for the umbilical cords, surgical instruments in a box containing scalpels, artery clamps, surgical forceps, curved scissors, and bandage scissors (Lister).
6. The surgical isolator is connected tightly with the transport isolator by a spacious plastic tunnel of 30 cm × 180 cm in size (**Fig. 1**). The entire system is stabilized by a low-pressure fan.

2.3.2. Transport Isolator

1. Reusable oval V4A steel sump (100 cm × 78 cm × 5 cm) covered with a PVC film balloon (90 cm × 100 cm × 80 cm) that is fixed to the sump by PVC and fiberglass adhesive tapes.
2. A high-grade-steel grating with narrow spaces, fixed 5 cm above the bottom of the sump. The grating is covered with towels.
3. At the top of the film balloon, a steel eye is installed for weighing the piglets inside the isolator at the outside beam scale, using a hammock.
4. Two pairs of puncture safe neoprene gauntlets at each side of the film balloon.
5. Air ducts and connection to the forced-ventilation system.
6. Contents of the transport isolator: rectangular piece of PVC film with four holes as a weighing hammock for the piglets, towels, and clamps for the umbilical cords.

2.3.3. Rearing Isolators

1. Reusable oval V4A steel sump (100 cm × 78 cm × 5 cm) with a PVC film balloon (100 cm × 100 cm × 80 cm) that is fixed to the sump by PVC adhesive tapes and fiberglass adhesive tapes.
2. Two pairs of neoprene gauntlets, one at each side of the film balloon.
3. The stainless-steel cage (60 cm × 40 cm × 40 cm), made from V4A wire netting (inside width 1 cm), is put inside the oval sump. It has a feces sump (66 cm × 50 cm × 9 cm) underneath and a punched steel sheet (perforation diameter–0.8 cm, distance between perforations-1 cm) as a floor, about 5 cm above the bottom of the feces sump. The side walls can be folded up along side (**Fig. 2**). The cage is

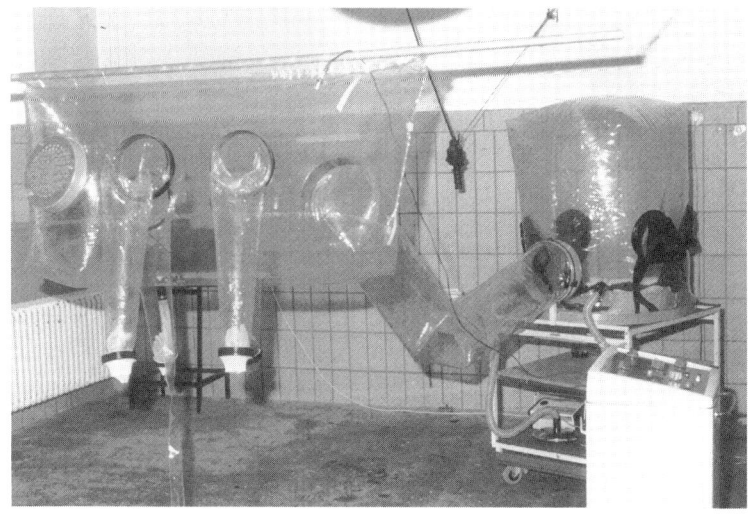

Fig. 1. Surgical isolator and transport isolator connected with a PVC tunnel.

Fig. 2. Animal cage with side wall folded up.

lengthwise divided into two halves by a steel plate. At the short sides, feeding plates (diameter–10 cm, depth–1 cm, capacity–50 mL) with central milk outlets and connections to the milk ducts are installed about 5 cm above the floor (**Fig. 3**).

Fig. 3. Feeding plate with milk outlet in the middle.

The ceiling consists of a steel pan which will be used for keeping instruments and sample tubes as well as for handling of the piglets.

4. A spacious PVC tunnel (130 cm × 30 cm) is welded to each isolator cover. It is closed by two large (50 cm) rubber covered metal clamps, leaving a sluice of about 40 cm between each other, which is filled with 400 mL of 5% peracetic acid/detergent solution as disinfectant (*see* **Subheading 2.6.**).

5. Air ducts and connections to the forced ventilation system (**Fig. 4**).

6. Contents of one rearing isolator: towels, about 12 serum tubes (2-mL capacity), 12 EDTA tubes for anticoagulated blood (1-mL capacity), 12 heparine tubes for anticoagulated blood (1-mL capacity), 12 tubes with 0.2 mL of 0.1 *M* sodium citrate (2-mL capacity) for anticoagulated blood that has to be taken in after γ-irradiation, about 6 sample swabs, 12 urine sample tubes, 30 syringes (5 mL), 30 needles (0.8 mm × 40 mm), 1 feeding tube with 2 lateral eyes (length about 50 cm, diameter–2 × 3 mm; B. Braun Melsungen, Melsungen, Germany), 1 curved olive-headed probe (2 mm × 80 mm), a pair of curved scissors, and about 6 small plastic bags plus tie strings.

2.4. Feeding (5,13)

1. Flexible silicone tubes as milk ducts (diameter: 0.4 cm × 0.9 cm).
2. Plastic tube couplings.
3. Peristaltic tube pumps, one for two isolators.
4. Stainless-steel screw clamps for blocking the milk ducts if necessary.
5. Automatic timer.

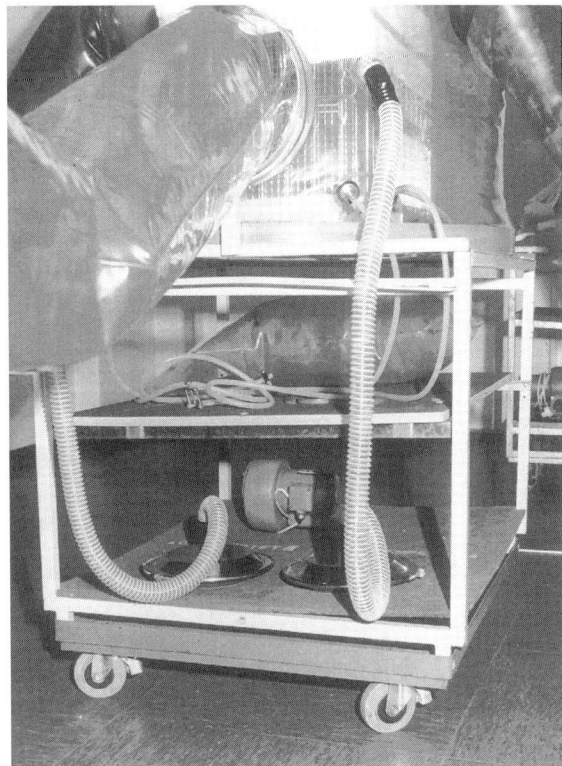

Fig. 4. Forced ventilation system with an electric low pressure fan underneath the isolator.

6. Four autoclavable glass bottles with spigots, for double distilled water (ddH$_2$O).
7. Bags for milk substitute (dry content: 1550 g) consisting of an inner polyethylene bag and an outer PVC bag (0.3 mm PVC, 50 cm × 40 cm).
8. In **Table 1** the ingredients for a 10-kg piglet diet, made from a milk substitute premix, are listed.

In every rearing isolator, two sterile flexible silicone tubes are connected to the central milk outlets of the two feeding plates. They converge before reaching the peristaltic pump and are leading into a bag with milk substitute underneath the isolator. The milk pumps are controlled by a timer that allows automated feeding of the animals at fixed time-points (*see* **Fig. 5A,B**).

2.5. Distribution System

1. Unbroken steel pipe (diameter 30 cm, length 100 cm) to connect the transport isolator to the distribution pipe.

Table 1
Ingredients of the Milk Diet for Gnotobiotic Piglets

Vitamin A	420,000 IU
Vitamin D_3	43,000 IU
Vitamin E acetat	1.5 g
Vitamin B_1	100.0 mg
Vitamin B_2	150.0 mg
Calcium D panthotenic acid	150.0 mg
Nicotinic acid	300.0 mg
Vitamin B_6	75.0 mg
Folic acid	50.0 mg
Vitamin B_{12}	0.30 mg
Ascorbic acid	1.67 g
Biotine	1.67 mg
Lysinemonohydrochloride	19.00 g
Vitamin K_3	60.00 mg
MgO	1.05 g
$Fe_2SO_4 \times 7H_2O$	960.00 mg
$ZnSO_4 \times 7H_2O$	162.00 mg
$MnSO_4 \times 1H_2O$	105.00 mg
$CuSO_4 \times 5H_2O$	45.00 mg

Full cream milk powder has to be added to all these ingredients up to a weight of 100 g. This is the milk substitute premix (26% fat, at least 26% protein, at least 36.5% lactose).

Preparation of 10 kg milk diet (1 bag/U): 1450 g full cream milk powder and 100 g milk substitute premix are dissolved in 8450 g sterile ddH₂O.

2. Distribution steel pipe (30 cm × 150 cm) with seven holes at the side, connecting the distribution pipe to the rearing isolators. Two additional holes, which are needed to pass on the piglets, are located at the top of this pipe. They are covered with blind-ending PVC film sleeves (20 cm × 50 cm). Near the front end of the distribution pipe, another PVC sleeve is fixed, containing towels to absorb excessive peracetic acid and keeping it away from the piglets (**Fig. 6**).
3. Eight to nine connecting plastic tunnels (30 cm × 150 cm) between transport isolator, unbroken steel pipe, distribution pipe, and up to six rearing isolators.
4. The whole system is stabilized by low-pressure fans (**Fig. 7**).

2.6. Sterilization and Waste Disposal

1. 40% Peracetic acid in a safety container with dosage pump (Wofasteril®; Kesla Pharma Wolfen, Greppin, Germany).

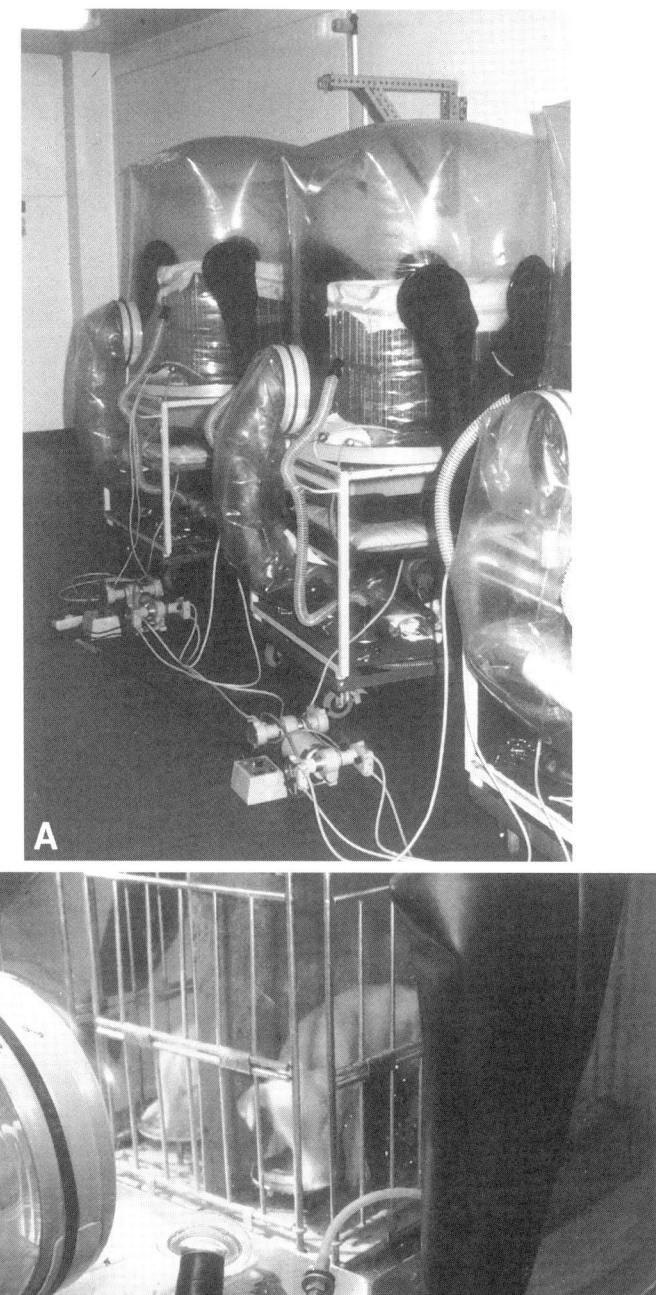

Fig. 5. (**A**) Peristaltic milk pumps on the floor of the animal laboratory. They are connected to the milk bags underneath the isolators. (**B**) Feeding of the piglets. Connections from left to right: PVC tunnel, air duct, milk duct.

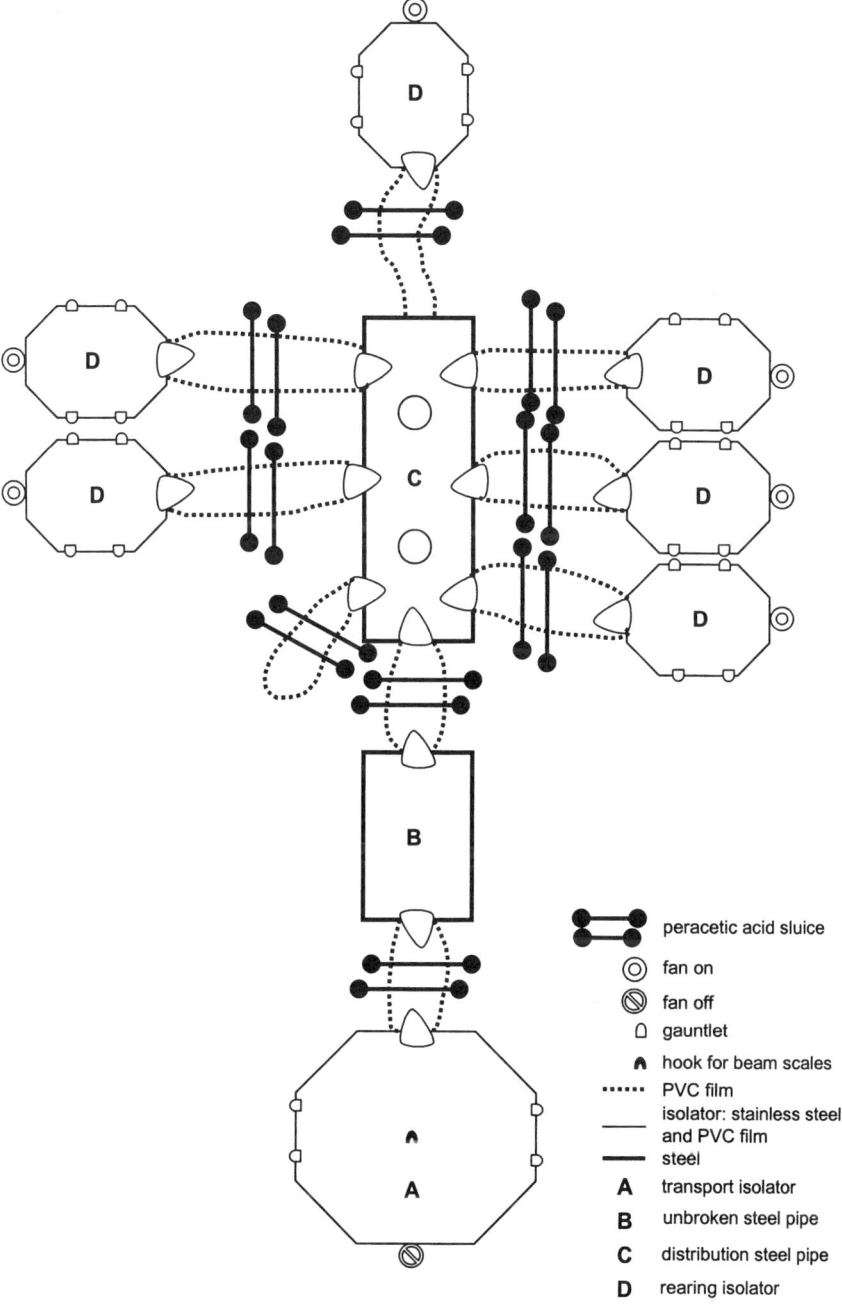

Fig. 6. Diagram of the piglet distribution line for an experiment with six rearing isolators.

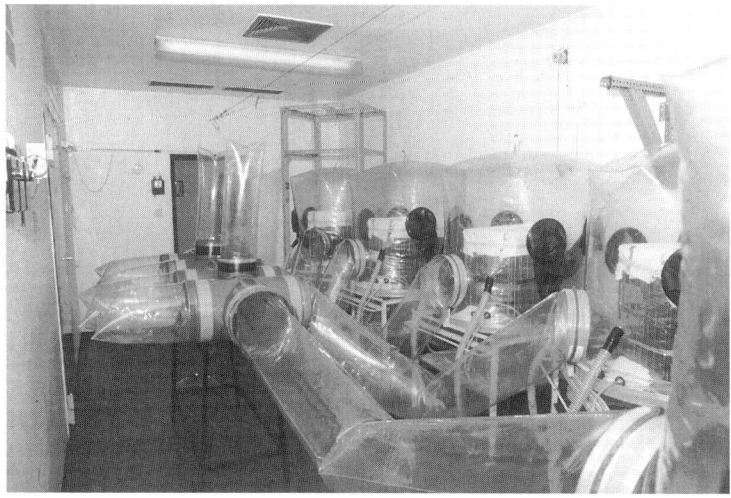

Fig. 7. Distribution system with four isolators connected to the distribution steel pipe.

2. Double distilled H_2O to prepare a 5% peracetic acid/detergent solution, ready to use.
3. Mixing tank (glass or stainless steel).
4. Pair of safety glasses.
5. Gas mask.
6. Acid-proof gauntlets.
7. Merckoquant® test strips (Merck, Darmstadt, Germany) to check peracetic acid concentrations.
8. Spray pistol connected to a compressor (5 atm) for chemical sterilization with peracetic acid.
9. 2 *M* Sodium hydroxide for neutralization.
10. High-grade-steel carcass pans that can be autoclaved.
11. Autoclavable plastic bags.
12. Heat sealing machine for waste bags.

2.7. Animals

1. Conventional sow of a German breeding program.
2. High health and condition state.
3. Fourth to sixth farrowing.
4. Only one insemination at a well-known time-point (*see* **Note 1**).

3. Methods
3.1. Preparation of Isolators and Diet

1. Start assembling all isolators and milk bags 3 mo before surgery.
2. Prepare the milk-substitute premix according to **Table 1**. The loss of vitamins by irradiation is taken into account by the composition of the diet *(13–15)* (*see* **Note 2**).

3. Twelve packages each containing 12 sample tubes with 0.2 mL of 0.1 *M* sodium citrate for anticoagulated blood have to be heat sealed into double plastic bags and sterilized by β-irradiation with at least 10 kGy.
4. Dissolve the milk substitute 10 d before surgery with 8450 mL sterile ddH$_2$O, connecting the diet bags with silicone tubes to the outlets of the water bottles (*see* **Note 3**).
5. Take the silicone tubes off the spigots and seal them at their open ends by a 5-cm peracetic acid sluice between two clamps.
6. Two days later, cut a separated piece of the milk duct and test the ready made diet for sterility by aerobic and anaerobic culture.
7. Set up the distribution line for the piglets from the transport isolator into the rearing isolators about 10 d before surgery (*see* **Note 4**).

3.2. Preparation of the Sow and Cesarean Section

1. Purchase the sow about 3 wk before the date of farrowing (d 114 of pregnancy).
2. Apply a progestational hormone (e.g., chlormadinonacetate, Gestafortin®) in a dose of 0.3 mg/kg body mass (bm) intramuscularly (im) at the 105th, 108th, and 112th d of gestation to prevent spontaneous farrowing (*see* **Note 5**).

113th d of Gestation

3. Implant a permanent catheter into an ear vein of the sow.
4. Shear, shave, and depilate the left flank of the sow.
5. Weigh the sow.

114th d of Gestation, Day of Surgery

6. Temperature in the operating room should be about 32°C.
7. Anesthetize the animal with a general anesthesia of 2 mg azaperone/kg bm im and 10–15 mg ketamine hydrochloride/kg bm iv (ear vein catheter) (*see* **Note 6**).
8. Inject 2% lidocaine hydrochloride through the foramen lumbosacrale, either 0.5 mL (subarachnoidally) or 0.7 mL (epidurally) per 10 cm crown rump length (*see* **Note 7**).
9. Wash the left flank of the sow with surgical soap using a brush.
10. Remove the grease of the skin with diethylether.
11. Disinfect the skin with alcoholic iodine solution.
12. Dry the skin completely by means of heat radiation.
13. In the meantime, clean the bottom of the surgical isolator with diethylether.
14. Apply contact glue onto the left flank of the sow and heat-dry for about 5 min.
15. The surgeon guides the isolator through his gauntlets when lowered onto the operating field.
16. Glue the bottom of the surgical isolator to the skin, avoiding wrinkles.
17. Now the assistant slips into the other pair of surgical gauntlets opposite to the surgeon (*see* **Note 8**).
18. PVC film and skin are cut with the electroknife to heat seal any spaces between film and skin.
19. Start the incision cranial and ventral to the tuber coxae and extend it ventrally to a point approx 5 cm dorsal to the cranial skin fold of the flank (paralumbar fossa incision).

20. Cut the cutaneous and the three abdominal muscles (external and internal oblique muscle of abdomen, transverse abdominal muscle) with a scalpel and a pair of bandage scissors.

21. Remove the subperitoneal fat with an artery clamp.

22. Open the peritoneum carefully avoiding injury of the gut.

23. Extort the closest uterine horn and make incisions of approx 8 cm parallel to nearly every piglet to guarantee a rapid delivery.

24. Gnotobiotic piglet serum can be collected from the umbilical cords before clamping them (*see* **Note 9**).

25. Transport piglets rapidly through the connection tunnel into the attached transport isolator.

26. Inject 40 mg pentobarbital–Na/kg bm into the ear vein catheter to euthanize the sow.

27. Separate the transport isolator from the surgical isolator in the middle of the connection tunnel by two large metal clamps. A 40-cm-wide sluice, filled with 5% peracetic acid/detergent solution, is left in between.

3.3. Handling of the Newborn Gnotobiotic Piglets

1. Take the transport isolator to the experimental laboratory (Biosafety Level 2, room temperature at least 32°C).

2. Rub the piglets with towels to stimulate their spontaneous breathing by massage, if necessary for several hours.

3. Disinfect the unbroken steel pipe with 5% peracetic acid in the meantime and connect it with the PVC tunnel of the transport isolator. The plastic tunnels to the distribution pipe and to the transport isolator are tightly clamped, avoiding airflow toward the piglets.

4. Leave peracetic acid in the system for at least 1 h.

5. Absorb excess peracetic acid inside the unbroken steel pipe with towels taken out of the blind-ending plastic sleeve at the distribution steel pipe.

6. Turn off the fan at the transport isolator.

7. Turn on all fans at the rearing isolators.

8. Weigh every piglet in the hammock with the beam scale at the ceiling of the laboratory.

9. Open the connection between transport isolator and unbroken steel pipe and take the piglets to the rearing isolators against the airflow.

10. Separate each rearing isolator from the distribution channel, leaving a PVC tunnel of about 150 cm. A permanent sluice, filled with 5% peracetic acid, is formed between two large metal clamps.

11. Piglets are kept in pairs, but separated from each other, for the duration of the experiment.

12. It is very important that piglets take in milk substitute on the first day of living. Therefore a small amount of diet has to be in the feeding plates all the time (*see* **Note 10**).

13. The room temperature of 32°C on the first days of life can be lowered to 30°C at the end of the first week.

3.4. Oral Infection and Clinical Monitoring

3.4.1. Sample Collection During the Experiment

1. Prepare needles, syringes, and sample tubes for quick availability.

2. Blood and urine samples are taken immediately before infection and about every second day thereafter. Every manipulation should be performed as quickly and skillfully as possible because the piglets are highly susceptible to stress (*see* **Notes 11** and **12**).
3. A maximum of 4 mL blood should be drawn in 1 d. This amount is divided into the following three samples: 0.5–1 mL EDTA blood, 2 mL citrate blood, and 1 mL serum or heparin blood.
4. Gently shake every tube immediately after the sample is taken.
5. Put all sample tubes into a plastic bag that is tied with a string.
6. Channel the bag out through the peracetic acid sluice.

3.4.2. Oral Infection

1. Grow EHEC strains overnight at 37°C in Luria–Bertani (LB) broth. On the day of surgery, inoculate fresh cultures at a 1:100 dilution in LB broth and grow them for another 3–4 h at 37°C with vigorous agitation. OD_{600} should reach 1.0, which equals about 1×10^9 CFUs/mL.
2. Wash the bacteria once to remove any toxin released into the supernatant and then dilute them to concentrations according to the experimental design. Apply the bacterial inoculum in 5 mL LB broth.
3. Ten to 12 h after cesarean section, piglets can be infected perorally with washed viable bacteria of either O157 or non-O157 EHEC wild-type strains. An apathogenic *E. coli* strain (e.g., *E. coli* Nissle 1917) should be used as negative control (*see* **Note 13**).
4. The inoculum is given either via a feeding tube directly into the stomach or with a curved olive-headed probe behind the root of tongue. Both routes resemble the natural route of infection (*see* **Note 14**).

3.4.3. Clinical Monitoring and Euthanasia

1. Perform a visual examination of all piglets six times a day. Critically ill animals will thus be detected early. First clinical symptoms may appear 6–12 h after infection (*see* **Note 15**).
2. Give a daily average score to each animal, based on the incidence and degree of diarrhea, locomotory disorder, and lethargy. The score ranges are 0 (normal), 1 (slight increase), and 2 (severe increase).
3. Pigs with a clinical score of 2 are euthanized by intravenous injection of pento-barbital–Na after taking the last blood, serum, and urine samples.
4. Remove dead piglets out of the isolator into the plastic tunnel and put them between two clamps.
5. Cut the whole bundle consisting of the piglet inside the tunnel and the two clamps and weigh it.
6. Unwrap the piglet under the S2 laminar airflow workbench and put it onto a dissection tray.
7. Weigh the rest of the bundle (PVC film, metal clamps) again.

3.4.3.1. Postmortem Examination

1. Open the body cavity of the piglets at the linea alba after skinning breast and abdomen and removing hind and fore limbs.
2. Remove the spleen from its mesentery and the mesentery from stomach and intestine.
3. Ligate the rectum and extirpate the entire intestines.
4. Take out the stomach after the esophagus is ligated and the stomach diaphragm ligament and the stomach liver ligament are cut.
5. Extirpate the liver by cutting the liver ligaments and the vena cava.
6. Extirpate the kidneys and remove their capsules.
7. Cut the ribs at the bone cartilage connection with a bone saw.
8. Remove all thoracic organs after the connections between sternum and pericardium are released and the aorta is ligated and cut.
9. Cut the intestines open for inspection of their mucosal surface and to take samples for microbiology.
10. Open the stomach and inspect the mucosa.
11. Separate the head.
12. Open the cranial cavity by a cross-section behind the eyes.
13. Position the skull on its cranium so that the brain comes out after the cranial nerves are cut.

3.5. Collection of Samples for Microbiological and Histological Examination

3.5.1. Microbiology

1. Take rectal sample swabs throughout the experiment to monitor the uniformity of the bacterial flora.
2. At the end of the experiment, a few milliliters of the gut contents from the small and the large intestines (spiral colon) are taken for bacterial counts.

3.5.2. Histology (Light Microscopy, 4% Formalin Buffered at pH 7.0)

1. Segments of jejunum and colon and the triangle consisting of ileum, cecum, and ascending colon spread out onto a piece of styrofoam, fastened with pins, and fixed in formalin.
2. Sagittal sections and wedge-shaped sections (renal cortex, renal medulla, renal pelvis) from both kidneys.
3. Sections of brainstem, cerebellum, and cerebrum.
4. Sections of liver, pancreas, and spleen.
5. Sections of heart muscle and lung.

3.5.3. Immunohistochemistry (Liquid Nitrogen) (see **Note 16**)

1. Sections of cerebellum.
2. Sections of both kidneys.
3. Section of large intestine.

3.5.4. Electron Microscopy (2.5% Glutaraldehyde)

1. Section of cerebellum.
2. Sections of both kidneys.
3. Section of large intestine.

3.6. Carcass Disposal and Chemical Sterilization of the Rearing Isolators

1. Carcasses in steel pans are double bagged into autoclavable plastic bags and heat sealed.
2. Autoclave them at a core temperature of 121°C for 20 min.
3. Sterile carcasses have to be processed by a skinnery.
4. Autoclave all the waste from the S2 laboratories in double plastic bags at 121°C for 20 min and dispose of it at the garbage disposal.
5. Disinfect dissection instruments, dissection trays, chopping boards, and bone saw in a disinfection bath before autoclaving them at 121°C for at least 20 min.
6. Put on protective clothing (gas mask, acid-proof gloves).
7. Prepare a fresh dilution of the 5% peracetic acid/detergent mixture.
8. Pump this disinfectant through the milk ducts into the feces sump of the rearing isolators, up to a level of about 3 cm (about 10 L).
9. Peracetic acid must also be filled into both air ducts.
10. Inside the rearing isolators, peracetic acid is dispersed with a spray pistol while the fan is running (2.5 mL/m^3).
11. Take a sample swab for sterility control 1 h after the procedure is finished.
12. Let peracetic acid take effect on the isolator contents until the next day.
13. Check peracetic acid concentration by Merckoquant test strips (at least 1%).
14. Fill the isolator with water up to the level of the PVC tunnel, when the sterility check is negative.
15. Neutralize peracetic acid to pH 7 with 2 M NaOH.
16. Dismantle the isolators and clean them.

3.7. Collection of Data

1. Collect feces directly from the small and the large intestines for bacterial counting and detection of Shiga toxin 1 or 2 (*see* **Note 17**).
2. Collect EDTA blood for complete and differential blood cell count, thrombocyte cell count, and observation of morphological changes in red blood cells such as anisocytosis, polychromasia, poikilocytosis, and fragmentocytes.
3. Take serum or heparin blood for determination of total protein, urea, creatinine, glutamate dehydrogenase activity, lactate dehydrogenase activity, sodium, and potassium.
4. Urine analysis includes semiquantitative measurement of pH, blood, number of leucocytes, protein content, concentration of glucose, and urobilinogen. Nitrite, ketone bodies, and bilirubin are measured qualitatively (Combur 9® test strips; Roche Molecular Biochemicals, Mannheim, Germany).

5. When the concentrations of creatinine, sodium, and potassium in the urine have been determined, the glomerular filtration rate (GFR) and the fractional excretions (FE) of water, sodium and potassium are calculated *(16)* (*see* **Note 18**).
6. Assess urinary sediments after centrifugation at 300*g* for 10 min.
7. Freeze aliquots of urine supernatants at –80°C for further investigations.
8. Take extra serum samples for serological assays and freeze them at –80°C.

4. Notes

1. Hysterotomy has to take place on the 114th d of gestation. Therefore, it is absolutely necessary to know the exact date of mating.
2. During a 4-wk trial with 12 piglets, about 30 bags with 10 kg milk substitute each will be used. About 1 mo before the experiment, isolators and diet bags have to be transported to a company performing γ-irradiation with [60]Co (isolators at least 25 kGy, diet at least 50 kGy).
3. Dry-milk diet powder must be dissolved with the correct amount of water (be careful about volume loss through autoclaving). If the concentration of the milk substitute is too high, there is a risk of sodium chloride poisoning.
4. All inner surfaces of the distribution line have to be sterilized chemically with 5% peracetic acid for 1 d and must then be aerated with a fan from the isolator overpressure system for the remaining days. It is important that there be no peracetic acid left at the day of surgery, because it will burn the respiratory tract of the piglets.
5. If piglets are delivered too early, they will show weakness and signs of immaturity (long soft claws, depressed breathing).
6. General anesthesia has to be as superficial as possible and surgery has to be performed as quickly as possible because of the depressive effect on the piglets caused by the anaesthetics. A repeated application of ketamine hydrochlorid should be avoided; however, if necessary, it can be used for a prolongation of general anesthesia. Analgesia should be achieved by local anesthesia through the foramen lumbosacrale.
7. For local anesthesia, the sow has to be positioned in ventral recumbency with the hind legs stretched and spread out forward as soon as possible. The sow should be lifted onto the operating table not later than 10 min after injection, lying on the right flank and tied up with ropes at the fore and hind limbs as well as at the upper jaw.
8. The assistant must remove blood and condensation from the inside wall of the surgical isolator that limit the vision of the surgeon. During the surgery, he also has to take sample swabs from the following areas: (1) operation field before the skin is cut; (2) skin and subcutaneous tissue after cutting; (3) abdominal cavity; (4) uterus after the piglets have been removed; (5) operation field at the end of surgery.
9. Bleeding out of the umbilical cords reduces the viability of the piglets.
10. If all piglets are drinking well on their own they will be fed automatically every 2 h. The peristaltic pumps are controlled by a timer. The daily amount of diet will be slowly increased from 200 mL at d 4 to 500 mL around d 21. At night, feeding times have to be linked to illumination. The automatic feeding system must be controlled every 4 h, because milk ducts can tear inside the peristaltic pumps and

must be closed before air can enter the affected isolator.

11. For taking blood samples, the piglets have to be set onto the steel top of the rearing isolator in dorsal recumbency. An assistant presses the head firmly to the ground with one hand and stretches the fore limbs tightly to the piglet's body with the other hand. In this position blood can be taken by puncture of the vena cava cranialis. Injury of the pericardium must be avoided. The site of puncture should be compressed for a few seconds. To draw urine samples, the piglet has to be held headfirst at the hind limbs, without reaching the ground. Cystocentesis is performed by pricking the paramedian between the last and next to last pair of teats in the craniodorsomedial direction.

12. Piglets must be handled with care because they might bite into the gauntlets and can destroy the closed system of the isolator.

13. Usually, the infectious doses of EHEC, given perorally, range from 5×10^8 to 5×10^{10} organisms. However, virulent O157 EHEC strains may cause lethal disease even in numbers as low as 10^4 bacteria. At the day of the surgery, the number of bacteria in the LB broth culture can only be estimated from the OD_{600}. To obtain an exact CFU count, a 10-fold dilution series has to be plated, which is analyzed the following day.

14. Peroral infection by feeding tubes takes the bacterial suspension directly down into the stomach. The feeding tube has to be led carefully in the middle of the tongue, making the piglet swallow; otherwise, there is a risk for pushing the tube into the trachea. When using the olive-headed probe for oral infection, the bacterial suspension must be applied slowly enough for the piglets to swallow in order to control that they receive the calculated infectious dose.

15. The first clinical symptom of EHEC infection is a yellowish, watery diarrhea (not to be mistaken with the physiological brown liquid feces of gnotobiotic piglets). Further symptoms are locomotory disorders especially in the hind limbs, swaying back, and a dog-sitting posture. Finally, piglets show apathy and lateral recumbency.

16. A wide receptacle with about 300 mL of liquid nitrogen should be used to snap-freeze tissue samples for immunohistochemistry. Sample pieces are put into 2-mL cryo tubes and dropped directly into the liquid nitrogen. The frozen specimens can be stored in a nitrogen container or at −80°C in a freezer.

17. Detection of Shiga toxins in the stool samples is performed without any further enrichment, using an Stx enzyme-linked immunosorbent assay (ELISA) (e.g., Ridascreen® Verotoxin, R-Biopharm, Darmstadt, Germany). Samples should be stored at 4°C no more than a week, although the toxin might be stable for longer periods. For bacterial counting, a 10-fold dilution series down to 10^{-8} is prepared in LB medium from 1 g or 1000 µL of the gut contents. Starting with the highest dilution, 10 µL of each dilution step are spotted counterclockwise onto a single blood and sorbit MacConkey (SMAC) agar plate and incubated at 37°C overnight. The next day, numbers are taken from all dilutions where individual colonies can be discriminated and the bacterial colonization/g stool is calculated. The purity of the stool cultures is also assessed on the solid growth media. Additionally, reisolated strains can be tested for genotypical and phenotypical changes by poly-

merase chain reaction, DNA sequencing, serology, or biochemical profiles.

18. The GFR is determined using the endogenous marker creatinine, which is exclusively eliminated by glomerular filtration. Creatinine clearance is highly correlated to body weight and can be estimated assuming a constant endogenous creatinine production and excretion. The endogenous creatinine production and excretion (E_{cr}) is calculated for individual piglets by the following regression equation: $E_{cr} = 83.6 \times \log$ body weight $+ 86.1$ *(16)*. The GFR is now calculated by dividing E_{cr} through the plasma creatinine concentration. Subsequently, the FE of electrolytes can be evaluated. For example, FE-Na = (Pl-crea/U-crea)(U-Na/Pl-Na). Pl-crea = plasma creatinine concentration; U-crea = urine creatinine concentration; U-Na = urine sodium concentration; Pl-Na = plasma sodium concentration.

References

1. Küster, E. (1912) Die keimfreie Züchtung von Säugetieren und ihre Bedeutung für die Erforschung der Körperfunktionen. *Zbl. Bakteriol.* **54,** 55.
2. Reyniers, J. A. (1941) Apparatus for a method of maintaining and working with biological specimens in a germfree controlled environment. US Patent 2244082.
3. Trexler, P. C. (1959) The use of plastics in the design of isolator systems. *Ann. NY Acad. Sci.* **78,** 29.
4. Plonait, H., Bickhardt, K., and Bähr, K.-H. (1966) Versuche zur Gewinnung gnotobiotischer Ferkel mit dem Isolator Hannover I. *Dtsch. Tierärztl. Wochenschr.* **73,** 539–543.
5. Bähr, K.-H., Richter, L., and Plonait, H. (1968) Versuche zur Gewinnung spezifisch pathogen-freier Ferkel mit dem Isolator Hannover II. *Dtsch. Tierärztl. Wochenschr.* **75,** 55–64.
6. Marques, L.R.M., Peiris, J.S.M., Cryz, S.J., and O'Brien, A.D. (1987) *Escherichia coli* strains isolated from pigs with edema disease produce a variant of Shiga like toxin II. *FEMS Microbiol. Lett.* **44,** 33–38.
7. Gyles, C. L. (1992) *Escherichia coli* cytotoxins and enterotoxins. *Can. J. Microbiol.* **38,** 734–746.
8. MacLeod, D. L., Gyles, C. L., and Wilcock, B. P. (1991) Reproduction of edema disease of swine with purified Shiga like toxin II variant. *Vet. Pathol.* **28,** 66–73.
9. Tzipori, S., Wachsmuth, K. I., Chapman, C., Birner, R., Brittingham, J., Jackson, C., et al. (1986) The pathogenesis of haemorrhagic colitis caused by *Escherichia coli* O157:H7 in gnotobiotic piglets. *J. Infect. Dis.* **154,** 712–716.
10. Tzipori, S., Wachsmuth, K. I., Smithers, J., and Jackson, C. (1988) Studies in gnotobiotic piglets on non-0157:H7 *Escherichia coli* serotypes isolated from patients with hemorrhagic colitis. *Gastroenterology* **94,** 590–597.
11. Tzipori, S., Gunzer, F., Donnenberg, M. S., de Montigny, L., Kaper, J. B., and Donohue-Rolfe, A. (1995) The role of the eaeA gene in diarrhea and neurological complications in a gnotobiotic piglet model of enterohemorrhagic *Escherichia coli* infection. *Infect. Immun.* **63,** 3621–3627.
12. Donohue-Rolfe, A., Kondova, I., Oswald, S., Hutto, D., and Tzipori, S. (2000) *Escherichia coli* O157:H7 strains that express Shiga toxin (Stx) 2 alone are more

neurotropic for gnotobiotic piglets than are isotypes producing only Stx1 or both Stx1 and Stx2. *J. Infect. Dis.* **181**, 1825–1829.

13. Waldmann, K. H. (1988) Gnotobiotische Gewinnung und Haltung von Ferkeln der Rasse Göttinger Miniaturschwein. *Tierärztl. Prax.* **3(Suppl.),** 84–92.

14. Ley, F. J. (1976) Radiation sterilization of diets. *J. Inst. Anim. Tech.* **26,** 87.

15. Travnicek, J. and Mandel, L. (1979) Gnotobiotic techniques. *Folia Microbiol.* **24,** 6–10.

16. Waldmann, K. H., Wendt, M., and Bickhardt, K. (1991) Kreatinin-Clearance als Grundlage klinischer Nierenfunktionsbestimmung beim Schwein. *Tierärztl. Prax.* **19,** 373–380.

25

Bovine *Escherichia coli* O157:H7 Infection Model

Evelyn A. Dean-Nystrom

1. Introduction

Cattle are a major source of Shiga toxin-producing *Escherichia coli* (STEC) O157:H7 and other STEC that cause serious food-borne diseases in humans. Most STEC-infected cattle are asymptomatic carriers of STEC. We have developed bovine STEC-infection models to identify the bacterial and host factors that are integral to intestinal colonization and shedding of STEC by cattle. Although STEC usually does not cause disease in cattle, experimentally infected colostrum-deprived, neonatal (<12 h old) calves develop diarrhea and enterocolitis by 18 h after inoculation with 10^{10} colony-forming units (CFUs) of STEC bacteria. STEC-infected calves become colonized with bacteria (i.e., have $\geq 10^6$ CFUs of inoculum bacteria/g of intestinal tissue or feces) and have attaching and effacing lesions in both the small and large intestines *(1,2)*. In some calves, the disease induced by STEC may be fatal.

This chapter describes the neonatal calf STEC-infection model we are using to study the mechanisms of STEC colonization in cattle. The focus is on methods used to infect neonatal calves with STEC O157:H7, observations for clinical signs of disease, necropsy procedures and collection of samples, bacteriological counts, microscopic examination of tissues, and immunohistochemical detection of O157 antigen in formalin-fixed tissues. The neonatal calf model provides information relevant for determining where and how STEC colonize the bovine gastrointestinal tract. This information will facilitate identification of methods for preventing infection of cattle with STEC and, thus, reduce the incidence of STEC infections in humans.

From: *Methods in Molecular Medicine, vol. 73: E. coli: Shiga Toxin Methods and Protocols*
Edited by: D. Philpott and F. Ebel © Humana Press Inc., Totowa, NJ

2. Materials

2.1. Neonatal Calves

1. Neonatal calves (<12 h old), removed from their dams before ingestion of colostrum and housed individually in pens on concrete floors with sanitized wood chips (*see* **Note 1**).
2. Milk replacer without antibiotics (Scooter Feeds from Arbie Mineral Feed Co., Inc., Marshalltown, IA or Premium Milk Replacer from Farm and Country, Des Moines, IA), prepared and administered according to the manufacturer's directions (*see* **Note 2**).
3. Oral electrolytes or lactated Ringer's solution with bicarbonate, as needed for treating dehydration (*see* **Note 3**).

2.2. Bacterial Inocula

1. *E. coli* inoculum strains are described in **Table 1** (*see* **Note 4**).
2. Broth media:
   a. Trypticase soy broth (TSB; BBL, Cockeysville, MD) prepared according to manufacturer's directions, autoclaved (121°C for 15 min), cooled, and stored at 4°C.
   b. Half-strength trypticase soy broth with 20% glycerol (TSB-glycerol). Mix 250 mL TSB + 100 mL glycerol + 150 mL distilled water (dH$_2$O), autoclave (121°C, 15 min) and store at room temperature.

2.3. Euthanasia, Necropsy, and Collection of Samples

1. Euthanasia: sodium pentobarbital (26% pentobarbital solution).
2. Necropsy supplies, including surgical drapes, sterile surgical instruments, disposable 150-mm Petri dishes, tared sterile 50-mL disposable tubes for tissues for bacteriological counts, and dry ice for freezing samples.
3. Histology samples: one jar containing 120 mL of 10% neutral buffered formalin for each tissue.
4. Electron microscopy (EM) samples: one vial containing 15–20 mL of 2.5% glutaraldehyde in 0.1 M sodium cacodylate buffer for each tissue, kept cold in bucket of wet ice.

2.4. Histologic Studies

1. 10% Neutral-buffered formalin.
2. Harris hematoxylin stain (Shandon, Inc., Pittsburgh, PA).
3. Alcoholic eosin stain (Shandon, Inc.).

2.5. Electron Microscopy Studies

1. Glutaraldehyde, 2.5% in 0.1 M sodium cacodylate buffer, pH 7.4.
2. Eponate 12 Resin (Ted Pella, Inc., Redding, CA).
3. Nickel grids.
4. Lead and uranyl acetate stain: 5% uranyl acetate in methanol and Reynolds lead citrate (*7*).

Table 1
E. coli Inoculum Strains

Strain	Serotype	Antibiotic resistance marker	*eae* gene	Stx type	Ref.
86-24	O157:H7	Streptomycin	+	Stx2	*3*
933	O157:H7	Streptomycin	+	Stx1, Stx2	*5*
3081	O157:H7	Kanamycin, ampicillin	+	Stx1, Stx2	*1,6*
123	O43:H28	Nalidixic acid	−	None	*4*

5. Desiccator for storing grids until ready to view.
6. Transmission electron microscope.

2.6. Bacteriological Counts

1. Media: MacConkey Agar (Becton Dickinson, Franklin Lakes, NJ) and Oxoid No. 3 Sorbitol–MacConkey agar (Oxoid, Inc., Ogdensburg, NY).
2. Antibiotic stock solutions: Stock solutions of antibiotics (Sigma, St. Louis, MO) are prepared as follows, sterilized by filtration (0.22-μm syringe filter), aliquoted in 500-μL vol and stored at −20°C.
   a. Ampicillin: 100 mg/mL; 1 g in final volume of 10 mL dH_2O.
   b. Kanamycin monosulfate: 100 mg/mL; 1 g in final volume of 10 mL dH_2O.
   c. Nalidixic acid: 20 mg/mL; 0.200 g in final volume of 10 mL of 0.2 M NaOH.
   d. Streptomycin 100 mg/mL; 1 g in final volume of 10 mL H_2O.
3. Tissue homogenization, serial dilutions and bacteriologic counts:
   a. Tissue homogenizer: Cyclone/Tempest IQ^2 Mechanical Homogenizer, 100-mL stainless-steel flasks, and macro shaft/open blade assemblies with aerosol-free caps (VirTis, Gardiner, NY).
   b. Diluent: sterile 0.3% peptone in water: 3 g of peptone (Difco, Detroit, MI) dissolved in 1 L dH_2O, aliquoted in 100-mL vol, autoclaved (121°C, 15 min), cooled to room temperature, and then stored at 4°C.
   c. Sterile disposable tubes for dilutions and disposable spreaders (Marsh BioProducts, Rochester, NY).
   d. Illuminated colony counter with magnifying lens.
   e. O157 Latex Agglutination Kit (Oxoid, Inc.).

2.7. Immunoperoxidase Staining with Anti-O157:H7

1. Paraffin-embedded, formalin-fixed tissue sections (4 μm thick) on Probe-On I ™ slides (Fisher Scientific, Pittsburgh, PA).
2. MicroProbe Manual Staining System (Fisher Scientific) or Techmate 500 Automated Pathology Slide Staining Workstation (BioTek Solutions, Inc., Santa Barbara, CA). *See* manufacturers' manuals to determine the appropriate slide holders, staining buckets, and reagent wells for each system.
3. Removal of paraffin:
   a. 85°C oven or incubator.

    b. Deparaffinization solutions. Use fresh aliquots of each stock solution for each assay. Prepare solutions in chemical fume hood and store in tightly capped bottles at room temperature. Follow safety regulations for disposal.

       (1) 3 : 1 Clearant: 375 mL HemoDe (Fisher Scientific) plus 125-mL xylene.

       (2) 1 : 1 Clearant: 150 mL HemoDe plus 150-mL xylene.

       (3) 100% and 95% ethanol solutions.

  4. Immunoperoxidase assay for O157:H7:

    a. Tris-buffered saline (TBS): 0.01 M Tris-HCl, 154 mM NaCl, pH 7.5, autoclaved, cooled to room temperature, and then stored at 4°C.

    b. Tris-buffered saline with 0.25% Tween-20 (TBS-T$_{20}$): Add 1.25 mL of Tween-20 (polyoxyethylenesorbitan monolaurate, Sigma) to 500 mL of sterile TBS after it has cooled to room temperature. Store at 4°C.

    c. 0.3% Hydrogen peroxide in methanol: Prepare fresh by adding 250 µL of 30% hydrogen peroxide (Sigma) to 25 mL dH$_2$O.

    d. 5% NRS (normal rabbit serum). Add 2 mL of normal rabbit serum (Vector Laboratories, Burlingame, CA) to 38 mL TBS-T$_{20}$ and sterilize by filtration (0.45-µm syringe filter [Millipore]).

    e. Primary antibody: Goat anti-O157:H7 and goat anti-*Salmonella* (Kirkegaard & Perry Laboratories, Inc., Gaithersburg, MD) diluted 1:20,000. Prepare a 1:50 stock solution (20 µL of antibody plus 980 µL of TBS) and store at 4°C. Prepare fresh 1:20,000 dilution on day of assay by adding 25 µL of the 1:50 stock to 10 mL of 5% NRS.

    f. Biotinylated antibody: Biotinylated rabbit anti-goat IgG (H&L) (Vector Laboratories) diluted 1:800 by adding 12.5 µL to 10 mL 5% NRS.

    g. Avidin–biotin complex (ABC). Prepared from BioMeda ABC kit (BioMeda Corp., Foster City, CA), according to manufacturer's directions.

    h. Enhancer (0.075% Brij in dH$_2$O). This is a 1:400 dilution of 30% Brij (BioMeda Corp.). A stock solution containing 0.75% Brij solution is prepared by adding 250 µL to 9.75 mL dH$_2$O and storing at room temperature. Enhancer is freshly prepared for each assay by adding 2 mL of the stock solution to 18 mL dH$_2$O.

    i. DAB (diaminobenzidine HCl) prepared from DAB kit (Kirkegaard & Perry) as directed by manufacturer. Note that this is a carcinogen!

    j. Tris-HCl (0.1 M Tris-HCl, pH 7.6).

    k. Gill's No. 2 hematoxylin solution (Sigma). Note manufacturer's safety warnings! This solution can be reused until it develops a metallic sheen.

    l. Permount/xylene.

3. Methods

3.1. Preparation of Stock Inocula

1. Subculture *E. coli* into TSB and incubate at 37°C (without agitation) for 16–18 h to obtain approx 10^9 CFUs of inoculum bacteria/mL of culture.

2. Centrifuge this culture at 10,000g for 20 min. Suspend the bacterial pellet in TSB–glycerol (to 1/10 of the original culture volume) to obtain an approximate concentration of 10^{10} CFUs/mL.

3. Store in 1-mL aliquots at –80°C until used.
4. Determine exact bacterial concentrations by doing bacterial colony counts on thawed aliquots of each batch of stock inocula.

3.2. Inoculation of Calves

1. Before inoculating calf, measure and record temperature (rectal). Collect a preinoculation fecal sample, place it in a sterile tared container, and store at –80°C until bacteriological counts are performed.
2. Prepare milk replacer according to manufacturer's instructions.
3. Add 1 mL of stock inoculum (~ 10^{10} CFUs of inoculum bacteria) to approx 200 mL of milk replacer in a milk bottle and feed this to a <12-h-old colostrum-deprived calf. Add approx 200 mL additional milk replacer to rinse bottle and feed to calf. This ensures that a calf receives the total inoculum dose, even if it does not ingest all of the remaining milk replacer. Finally, feed the calf as much of the remaining milk replacer as it will take. Note the total volume of milk replacer ingested.

3.3. Feeding calves

Feed calf with milk replacer at 8 h intervals after inoculation and note volume of milk replacer ingested at each feeding.

3.4. Observing Calves

1. Observe calf every 4–8 h after inoculation for signs of clinical disease.
2. Record specific observations of appetite, responsiveness, temperature (rectal), fecal appearance and consistency, and hydration status (checked by skin turgor) (*see* **Note 3**).
3. Collect sample of feces in sterile tared containers every 24 h postinoculation and store at –80°C until processed.
4. Euthanize (*see* **Subheading 3.5.**) any calf that is moribund and unable to rise or is severely dehydrated (checked by skin turgor) and does not respond to fluid therapy within 1 d. Perform necropsy immediately after euthanasia of calf (*see* **Subheading 3.6.** and **Note 5**).

3.5. Euthanasia

Euthanize calves at 1–3 d postinoculation or when down (*see* **Subheading 3.4.**) by intravenous administration of an overdose of sodium pentobarbital (10 mL of 26% solution/100 lb body weight).

3.6. Necropsy

1. Aseptically collect tissue samples from mid-cecum, ileum, spiral colon, and distal colon (at pelvic inlet) for bacteriological counts. Collect tissues for histology and electron microscopy from same sections after samples for bacteriology have been collected. To prevent leakage and cross-contamination when intestinal segments are cut, locate and tie off approx 7- to10-cm-long sections of mid-cecum,

ileum (approx 30 cm from ileocecal valve), and spiral colon. Cotton cord (16-ply) is suitable for ligation of intestinal sections. The distal colon does not have to be tied off if it is the last tissue collected for culture. Position two ties close together in the mid-cecum and at each end of the sections of ileum and spiral colon to be collected. Cutting between these two ties reduces contamination of other tissues with intestinal contents that leak from the cut sections. Place each tissue section in a sterile Petri dish for the collection of the following samples:

   a. For bacteriological counts: aseptically excise a section of each tissue (approx 1–2 g) and place in a tared sterile 50-mL conical disposable tube, place on dry ice for transport to the laboratory, and store at –80°C until processed. Additional intestinal segments and tissues from internal organs are not routinely collected for bacteriological counts, but can be collected aseptically if bacteriological cultures are to be performed. The next two steps do not need to be done aseptically.

   b. For histology and immunohistochemistry: remove an adjacent section (approx 2.5 × 3 cm) of each intestinal tissue, staple it flat onto a labeled index card (lumen side up), and submerge in formalin jar (*see* **Note 6**). Keep formalin-fixed tissues at room temperature.

   c. For electron microscopy: cut an adjacent section of each tissue into small (approx 1 mm) pieces (*see* **Note 7**) and immediately place in ice-cold 2.5% glutaraldehyde in glass vials. Immediately place the glass vials containing tissues on ice for transport to the laboratory and keep at 4°C until processed.

2. Additional intestinal segments (e.g., ascending and descending colon, ileocecal valve) and tissues from internal organs are not routinely collected for bacteriological culture, but should also be collected aseptically if such cultures are to be performed. Collect any tissues containing macroscopic lesions or tissues needed to identify confounding infections in formalin for histopathological examination.

3.7. Histological Studies

1. Within 48 h of necropsy, trim selected portions of the formalin-fixed tissues and place in cassettes for routine processing for histopathological examination. Routine processing includes alcohol degradation, paraffin embedding, preparation of 4-μm sections, and staining with hematoxylin and eosin (H&E) using Harris hematoxylin and alcoholic eosin stains. Additional sections are prepared on ProbeOn+(slides for immunoperoxidase staining (*see* **Subheading 3.9.**) at the same time as the ones for H&E staining.

2. Code stained tissue slides (to prevent subjective bias) and examine by the use of light microscopy. Score attaching and effacing lesions: 0, negative; 1+, rare (negative by H&E, but detected by HRPO (*see* **Subheading 3.9.**, **step 3**); 2+, < 10% of the villi or surface epithelium affected; 3+, 10–50% of the villi or surface epithelium affected; 4+, > 50% of the villi or surface epithelium affected.

3.8. Bacteriological Counts

1. Preparation of media: Prepare media according to manufacturer's directions, autoclave (121°C, 15 min), and cool to 60°C in a water bath before adding antibiotics and pouring plates. Add antibiotics (500 μL of stock antibiotics/500 mL of

media; *see* **Table 1** for selecting antibiotics) to media immediately before pouring plates. Store all plates at room temperature in the dark and use within 1 wk of preparation.

2. Homogenization of samples:
   a. Record tissue weight (weight of tared tube containing tissue minus the tube weight).
   b. Add peptone buffer to tissue or feces to a final volume of 30 mL, vortex, and transfer tissue and buffer into a steel flask. Follow manufacturer's directions for assembling the blade and cap on the flask and for operating the homogenizer.
   c. Homogenize sample for 20 s at a setting of 30,000 rpm.

3. Serial dilutions of samples:
   a. Make 10-fold serial dilutions of the homogenate (100 µL homogenate + 900 µL of peptone buffer) from 10^{-1} to 10^{-4}. If homogenate is viscous, use 1 mL homogenate + 9 mL peptone buffer for first dilution.
   b. Transfer 100 µL of each dilution to two duplicate plates and spread evenly with disposable spreader.
   c. Incubate at 37°C overnight.

4. Bacteriological counts (*see* **Note 8**):
   a. Count appropriate colonies (sorbitol-negative O157:H7 forms colorless colonies on SMAC; lactose-positive coliforms form pink colonies on MacConkey), based on characteristics of the inoculum strain. Confirm sorbitol-negative colonies as O157:H7 by using the Oxoid O157:H7 latex agglutination kit (according to manufacturer's directions).
   b. Calculate the number of colony-forming units (CFUs) of inoculum bacteria per gram of tissue or feces using the following formula. The number of colonies/0.2 mL is the total number of colonies on both plates of the lowest dilution with 20–200 colonies of interest per plate.

$$\text{CFUs/g of tissue (or feces)} = (\text{no. of colonies})/0.2 \text{ mL} \times \text{Dilution factor} \times (30 \text{ mL final volume})/\text{Wt. of tissue (or feces) (in g)}$$

3.9. Immunoperoxidase Staining with Anti-O157:H7

1. Principles of microprobe staining technique: Probe-On+ slides are sandwiched together so that tissue sections face inside and reagents are drawn into the sandwich via capillary action. Addition of detergents (0.25% Tween-20 or 0.075% Brij) to reagents facilitates filling and emptying reagents via capillary action.

2. Protocol for removal of paraffin, tissue rehydration, and blocking endogenous peroxidase (manual slide holder works best for this):
   a. Heat slides for 15 min at 85°C. Remove from incubator.
   b. Dip in 3 : 1 clearant for 1 min, incubate at 37°C for 5 min, blot (use organic blotting pad until introduction of TBS-T$_{20}$, **step 7**) and repeat once.
   c. Dip in 1 : 1 clearant for 30 s, blot, and repeat once.
   d. Dip in 100% ethanol for 10 s, blot, and repeat once.
   e. Dip in 95% alcohol for 2 min, blot, and repeat once.
   f. Dip in 0.3% hydrogen peroxide in methanol, incubate for 30 min, and blot.

g. Dip in TBS-T_{20} for 30 s, blot on aqueous blotting pad, and repeat once. Dip again in TBS-T_{20}, and leave in TBS-T_{20} for next step.

h. Use same slide holder if doing manual procedure. If using automated staining system, transfer the slides into an appropriate slide holder at this point and blot.

3. Horseradish peroxidase (HRPO) immunoassay protocol (perform all incubations at room temperature):

   a. Dip in 5% NRS and incubate for 20 min; blot (use aqueous blotting pad for all blots in HRPO assay).

   b. Dip in primary antibody (anti-O157:H7 or anti-*Salmonella*) and incubate for 2 h. (If using the manual procedure, this incubation can be done overnight at 4°C).

   c. Blot and rinse in TBS-T_{20} for 30 s; blot and place in TBS-T for 30 min; blot.

   d. Dip in biotinylated antibody for 30 s, blot, dip in biotinylated antibody again, and incubate for 30 min; blot. Rinse in TBS-T_{20} for 30 s; blot and repeat twice.

   e. Dip in TBS for 30 s and blot.

   f. Dip in ABC and incubate for 30 min; blot.

   g. Dip in enhancer for 30 s; blot and repeat twice.

   h. Dip in DAB for 2.5 min; blot and repeat once.

   i. Dip in dH_2O for 30 s; blot and repeat once.

   j. Dip in Gill's hematoxylin for 1 min; blot.

   k. Dip in dH_2O for 30 s; blot and repeat once.

   l. Dip in TBS-T_{20} for 30 s; blot and repeat once but do not blot until ready for dehydration.

4. Dehydration:

   a. If necessary, transfer slides into manual slide holder; blot.

   b. Dip in 95% ethanol for 10 s; blot and repeat twice.

   c. Dip in 100% ethanol for 10 s; blot and repeat once.

   d. Dip in 1:1 clearant for 10 s; blot and repeat once.

   e. Dip in 3:1 clearant for 10 s; blot and repeat twice.

   f. Mount coverslip slides using 1:1 Permount/xylene.

5. Code HRPO-stained tissue slides and use light microscopy to examine for the presence of attaching and effacing bacteria that are stained brown in slides incubated with anti-O157, but are not stained with anti-*Salmonella* (*see* **Fig. 1** and **Note 9**).

4. Notes

1. Because colostrum-deprived calves are highly susceptible to microbial infections, precautions must be taken to prevent the introduction of other infectious agents. These precautions include decontamination of pens before an experiment begins and changing clothes, boots, and gloves when entering animal biocontainment areas.

2. Milk replacer must not contain antibiotics that interfere with recovery of *E. coli* inoculum. It is a good idea to test whether milk replacer inhibits the growth of the inoculum bacteria in vitro before beginning calf experiments.

3. STEC O157:H7 can cause severe, fatal disease in some calves. Rehydration therapy may be required for some calves, especially if duration of the experiment is longer than 2 d.

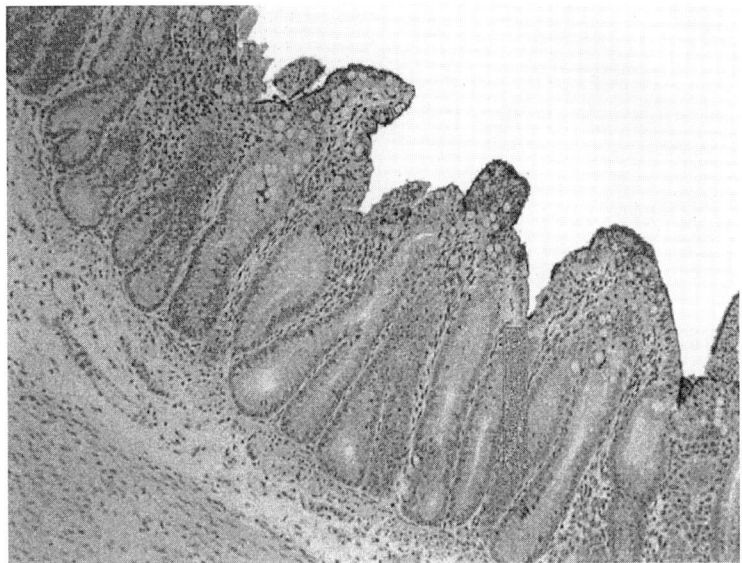

Fig. 1. Horseradish-peroxidase-stained section of spiral colon from a neonatal calf 18 h after inoculation with STEC O157:H7 strain 3081. Immunostained areas (dark patches) on the epithelium illustrate colonies of attaching and effacing STEC O157:H7.

4. *Escherichia coli* O157:H7 and other STEC are human pathogens and must be handled with caution. Perform all experiments under strict BL-2 containment using special precautions. Use biological safety cabinets and biocontainment centrifuges and rotors for all laboratory procedures. Always wear protective clothing (boots, coveralls, gloves, hairnet, and face shields [8710 respiration masks]) when inoculating and working with infected animals. Always disinfect boots and gloves before entering and leaving animal rooms. Remove all contaminated clothing and shower before leaving barns. Sterilize or decontaminate all contaminated clothing, equipment, and supplies at the end of the experiment. Surface decontaminate and double-bag all samples for transfer from barns to laboratory.

5. Perform complete necropsies to identify any confounding infections or diseases that may be responsible for clinical signs and pathological changes that are not associated with the experimental infection.

6. Use staples to attach opened intestinal tissues, lumen side up, onto labeled index cards cut to fit in jars of formalin. This improves tissue fixation and morphology.

7. Use a single-edged razor blade to cut tissues into small (approx 1 mm) pieces before placing them into cold glutaraldehyde. A sheet of dental wax is a useful cutting surface for this. To reduce the time and expense of collecting glutaraldehyde-fixed tissues for all samples, selected formalin-fixed tissues can be refixed in glutaraldehyde for EM.

8. High numbers of other bacteria can affect the sensitivity of detection and the quantitation of the inoculum bacteria. Using inoculum strains with antibiotic markers is helpful, but bacterial recovery on selective media containing antibiotics may be less efficient than on nonselective media and result in underestimating the number of bacteria.

9. Some calves may be infected with attaching and effacing bacteria that are not O157:H7. These can be differentiated from the STEC O157-positive inoculum strain by the immunoperoxidase assay for O157. Non-O157 attaching and effacing bacteria will not stain with anti-O157.

References

1. Dean-Nystrom, E. A., Bosworth, B. T., Cray, W. C., Jr., and Moon, H. W. (1997) Pathogenicity of *Escherichia coli* O157:H7 in the intestines of neonatal calves. *Infect. Immun.* **65,** 1842–1848.

2. Dean-Nystrom, E. A., Bosworth, B. T., Moon, H. W., and O'Brien, A. D. (1998) *Escherichia coli* O157:H7 requires intimin for enteropathogenicity in calves. *Infect. Immun.* **66,** 4560–4563.

3. McKee, M. L., Melton-Celsa, A. R., Moxley, R. A., Francis, D. H., and O'Brien, A. D. (1995) Enterohemorrhagic *Escherichia coli* O157:H7 requires intimin to colonize the gnotobiotic pig intestine and to adhere to HEp-2 cells. *Infect. Immun.* **63,** 3739–3744.

4. Moon, H. W., Sorensen, D. K., and Sautter, J. H. (1968) Experimental enteric colibacillosis in piglets. *Can. J. Comp. Med.* **32,** 493–497.

5. O'Brien, A. D., Melton, A. R., Schmitt, C. K., McKee, M. L., Batts, M. L, and Griffin, D. E. (1993) Profile of *Escherichia coli* O157:H7 pathogen responsible for hamburger-borne outbreak of hemorrhagic colitis and hemolytic uremic syndrome in Washington. *J. Clin. Microbiol.* **31,** 2799–2801.

6. Thomas, L. A., Reymann, R. A., Moon, H. W., Schneider, R. A., Cummins, D. R., Beckman, M. G., et al. (1992) Characterization of serotypes O157:H7 and O157:NM *Escherichia coli* isolated from dairy heifer feces, in Proceedings of the Annual Meeting of the American Association of Veterinary Laboratory Diagnosticians, p. 83.

7. Reynolds, E. S. (1963) The use of lead citrate at high pH as an electron-opaque stain in electron-opaque stain in electron microscopy. *J. Cell. Biol.* **17,** 208–212.

Index

A

A/E lesion, *see* Attaching and effacing
　lesion
AFLP, *see* Amplified fragment length
　polymorphism
Alabama rot, *see* Greyhound, Shiga
　toxigenic *Escherichia coli* infection
Amplified fragment length
　polymorphism (AFLP), Shiga
　toxigenic *Escherichia coli* typing,
　advantages and limitations, 61
　interpretation, 61
　materials, 57, 58
　primer set selection, 61
　principles, 56
　steps, 60
Apoptosis,
　assay overview, 231, 232
　Bcl-2 Western blot analysis,
　　235–239
　biochemical features, 231
　caspase inhibitor studies, 236, 239
　dysregulation in disease, 234
　enzyme-linked immunosorbent
　　assay, 235, 237, 240
　morphologic assessment,
　　fluorescent dye staining, 235, 237,
　　　239, 240
　　materials, 235
　　transmission electron microscopy,
　　　235–237, 239
　morphologic features, 229, 231
　Shiga toxin induction,
　　endothelial cells, 245, 246
　　overview, 172, 173

　peripheral blood mononuclear cell
　　TUNEL assay, 281, 286, 287
　signaling,
　　Bcl-2, 232, 233
　　caspases, 231, 232
　　gene inactivation studies, 234
　　inhibitors, 233, 234
　　transient transfection of death
　　　genes, 236, 239
Attaching and effacing (A/E) lesion,
　bacteria type distribution, 137
　characteristics, 93
　formation, 137
　locus of enterocyte effacement,
　　92–94, 137
　microscopy, *see* specific methods

B

Baboon models,
　advantages, 302
　inoculation route, 292, 302
　materials, 298
　outcomes, 296
BAEC, *see* Bovine aortic endothelial
　cell
B-cell, *see* Lymphocyte
Bcl-2,
　apoptosis signaling, 232, 233
　Western blot analysis of Shiga toxin-
　　induced apoptosis, 235–239
Bovine aortic endothelial cell (BAEC),
　isolation and culture, 251, 252
Bovine model, *see* Calf, Shiga
　toxigenic *Escherichia coli*
　infection

C

Calf, Shiga toxigenic *Escherichia coli* infection,
  clinical significance, 75
  diagnosis,
    culture, 80, 84, 85
    eae polymerase chain reaction, 81, 85, 86
    gel electrophoresis of amplification products, 82, 86
    materials, 78–80, 82, 84
    STEC multiplex polymerase chain reaction, 80, 81, 85, 86
  model,
    advantages, 329
    bacteriological counts, 331, 334, 335, 338
    carriers, 329
    clinical features, 329
    euthanasia, 330, 333
    feeding of calves, 333
    histologic studies, 330, 334
    immunoperoxidase staining with anti-O157:H7, 331, 332, 335, 336, 338
    inoculation route, 292
    inoculum,
      inoculation, 333
      preparation, 332, 333
      strains, 330, 331
    materials, 330–332, 336, 337
    necropsy, 330, 333, 334, 337
    observation of calves, 333, 336, 337
  pathogenesis, 77, 78
Caspase,
  apoptosis signaling, 231, 232
  inhibitor studies of Shiga toxin-induced apoptosis, 236, 239
CD19, Gb3 signal transduction, 170–172
CD77, *see* Gb3

Cellular microbiology,
  scope of field, 91, 125
  Shiga toxigenic *Escherichia coli* infection, 91–95
Chicken models,
  inoculation route, 292, 296
  outcomes, 296
Colony hybridization,
  food Shiga toxigenic *Escherichia coli* assay,
    digoxigenin labeling of probes, 71, 73
    hybridization, 72–74
    materials, 68, 70
    Shiga toxin gene detection, 12

D

Dendritic cell, Shiga toxin antigen presentation, 177

E

E-Hly, *see* Hemolysin
eae, polymerase chain reaction detection, 14, 15, 71, 73
Efa-pathogenicity island, features, 100, 101
ELISA, *see* Enzyme-linked immunosorbent assay
Endoplasmic reticulum (ER),
  KDEL tagging for targeting, 212, 213, 222
  pH measurement using FITC-labeled Shiga toxin B-subunit,
    digital fluorescence microscope set-up, 222–224, 226, 227
    materials, 224, 227
    overview, 221, 222
    pH determination, 225-227
    StxB labeling, 224, 225
Endothelial cell,
  commercial sources, 252, 258

culture study advantages and
limitations, 243, 244
freezing and thawing, 252, 253
Gb3 expression, 244
glycolipid receptor extraction and
separation,
extraction, 255, 256, 258
materials, 248
thin-layer chromatography, 256,
257
injury in hemolytic uremic
syndrome, 4
isolation and culture,
bovine aortic endothelial cells,
251, 252
glomerular microvascular
endothelial cells, 249–251
human umbilical vein endothelial
cells, 248, 249, 257, 258
materials, 246–248, 257
media, 247, 257
Shiga toxin binding,
apoptosis induction, 245, 246
binding conditions and Scatchard
analysis, 254, 255, 258
cytotoxicity assays, 253, 254, 258
effects on cells, 244, 245
molecular studies, 257, 259
overview, 244
Enzyme-linked immunosorbent assay
(ELISA),
lipopolysaccharide, 33-35, 38
Shiga toxin detection,
kits, 11
principles, 11
sensitivity, 11
Shiga toxin-induced apoptosis, 235,
237, 240
ER, *see* Endoplasmic reticulum
Escherichia coli O157, *see also* O157
antigen,
agar for isolation, 16
culture on SMAC, 15, 16

diagnosis, *see* specific assays
glucuronidase deficiency in strain
H7, 16
immunomagnetic separation, 18
non-O157 strains,
clinical features of disease, 27
culture, 17, 18
Esp proteins, functions, 94, 95

F

fliCh7, polymerase chain reaction
detection, 14
Flow cytometry, Gb3 detection on
peripheral blood mononuclear
cells, 284, 287, 288
Food,
sources of infection, 1, 2, 67, 263
Shiga toxigenic *Escherichia coli*
assays,
colony hybridization assay,
digoxigenin labeling of
probes, 71, 73
hybridization, 72–74
materials, 68, 70
immunomagnetic separation, 68,
70, 72, 73
materials, 68–70
media, 68
polymerase chain reaction,
eae, 71, 73
materials, 68
primers, 69
Shiga toxin genes, 71, 73

G

Gb3,
affinity chromatography for Stx1
purification, 187, 190–193
cell-type distribution, 169, 170, 175
endothelial cells,
expression, 244

extraction and separation,
  extraction, 255, 256, 258
  materials, 248
  thin-layer chromatography,
    256, 257
functions, 197
hemolytic uremic syndrome risk
  factor, 174–176
immunohistochemical staining using
  Stx1, 198–200, 203
isoforms, 277
isolation,
  cell extraction, 200, 201, 204, 205
  materials, 199
  saponification, 201, 202, 205, 206
  silica gel chromatography, 202,
    206
  tissue extraction, 201, 205
MDR regulation of synthesis, 173–175
peripheral blood mononuclear cell,
  binding assay, 285, 288
  detection, 284, 287, 288
Shiga toxin binding,
  affinity, 165, 166
  analog binding, 168
  antineoplastic activity, 173, 174
  binding sites, 166, 167
  crystal structure, 167
  fluorescence resonance energy
    transfer studies, 167
  retrograde transport, 168–170
  site-directed mutagenesis studies,
    168
signal transduction, 170–173
thin-layer chromatography overlay
  assay, 199, 200, 202, 203, 206
tumorigenesis role, 173, 174, 197
Globotriaosylceramide, *see* Gb3
Glomerular microvascular endothelial
  cell (GMVEC), isolation and
  culture, 249–251
GMVEC, *see* Glomerular microvascular
  endothelial cell

Golgi complex, pH measurement
  using FITC-labeled Shiga toxin
  B-subunit,
  digital fluorescence microscope set-
    up, 222–224, 226, 227
  materials, 224, 227
  overview, 176, 177, 221, 222
  pH determination, 225–227
  StxB labeling, 224, 225
Granulocyte, Shiga toxin binding, 244
Greyhound, Shiga toxigenic
    Escherichia coli infection,
  clinical significance, 75
  diagnosis,
    culture, 80, 84, 85
    materials, 78–80, 82, 84
  model,
    advantages, 302
    inoculation route, 292
    outcomes, 297
  pathogenesis, 78

H

Hemagglutination assay,
  lipopolysaccharide, 35, 39
Hemolysin,
  E-Hly,
    detection on blood agar plates,
      culture, 155
      erythrocyte preparation, 154,
        155, 161
      materials, 153, 160
    gene, 152
    hemolytic activity assay,
      extracellular protein
        precipitation, 155
      incubation conditions, 156
      materials, 153
      overview, 152
    overproduction, 152
    pore-forming activity assay,
      data analysis, 157, 159

lipid bilayer technique, 156, 157, 160
materials, 153, 154
overview, 153
purification of E-Hly, 156
recording, 157, 161, 162
α-hemolysin export, 152
hemolytic *Escherichia coli* strains, 151, 152
history of study, 151, 152
mechanism of action, 92
structure, 92
Hemolytic enterocyte effacement, epidemiology, 2
Hemolytic uremic syndrome (HUS),
clinical features, 2, 3, 291
diagnosis, 27
epidemiology, 1, 2
Escherichia coli,
infectious dose, 3
serotypes, 2
Gb3 risk factor, 174–176
monocyte role, 170
morbidity and mortality, 1
onset, 9
pathogenesis, 3, 4
sources of infection, 1, 2
treatment, 1
High pathogenicity island (HPI), features, 100
HPI, *see* High pathogenicity island
Human umbilical vein endothelial cell (HUVEC), isolation and culture, 248, 249, 257, 258
HUS, *see* Hemolytic uremic syndrome
HUVEC, *see* Human umbilical vein endothelial cell

I

IECs, *see* Intestinal epithelial cells
IFNAR1, Gb3 signal transduction, 170, 172

Immunoelectron microscopy, *see* Scanning electron microscopy; Transmission electron microscopy
Immunofluorescence microscopy,
attaching and effacing lesions, 138
infection of cell monolayers, 140
materials, 138, 139, 146–148
Immunomagnetic separation (IMS),
food Shiga toxigenic *Escherichia coli* assay, 68, 70, 72, 73
Shiga toxigenic *Escherichia coli*, 18
IMS, *see* Immunomagnetic separation
Intestinal epithelial cells (IECs), Shiga toxin translocation,
assays,
cell lines, 265, 271
collagen-coated transwell filers, 265, 267, 271
insulin control, 268, 272
materials, 265, 271
migration conditions, 268, 272
tissue culture, 267, 272
mechanisms, 264
polymorphonuclear leukocyte transmigration effect assays,
inverted monolayer growth, 266, 268
materials, 266
migration conditions, 270
myeloperoxidase assay, 266, 270
polymorphonuclear leukocyte isolation, 266, 269, 270
Intimin,
antibody detection in serum, 20
function, 67
gene, *see eae*
immune response, 276
Island probing, *see* Pathogencity island
Isogenic deletion mutant, Shiga toxigenic *Escherichia coli* generation,
allelic exchange by homologous recombination,

integrated plasmid excision,
        122, 123
integration of knockout vector
        into target gene, 122, 123
conjugation of knockout vector,
        120, 121
electroporation of knockout vector,
        121–123
knockout vector construction,
        chromosomal DNA isolation, 118,
        119, 123
        deletion allele cloning, 119
materials, 117, 118, 123
principles, 113, 116, 117
suicide vector systems, 117
transformation of knockout vector
        into K-12 strains,
        competent cell preparation, 119,
        120, 123
        transformation, 120

L

LEE, *see* Locus of enterocyte
        effacement
Lipopolysaccharide (LPS),
        antibody detection in serum, 19, 21, 29
        enzyme-linked immunosorbent
                assay, 33-35, 38
        extraction, 32, 38
        gel electrophoresis, 32, 33, 38
        hemagglutination assay, 35, 39
        non-O157:H7 strains, 29
        Western blotting with serum
                antibodies, 33, 38
Locus of enterocyte effacement (LEE),
        see also Pathogencity island,
        attaching and effacing lesion
                pathology, 92–94
        effector proteins, 94, 95
        expression control, 94
        operons, 94, 95
        pathogenicity islands, 93

structure, 100
translocation protein genes, 94
LPS, *see* Lipopolysaccharide
Lymphocyte,
        lymphostatin effects, 275, 276
        Shiga toxin effects,
                blast cell composition analysis,
                        283, 284, 287
                immunosuppression, 275
                materials for assays, 277, 278, 285
                morphology analysis, 281, 282, 287
                overview of assays, 276, 277

M

MDR, regulation of Gb3 synthesis,
        173–175
Microscopy, *see* specific methods
Monoclonal antibody, Shiga toxigenic
        Escherichia coli antigens,
        advantages, 126
        generation,
                antigen preparation, 126, 128,
                        129, 133
                feeder cell isolation, 126, 127,
                        129, 133
                fusion, 127, 128, 130–132, 134
                hybridoma storage, 128, 132, 134
                immunization, 127, 129, 130,
                        133, 134
                materials, 126–128
                myeloma cell culture, 127, 130, 134
                screening, 132, 134
                subcloning, 128, 132, 133
        polyclonal antibody advantages and
                limitations, 125
Mononuclear cell, *see* Lymphocyte;
        Peripheral blood mononuclear cell
Mouse models,
        advantages, 302
        inoculation route, 292-294
        intragastric inoculation,
                C3H mice, 300

CD-1 mice treated with
ciprofloxacin, 299, 300, 302
str- and mitomycin C-treated
mice, 299, 302
malnourished mouse model, 300
materials, 297, 298
outcomes, 293, 294
str-treated, orally infected CD-1
mice, 299
types, 293, 294

N

Neutralization assay, Shiga toxins, 28,
35, 36, 39, 40

O

O157 antigen,
antibody detection in serum, 30
direct detection in fecal
samples, 17
immunoperoxidase staining
with anti-O157:H7 in calf model,
331, 332, 335, 336, 338

P

PAI, *see* Pathogencity island
Pathogencity island (PAI), *see also*
specific islands,
definitive characteristics, 99
detection,
island probing,
deletion frequency
determination, 109–111
DNA manipulation, 108
mutagenesis plasmid
construction for allelic
exchange, 108–111
overview, 103–105
sacB-sacR insertion, 109, 111
materials, 105, 106

principles, 101–103, 105
subtractive hybridization,
adaptor ligation, 107, 110
hybridization, 107, 110
polymerase chain reaction,
107, 108, 110
restriction of driver and
tester DNA, 107
transfer RNA screening, 105–107,
110
loss, 103, 105
transfer RNA genes as integration
sites, 101
types, 99–101
PBMC, *see* Peripheral blood
mononuclear cell
PCR, *see* Polymerase chain reaction
Peripheral blood mononuclear cell
(PBMC), Shiga toxin effects,
Gb3,
binding assay, 285, 288
detection, 284, 287, 288
overview of assays, 276, 277
viability assays,
materials, 277, 278, 285
MTT assay, 279, 286
peripheral blood mononuclear cell
preparation, 278, 285, 286
propidium iodide uptake, 280,
286
toxin dilution on microtiter plates,
278, 279, 286
TUNEL assay, 281, 286, 287
PFGE, *see* Pulsed-field gel
electrophoresis
Pig, Shiga toxigenic *Escherichia coli*
infection,
clinical significance, 75
diagnosis,
culture, 80, 84, 85
gel electrophoresis of amplification
products, 82, 86
materials, 78–80, 82, 84

STEC multiplex polymerase chain
    reaction, 81, 82, 85, 86
model using gnotobiotic piglets,
    breeding, 318, 324
    clinical features, 308
    clinical monitoring and
        euthanasia, 321, 325
    data collection, 323-326
    facilities, 308–310
    feeding and distribution system,
        313–315, 318, 319, 324
    inoculation route, 292, 308
    isolators,
        assembly, 318, 319
        materials for preparation,
            310, 311, 313
        rearing isolator, 311–313
        surgical isolator, 311
        transport isolator, 311
    neonate handling, 320, 324
    oral infection, 321, 325
    postmortem examination, 322
    sample collection, 320, 321–325
    sow preparation and cesarean
        section, 319, 320, 324
    sterilization and waste disposal,
        315, 318, 323
    pathogenesis, 76, 77
    risk factors, 307, 308
PML, *see* Polymorphonuclear
    leukocyte
Polymerase chain reaction (PCR),
    eae detection, 14, 15
    fliCh7 detection, 14
    food assays,
        eae, 71, 73
        materials, 68
        primers, 69
        Shiga toxin genes, 71, 73
    multiplex assays, 15
    pathogencity island detection with
        subtractive hybridization, 107,
        108, 110

Shiga toxigenic *Escherichia coli*
    typing, *see* Amplified fragment
        length polymorphism; Random
        amplified polymorphic
        DNA-PCR
Shiga toxin gene detection,
    advantages, 14
    primers, 12, 13
    samples, 12, 13
    sensitivity, 13, 20, 21, 45
    time requirements, 13, 14
stool sample detection of Shiga
        toxigenic Escherichia coli,
    amplification reaction, 48, 52
    contamination prevention, 51, 52
    materials, 47
    multiplex assay, 47, 51–53
    primers, 49
    product detection, 48
    screening assay, 47
    serotype targets, 47
    Shiga toxin gene screening, 50–52
    specimen preparation, 47, 50, 52
    target sequences, 45–47
uidA mutation detection, 14
veterinary diagnostics, *see* Calf;
        Greyhound; Pig
Polymorphonuclear leukocyte (PML),
    transmigration effect on Shiga
        toxin translocation across
        intestinal epithelium,
    inverted monolayer growth, 266, 268
    materials, 266
    migration conditions, 270
    myeloperoxidase assay, 266, 270
    polymorphonuclear leukocyte
        isolation, 266, 269, 270
Pulsed-field gel electrophoresis
        (PFGE), Shiga toxigenic
        Escherichia coli typing,
    advantages and limitations, 61
    bacteria embedding in agarose, 58,
        61, 62

gel electrophoresis, 59, 62, 63
gel loading, 59, 62
lysis of samples, 58
materials, 56
principles, 55, 56
restriction digestion, 58, 62

R

Rabbit models,
  infection of infant rabbits, 300, 302
  inoculation,
    diarrheal pathogen, 301
    route, 292
    Stx1, 301
    Stx2,
      bolus, 301
      continuous infusion, 301
  materials, 298
  oral inoculation models, 295
  toxin administration routes, 295, 296
Random amplified polymorphic
    DNA-PCR (RAPD-PCR), Shiga
    toxigenic *Escherichia coli*
    typing,
  advantages and limitations, 61
  DNA preparation, 59, 60
  gel electrophoresis, 60
  materials, 57
  polymerase chain reaction, 60, 63
  principles, 56
RAPD-PCR, *see* Random amplified
    polymorphic DNA-PCR
Retrograde transport, StxB assays,
  C-terminal polylysine mutant,
    biotinylation, 213, 215, 218, 219
    generation, 210
  early endosome-to-Golgi transport
    kinetics, 210–212
  glycosylation assay, 215, 218, 219
  internalization,
    assay, 213, 214–217, 219
    kinetics, 210

intracellular trafficking of Shiga
    toxin, 168–170, 209–213
  iodination of protein, 214, 215,
    218, 219
  KDEL tagging for endoplasmic
    reticulum targeting, 212, 213
  materials, 213-215
  pH determination in organelles, *see*
    Endoplasmic reticulum; Golgi
    complex
  purification of protein, 213, 215, 218
  sulfation of protein,
    assay, 214, 217, 219
    overview, 210–212
Reverse passive latex agglutination
    (RPLA), Shiga toxin detection,
    11–12
RPLA, *see* Reverse passive latex
    agglutination

S

Scanning electron microscopy (SEM),
  attaching and effacing lesions, 137,
    138
  immunoelectron microscopy, 145,
    146, 148
  infection of cell monolayers, 140
  materials, 138–140, 148
SEM, *see* Scanning electron
    microscopy
Shiga toxigenic *Escherichia coli*
    (STEC), *see also Escherichia coli*
    O157,
  animal models, *see* Baboon models;
    Calf, Shiga toxigenic
    Escherichia coli infection;
    Chicken models; Greyhound,
    Shiga toxigenic *Escherichia
    coli* infection; Mouse models;
    Pig, Shiga toxigenic
    Escherichia coli infection;
    Rabbit models

assays, *see* specific assays
isogenic deletion mutants, *see* Isogenic
    deletion mutant, Shiga toxigenic
    Escherichia coli generation
microscopy, *see* specific methods
monoclonal antibody generation, *see*
    Monoclonal antibody, Shiga
    toxigenic *Escherichia coli*
    antigens
Shiga toxin,
    adsorption agents, 9
    antigen presentation studies, 177
    antineoplastic activity, 173, 174
    apoptosis induction, *see* Apoptosis
    detection, *see* specific assays
    DNA transfection application, 177
    endothelial cell interactions, *see*
        Endothelial cell
    expression stability in culture, 20
    intestinal epithelium interactions, *see*
        Intestinal epithelial cells
    intracellular trafficking, 168–170,
        209–213
    mechanism of action, 92
    mononuclear cell effects, *see*
        Lymphocyte; Peripheral blood
        mononuclear cell
    pH measurement in organelles using
        FITC-labeled B-subunit, 176,
        177, 221–227
    purification of Stx1
        chromatofocusing column, 189,
            190, 192
        dye affinity chromatography, 190
        Gb3 affinity chromatography,
            187, 190–193
        hydroxyapatite chromatography,
            189, 192
        materials, 187, 188
        overview, 187
        recombinant protein expression
            and harvesting, 188, 189,
            191, 192
    receptor, *see* Gb3
    retrograde transport, *see* Retrograde
        transport, StxB assays
    28S rRNA inactivation, 264
    serum antibodies, 19, 28
    subunits, 3, 92, 209, 264
    toxicity, 3
    types, 3, 264
STEC, *see* Shiga toxigenic *Escherichia
    coli*
Stool, detection of Shiga toxigenic
        Escherichia coli with polymerase
        chain reaction,
    amplification reaction, 48, 52
    contamination prevention, 51, 52
    materials, 47
    multiplex assay, 47, 51–53
    primers, 49
    product detection, 48
    screening assay, 47
    serotype targets, 47
    Shiga toxin gene screening, 50–52
    specimen preparation, 47, 50, 52
    target sequences, 45–47
    veterinary diagnostics, *see* Calf;
        Greyhound; Pig
Stx, *see* Shiga toxin

T

TAI, *see* Tellurite resistance- and
        adherence-conferring island
T-cell, *see* Lymphocyte
Tellurite resistance- and adherence-
        conferring island (TAI), features,
        101
TEM, *see* Transmission electron
        microscopy
Thin-layer chromatography (TLC),
    glycolipid receptors from endothelial
        cells, 256, 257
    overlay assay for Gb3, 199, 200,
        202, 203, 206

Thrombotic microangiopathy (TMA),
    pathogenesis, 263, 264
Tir,
    function, 94, 95
    phosphorylation, 94, 95
Tissue culture cytotoxicity assay, Shiga
        toxin detection,
    advantages and limitations, 10
    Vera cells, 10
TLC, *see* Thin-layer chromatography
TMA, *see* Thrombotic microangiopathy
Transmission electron microscopy
        (TEM),
    attaching and effacing lesions, 137,
        138, 141
    immunoelectron microscopy,
        post-embed staining of surface
            antigens, 144, 145, 148
        pre-embed staining of surface
            antigens, 143, 144
        immunonegative staining electron
            microscopy, 141–143, 148
        infection of cell monolayers, 140
        materials, 138–140, 148
        Shiga toxin-induced apoptosis,
            235–237, 239

W

Western blot,
    lipolysaccharide antibody detection,
        33, 38
    Shiga toxins, 28, 36–38, 40, 41